Biochar Production Engineering

Arpita Roy · Rouf Ahmad Bhat ·
Gowhar Hamid Dar
Editors

Biochar Production Engineering

Innovative Technology for Environmental Decontamination

Springer

Editors
Arpita Roy
Department of Biotechnology
Sharda University
Greater Noida, Uttar Pradesh, India

Rouf Ahmad Bhat
Department of School Education
Government of Jammu and Kashmir
Srinagar, Jammu and Kashmir, India

Gowhar Hamid Dar
Department of Environmental Science
Government Degree College Kulgam
Kulgam, Jammu and Kashmir, India

ISBN 978-3-031-92593-1 ISBN 978-3-031-92594-8 (eBook)
https://doi.org/10.1007/978-3-031-92594-8

This Springer imprint is published by the registered company Springer Nature Switzerland AG
The registered company address is: Gewerbestrasse 11, 6330 Cham, Switzerland

Preface

In the current development, environmental pollution is one of the major threats all over the world. Due to rapid urbanization and industrialization, various types of pollutants are coming up at an alarming rate, which leads to potential damage to the quality ecosystems. In the current situation, remediation of pollutants is one of the major challenges which our societies is facing around the globe. Increase in human population is also one of the major factors for pollution. This book entitled, *Biochar Production Engineering—Innovative Technology for Environmental Decontamination* provides broad spectrum potential alternatives for remediation of pollutants. Moreover, this book provides a comprehensive discussions on the compatibility of biochar production from wide-ranging raw materials and their application in the cleaning up of the contamination from different environs. In addition, the book discussed various eco-friendly, economical and aesthetically acceptable biochar-based decontamination technologies. The book is an excellent effort to bringing up biochar-based technologies for environmental decontamination in single bind. This volume will be enormously useful for a diverse group of readers vis-a-vis environmentalists, biotechnologists, microbiologists, nanotechnologists, environmental biotechnologists and those who are interested in clean eco-friendly technologies.

We are enormously thankful to all the authors who have contributed in this volume and to the Springer team for their substantial teamwork for printing up this volume.

Greater Noida, India — Dr. Arpita Roy
Srinagar, India — Dr. Rouf Ahmad Bhat
Kulgam, India — Dr. Gowhar Hamid Dar

About This Book

Biochar Production Engineering—Innovative Technology for Environmental Decontamination covers biochar production from different materials. The book discusses comprehensively the compatibility of biochar for cleaning up contamination from different environments. In addition, the book also reconnoiters various eco-friendly, economical, and aesthetically acceptable biochar-based decontamination technologies.

The book alluringly highlights the cutting-edge investigation in this field and provides intriguing data on the biochar application for the treatment of various kinds of pollutants to restore the ecosystem features and consequently reveal newfangled scopes of biochar application.

Key Features

- Provides various methods and strategies for biochar production from different raw materials and their physicochemical properties.
- Investigation in relation to compatibility of biochar as pollution decontamination tool.
- Address current trends and challenges related to the biochar production and economic aspects of biochar and its use in pollution remediation.

Contents

About the Editors

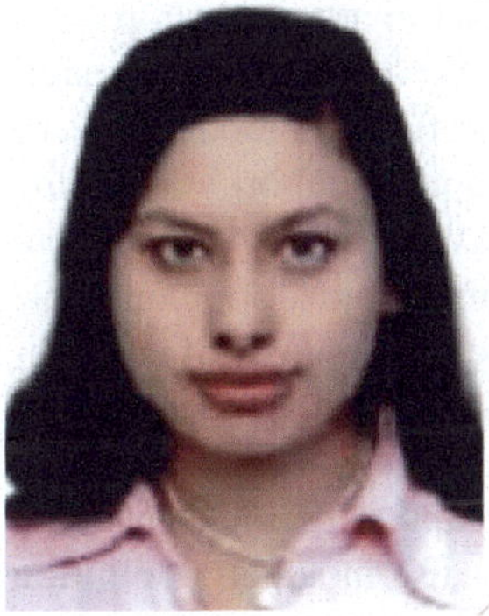

Dr. Arpita Roy (Ph.D.) is working as an Assistant Professor in the School of Bioscience and Technology (Department of Biotechnology) at Sharda University, Greater Noida. She obtained her M.Tech (Industrial Biotechnology) and Ph.D. (Biotechnology) from Delhi Technological University. Her areas of interest include Plant Biotechnology, Environmental Biotechnology, Green nanoparticle synthesis and Microbiology. She has published more than 100 papers in peer-reviewed journals (h-index 43; i-index 125; Total citation >6500), 4 patent published, 24 books chapters (Springer, Elsevier, CRC press, etc.) and edited ten books (Elsevier, Nova Publisher and Springer). She has qualified GATE, CSIR-UGC-NET and ICAR-ASRB NET. Enlisted in Top 2% scientists in the world for consecutive three years (Stanford University report published by Elsevier 2022, 2023 and 2024). Catalyst (MDPI), Highly cited paper in 2020–2021 in the section "Nanostructured Catalysts". Edited several special issues in different journals. She is also a recipient of "Commendable Research Excellence Award DTU 2019, 2020 and 2021" from Delhi Technological University, Delhi. She has presented more than 15 papers in national and international conferences.

Dr. Rouf Ahmad Bhat (Ph.D.) has pursued his doctorate at Sher-e-Kashmir University of Agricultural Sciences and Technology Kashmir (Division of Environmental Sciences) and presently working in the Department of School Education, Government of Jammu and Kashmir. Dr. Bhat has been teaching for the past nine years. He is an author of more than 55 research articles (h-index 34; i-index 59; total citation 4700) and 50 book chapters, and has published more than 50 books with international publishers (Springer, Elsevier, CRC Press Taylor and Francis, Apple Academic Press, John Wiley, IGI Global). He has his specialization in Toxicology, Phytochemistry, and Phytoremediation. Dr. Bhat has presented and participated in numerous state, national, and international conferences, seminars, workshops, and symposium. Besides, he has worked as an associate environmental expert in World Bank-funded Flood Recovery Project and also the environmental support staff in the Asian Development Bank (ADB) funded development projects. He has received many awards, appreciations, and recognition for his services in the field of environmental sciences. He has served as an editorial board member and a reviewer of reputed international journals. Dr. Bhat is still writing and experimenting with diverse capacities of plants for use in aquatic pollution remediation.

Dr. Gowhar Hamid Dar (Ph.D.) currently works as Assistant Professor in the Higher Education Department Government of Jammu and Kashmir and is currently posted in Government Degree College Kulgam where he serves as Head, Department of Environmental Science and Coordinator, Internal Quality Assurance Cell (IQAC). Before this he was posted at Government Sri Pratap College Srinagar, the oldest higher education institution of the UT of Jammu and Kashmir. He has a Ph.D. in Environmental Science with specialization in Environmental Microbiology (Fish Microbiology, Fish Pathology, Industrial Microbiology, Taxonomy and Limnology). He has been teaching postgraduate and graduate students for the past many years in Higher Education Department, Government of Jammu and Kashmir. He has more than 80 research articles (h-index 19; i-index 25; total citation >1219) in international and national journals of repute and a more than 20 books with international publishers (Springer, Elsevier, CRC Press Taylor and Francis, Apple Academic Press, John Wiley, IGI Global) to his credit. Moreover, he supervises a number of students for the completion of degrees (PG/Doctorate). He has been working on the Isolation, Identification and Characterization of microbes, their pathogenic behaviour and impact of pollution on development of diseases in fish fauna for the last several years. He has received many awards and appreciations for his services towards the science and development. Besides, he also acts as a member of various research and academic committees. Further, Dr. Dar is Principal Investigator and Co-Principal Investigator for different R&D projects sanctioned by Government of India and Government of Jammu and Kashmir.

Chapter 1
Fundamentals of Biochar Production

J. A. Ruley, B. M. Tomor, and Gowhar Rashid

Abstract The anticipated rise in population by 2050 has been greeted by several speculations about the potential threats to food insecurity and abrupt changes in climate. Collectively, their effects are feared to culminate into a human disaster and more so, if the intermediate determinants such as soil health have not been ensured. In this chapter, the discovery of biochar and the extant uses to which it can be put to offset the challenges of food insecurity and climate change to achieve a state of balance of nature is provided. The multifarious uses of biochar in realizing sustainable development are presented while also the multi-dimensional avenues through which it can be applied are appropriated. Observably, biochar is a multi-beneficial material, the reason why this chapter also provides a precise but concise overview about modifications that have been invented to scale up the usefulness of biochar for sustainable development.

Keywords Biochar · Climate change · Food insecurity · Soil health

1 Introduction

The term "biochar" is new in environmental and soil science literature. It was coined and first used by Peter Read in 2005 (Ernsting and Smolker 2009; Godbey 2016). Other sources quote 1998 as the first time, the term came in vogue to describe solid mass produced from biomass pyrolysis (Bapat and Manahan 1998). This followed paradigmatic shift in 1980s from biomass pyrolysis for energy and chemical resource

J. A. Ruley (✉)
Department of Agricultural Sciences, University of Juba, Juba, South Sudan
e-mail: janenajeb@yahoo.com

B. M. Tomor
Department of Wildlife, University of Juba, Juba, South Sudan

G. Rashid
Department of Medical Laboratory Technology, Amity Medical School, Amity University Haryana, Gurgaon, India

A. Roy et al. (eds.), *Biochar Production Engineering*,
https://doi.org/10.1007/978-3-031-92594-8_1

to sequestration of atmospheric carbon (Spokas et al. 2012). Biochar is portmanteau combining two words bio and char (Tan et al. 2023). It was loaned to English in the twentieth century from two Greek words βίος or bios, meaning "life" and "char" meaning charcoal produced by burning of biomass. The physical appearance of biochar may be confused for charcoal (Cyclic Carbon 2020; King 2013). However, biochar and charcoal are starkly different (Cyclic Carbon 2020). The slight similarity between the two is that biochar is charcoal produced from biomass that has undergone pyrolysis. Also, while charcoal is largely produced by burning living grass, shrubs and wood, biochar on top of the fore listed can be produced from dead plants (Breulmann et al. 2015; Khan et al. 2024; Stefanie 2018). Several other materials such as biogenic residues (such as agricultural residues, sewage sludge, and digestate from biogas production), food industry residues (such as brewer spent grain), or modern processes residues (such as ethanol production from straw, bagasse, and macroalgae) provide excellent raw materials for biochar are depicted in Fig. 1 (Breulmann et al. 2015; Roy and Bharadvaja 2021; Chelsea Green Publishing 2022; Hornung et al. 2021). The materials undergo processing the chief output biochar. Alongside biochar, syngas, and bio-oil are produced (Fig. 2).

Breulmann et al. (2015), Ilango and Lefebvre (2016) and Koulouri et al. (2024) suggested that besides, biochar can be got from human waste. Human waste is underutilized resource worldwide, rich in soil nutrients and carbon (Breulmann et al. 2015; Nicholas et al. 2024). Thus, production of biochar from these residues offsets environmental threats, further highlighting how important, biochar production is, in ridding environment of contaminants (Breulmann et al. 2015; Hornung et al. 2021; Nicholas et al. 2024) and acting as key solution to problems caused by poor human waste plaguing many developing countries (Breulmann et al. 2015). Also, biochar production is a clean-up strategy for farming zones where accumulation of crop residues is a nuisance (Billa et al. 2019; Gupta et al. 2020; Ruley 2024). Thirdly, biochar enhances safe storage of carbon, which is not possible with charcoal (Mishra et al. 2023; Shoudo et al. 2024; Stefanie 2018). Lastly, in terms of physical appearance, biochar is black, poked with numerous pores, light in weight, fine grained, and has larger surface area (Chen et al. 2023). Charcoal is not. Chemically, a large percentage of biochar (about 70%) is carbon while the rest (30%) comprises N, H_2, O_2, and few other elements. It should also be remembered that biochar chemistry is largely dependent on materials and approach used in its production (Stefanie 2018).

In recent years, the usefulness of biochar has gained great attention mainly from soil scientists and environmentalists (Burrow 2023). While the former group of scientists have vested interest in it as soil amendment for restoration of fertility, the latter category of professionals bestows great importance to it as a catalyst for addressing climate change due to its prowess in sequestration of carbon (King 2013). This among many reasons for prominence to be unveiled in this chapter accounts for why extant research and scholarship by environmentalists, agronomists, soil health experts, and biodiversity conservation is vested in it.

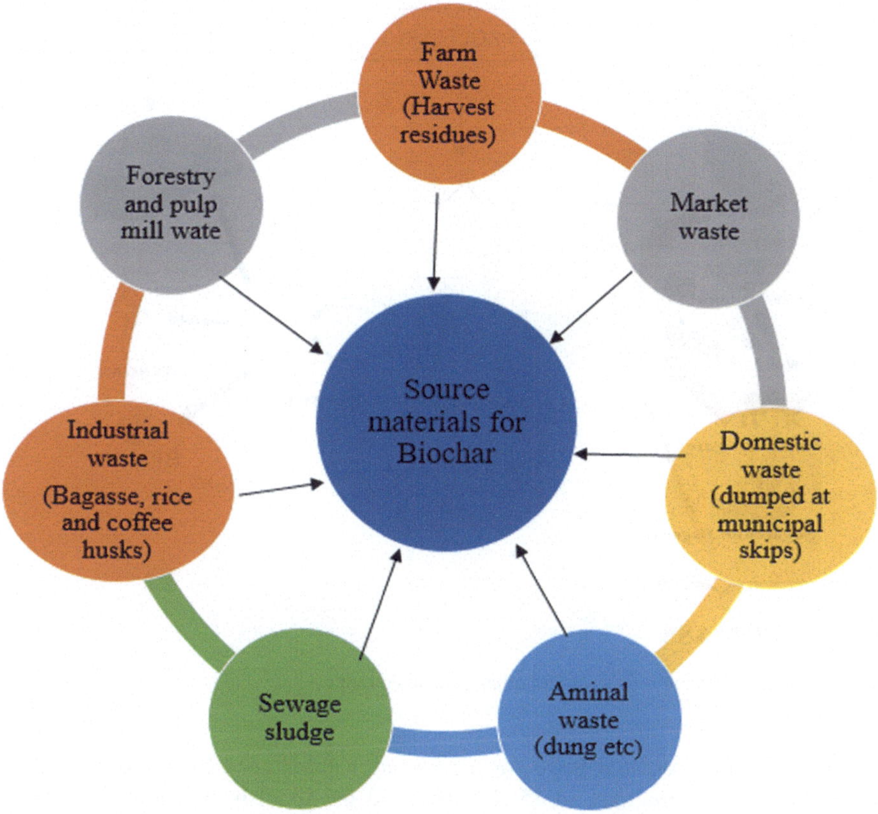

Fig. 1 Varied source materials for biochar production

2 History of Biochar

Biochar is not a novel innovation in science. It has a long and rich history (King 2013; Spokas et al. 2012). The Amazon Forest of Brazil is believed to have birthed use of biochar. King (2013) and United States Biochar Initiative (USBI) (2021) mention that over 2000 years ago, biochar was in use in this area. Over this period, archeological studies have indicated that natives of Amazon Forest were faced with the challenge of nutrient deficient soils. The landscape of the Amazon basin is covered by layers of Oxisols and Ultisols (Manuschevich et al. 2023; Silva et al. 2021; Zinck and Montero 2023). The two categories of soil are highly weathered, acidic in nature, and deficient in soil nutrients. According to Silva et al. (2021), much as they are habited by adapting plants tolerate the harsh conditions, crop species cannot do well leading to limited food production even in the face of intensive farming techniques. Habited by agrarian communities, soil fertility was restored by applying charcoal (also known as biochar) together with organic matter (King 2013). This innovation

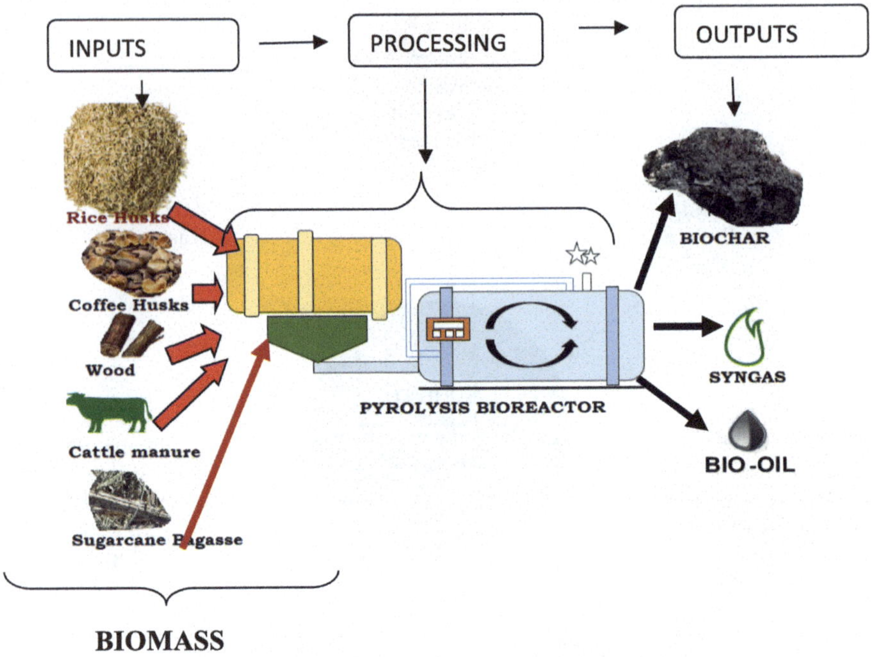

Fig. 2 Biochar production process and other associated outputs

was a eureka that restored hope for a population threatened by starvation. Until the breakdown of smallpox in 1561 introduced by Spanish expeditions (Castillo 2021), there was progressive increase in population that subsisted of agriculture in biochar rejuvenated soils (King 2013). During these early biochar days in Amazon Forest, indigenous people used slash and char technique of farming from which they produced biochar by igniting, burying, and smoldering biomass. Doing so, they created low-oxygen conditions that led to production of charcoal. The adoption of slash and char system was an upgrade of slash and burn method, also called shifting cultivation, whose net benefits were far less than what ensued under slash and char. Specifically, slash and char produced charcoal and sequestered over 50% carbon. Slash and burn produced ash and sequestered between 1 and 3% carbon. Besides the historical narrative, biochar has a global footprint. In this chapter, the authors address it as a ubiquitous solution and provide justifications why it is called a do-it-yourself adaptation of twenty-first century (King 2013).

3 Biochar Production Processes

Several processes or techniques are used to produce biochar (Manikandan et al. 2023; Safarian 2023). These have changed over time (USBI 2021). Unlike in the past centuries when a pile of wood logs would be built and covered with soil to prevent entry of air (oxygen) leading to low production of biochar; in present time, it is produced on a large-scale production using modern reactors. Discussed below are processes of producing biochar.

3.1 Hydrothermal Carbonization

Hydrothermal carbonization (hereafter, HTC) is the most common biochar production method from feedstock (Ercan et al. 2023; Hornung et al. 2021; Zhang et al. 2024). HTC is mainly applied when feedstock contains a high moisture content. Examples of feedstocks with high moisture content include sewage sludge and digestate. With this process, every one unit of solid feedstock is mixed with 6 units of water so that a suspension is obtained. Thereafter, the suspension is fed into a bioreactor that converts it into biomass at a temperature of 180–260 °C and pressure of 15–60 bar for 2–16 h (Hornung et al. 2021; Kostyniuk and Likozar 2024). The products from this process are biochar and the process water. The pitfall of this method is huge amount of process water. Equally, on environmental conservation grounds, not all pollutants in sewage sludge are destroyed such as pharmaceutical residues. As well, nitrogen compounds in sludge can inhibit efficiency of biochar production through this method.

3.2 Pyrolysis

According to Hornung et al. (2021), Li et al. (2023) and Wang et al. (2023a, b), pyrolysis is the commonest method of producing biochar. There are intra-differences within the process itself. These are threefold: fast pyrolysis, intermediate, and slow pyrolysis. Much as there are differences between the three as seen in the following discussion, the commonality between them is that the operations do not involve air or oxygen. In fast pyrolysis, the motivation is to produce more liquid. As a result, very high temperatures of about 1000 °C per second are applied. This causes vaporization (Hornung et al. 2021; Zhu et al. 2023). The cooled vapor deters further thermal decomposition. One condition that must be observed under this process is grounding down of feedstock into small and even very smaller particles so that effect of high heating rates is realized. In contrast, slow pyrolysis is applied to convert biomass into biochar. While biochar under this method is produced under a charcoal pile, at an industrial scale, retorts are used. The biomass is treated to a high temperature of about

Table 1 Categorization of the pyrolysis process

Type of pyrolysis	Requisite conditions	Resident time	Pretreatment requirements	Outputs (in %age)		
				Liquid	Biochar	Syngas
Fast	Very high temperatures above 1000 °C	Very short of about 1 s	Reducing the feedstock into smaller and microparticles	75	12	13
Intermediate	High temperatures (400–500 °C)	Short of about 10–20 s	Chopping of material to moderate size	50	20	30
Slow	Low temperatures (400 °C and below)	Long resident time of hours and even days	No need for size reduction of materials	30	35	35

400 °C for hours or days. This differentiates slow pyrolysis from fast pyrolysis where extremely high temperatures are involved with residence time even far shorter in seconds. As well, while chopping of materials into small pieces and minute particles is a required for fast pyrolysis, slow pyrolysis can involve feeding of whole logs in pile or retort (Hornung et al. 2021; Masek et al. 2024; Pereira et al. 2023). Lastly, intermediate pyrolysis falls between fast and slow pyrolysis. In this category, resident time ranges for few minutes while heating rates range between 200 and 300 °C/min. The pretreatment conditions of feedstock for intermediate pyrolysis are coarse shredding or chopping of material and limiting moisture content of materials to about 40 wt %. Preferably, process should be run at temperatures of 400–500 °C. The products from intermediate pyrolysis are lower liquid and higher biochar yield. Synthesis outputs of pyrolysed biochar are presented in Table 1.

3.3 Gasification

This process or technique is applied when the goal is to output synthesis gas and not biochar. The gas is produced at high temperatures ranging between 700 and 1400 °C (Adeniyi et al. 2024; Hornung et al. 2021). The outputs of this process are carbon monoxide, hydrogen, and carbon dioxide. As well, tar can be produced (Liu et al. 2019; Yue et al. 2023). Besides tar, are particulate matter which must be removed before harvested gas is turned into combustion engine (Ottani et al. 2023). One of the pitfalls of this method is that if air is used as gasification agent, nitrogen has a dilution effect on produced synthesis gas as well as slowing heating process (Aguado et al. 2023; Hornung et al. 2021).

4 Basic Properties of Biochar

4.1 *Physicochemical Properties of Biochar*

The physicochemical properties differ between the biochar products depending on certain factors (Li and Skelly 2023; Zheng et al. 2022). For example, temperature is a key factor that influences quality and type of output. To mention, at low temperature or slow pyrolysis, biochar solid is produced while fast pyrolysis yields liquid output. The basic physical characteristics of biochar are: it has a large surface area, is poked with several pores, and charges and exists in different functional groups. Further, compare to the raw materials, biochar is small as it shrinks during production process. These physical attributes affect soil chemistry such as pH, CEC, and surface adsorption. For example, addition of biochar in soil has a liming effect that checks pH levels (Bolan et al. 2023a, b; Rawat et al. 2019; Tusar et al. 2023). Further impact on soil chemistry is the observable increase in availability of C, N, Ca, Mg, K, and P (Rawat et al. 2019; Zheng et al. 2022).

4.2 *Nutritious Properties of Biochar*

Biochar is rich in essential plant nutrients since it contains K, Mg, K, Ca, and P (Ilay 2024; Mona et al. 2020; Zheng et al. 2022). These mineral nutrients are formed during pyrolysis (Bilias et al. 2023). Besides, dissolved organic materials are also precipitated during pyrolysis. For this reason, when biochar is applied in soil environment, it provides an excellent source of bioavailable nutrients to growing plants and resident rhizobacteria (Zheng et al. 2022). Despite this linkage, the viability of biochar is determined by raw materials used and the conditions under which it is produced (El-Bassi et al. 2021; de Figueiredo et al. 2021; Yang et al. 2019). Therefore, to realize substantial return on investment (ROI) in biochar, efforts should be made to obtain and use nutrient-rich raw materials and subject them to appropriate pyrolysis conditions if nutrient-rich biochar is to be obtained (Ghadirnezhad Shiade et al. 2024; Yang et al. 2024).

4.3 *Constancy Property of Biochar*

Biochar differs from other organic soil amendments. It stays in soil for a very long period (Azzizi et al. 2024; Zheng et al. 2022). This attribute was also raised by Lian and Xing (2017) who argue that in absolute and relative terms, biochar stays in soil for quite a long period of time eventually leading to improvement in capacity of soil to store carbon. Comporting evidence is provided by Williams et al. (2019) and Azizzi et al. (2024) that low rate of biochar mineralization accounts for over 100 years of carbon half-life. Further evidence of low rate of mineralization is reported by Agegnehu et al. (2017) and Drewer et al. (2023) whose studies established that carbon from biochar had half-life of over 1000 years. This long retention period improves physiology of plants such as germination, shoot, and root biomass and yield (Sanchez-Reinoso et al. 2020).

5 Biochar as a Panacea

The global population is projected to soar to about 10 billion people by 2050 from the current 7.6 billion people (in November 2024) (Aloo et al. 2021; International Institute of Sustainable Development [IISD] 2020; Mahawar et al. 2020). The accompanying begging question is, will nations feed this population? In providing possible answers to this question, maintenance of soil health and environmental conservation have been cited as sustenance factors. Incidentally, several studies (such as Elaoufi 2024; Ferreira et al. 2024; Stefanie 2018; Verma et al. 2022) have noted that soil degradation is unprecedented worldwide. If left unchecked, it might culminate in low agricultural productivity leading to massive starvation and death of many people. Aside from feared degradation of soils, is the twin bedfellow challenge of climate change? Like diminishing soils, climate change is feared to induce undesirable outcomes such as extinction of species of fauna and flora. As an antidote to these looming defects, extant experimental and non-experimental works have been undertaken by researchers both in the science and non-science fields to ascertain the extent to which application of biochar can redeem depleted soils and restore the lost fertility. Bekchanova et al. (2024), El Moussaoui et al. (2024) and Stefanie (2018) report that when protocols of use are followed, use of biochar causes an improvement in soil fertility. Specifically, it serves as a soil conditioner, improves soil water, checks acidity levels of soil, improves soil porosity, regulates leaching, stabilizes soil CEC, and lastly, it improves soil environment for survival, growth, and multiplication of microbial communities. On aspect of environmental conservation, biochar sequesters carbon and reduces emission of other GHGs while also preventing pollution of groundwater and decontamination of wastewater and other industrial effluents (Oni et al. 2019; Stefanie 2018; Wang et al. 2023a, b; Yang and Yang 2024).

In summary, use of biochar offers the following listed economic and environmental benefits (Bista et al. 2019; Bolan et al. 2023a, b; Campion et al. 2023; Ernsting and Smolker 2009; King 2013; Maroušek et al. 2023; Oni et al. 2019; Six 2014; USBI 2021),

1. One of the environmental benefits pertains to sequestration of carbon. Several studies in this area have unanimously demonstrated that use of biochar helps in retention of and or serves as sinks for carbon for very many years. As well, the introduction of biochar in the soil also helps to suppress the ability of the soil to emit greenhouse gases such as methane and nitrous oxide.

2. Use of biochar restores lost nutrients in the soil leading to restoration of impoverished soils that have become wastelands. In the soil, biochar also improves the physicochemical properties of the soil leading to nutrient balance and nutrient retention. It has a liming effect in the soil that helps to reduce the acidity of the soil, hence creating favorable conditions for survival of resident microorganisms and plants. Its application in the soil environment helps to reduce soil density and soil hardening while at the same time increasing soil aeration and CEC. On the third prong, biochar is an excellent bioremediator of heavy metals from contaminated soil, petroleum hydrocarbons (PHC), and cleansing of soils polluted by herbicides, pesticides, and inorganic fertilizers.
3. On the economic note, production of biochar engenders production of biofuels such as bio-oil and syngas, and these are known to provide energy sources for both domestic, industrial, transport, and aviation uses.
4. Application of biochar increases the vitality of crops leading to significant yields. Bista et al. (2019) for example, using the wheat as an example, established that the application of 0.5 to 135 tonnes of biochar translated into a significant increase in yields from – 29 to 324%.
5. The introduction of biochar in the soil also helps to trap the toxins as well as excess nitrogen in the soil consequently protecting the quality of the ground water as well as other open water sources that are liable to infiltration by runoff erosion.
6. Biochar production helps to reduce on the need and consequent use of fossil fuels and application of inorganic or fossil fuel-based fertilizers, hence protecting the ecosystem from the deleterious effects of both hydrocarbon and inorganic fertilizer pollution.
7. The entire process of biochar production translates into creation of jobs which improves household livelihoods, steers community, and regional development and extends those available jobs nationally, leading to economic growth and economic development. Besides, it supports business growth and development especially sellers of accessories and gear for pyrolysis, middlemen, and other auxiliary businesses such as transporters and stockists.

6 In-Depth Analysis of Composite Roles of Biochar in Soil and Water Ecosystems

6.1 Biochar as a Soil Conditioner

Poor methods of farming and other anthropogenic activities have resulted in problem soils (Yu et al. 2019). Biochar is a remedy for problem soils (Pandian et al. 2024; Razzaghi et al. 2020; Yu et al. 2019). Biochar as an organic soil amendment provides far superior attributes over other organic amendments that qualify it to be the best choice for soil conditioning (Agarwal et al. 2022; Cig et al. 2021; Das and Gosh 2020; Hossain et al. 2020; Kocsis et al. 2021; Neogi et al. 2021). Firstly, biochar has a long half-life in soil, hence posting long-term benefits to soil. Besides, it retains nutrients over a long period of time while also checking acidity levels of soil, improving moisture holding capacity, CEC, soil aeration, and supporting abundance of microbial flora. In soil with lower pH, controlled conditions under which biochar is made can be varied to produce acidic biochar which can help to increase soil pH to desired levels of both plants and microorganisms (Shetty and Prakash 2020; Wang et al. 2021). The additional conditioning role of biochar is that biochar pertains ridding soil of air-bone and soil-borne pathogens (Liu et al. 2024; Xu et al. 2024). This was proven by Bonanomi et al. (2017) using citrus wood biochar and Elmer and Pignatello (2011) using hardwood dust. More supporting evidence of soil-borne pathogen suppression is provided by Nerome et al. (2005) who used biochar derived from municipal waste to suppress tomato bacterial wilt. The calcium compounds in biochar explain its ability to control diseases caused by soil-borne pathogens (Kumari et al. 2024; Ogawa 2009).

In arid ecological zones where freshwater is scarcer, use of biochar provides opportunities for mitigating scotching effects of saline irrigation water for arable farming. Substantial evidence of this matter is reported by Liu et al. (2021) that biochar and compost improve physicochemical and biological properties of saline water irrigated soils leading to improvement in quality. Thus, biochar and compost enable plants to offset negative effects of imparted by saline water in soil (Feng et al. 2021; Mahmoud et al. 2019; Liang et al. 2021; Liu et al. 2021). Besides, in drought infested zones, biochar improves soil conditions such as nutrient uptake and gaseous exchange (Gao et al. 2024). As well, application of biochar improves water holding capacity of soil while it also improves stomatal conductance and improved uptake of minerals by plants. By doing so, plants can hardly be scotched by xeric conditions, leading to survival and improved productivity in case they are crop plants (Bala et al. 2019; Mansoor et al. 2021). Thus, in saline and drought affected zones, Abd El-Mageed et al. (2020) confirmed that biochar increases crop tolerance to deleterious conditions.

In weathered soils, application of biochar from locally available materials restores fertility, making soils fertile for crop growth. Evidence is of this matter is traceable in works of Berek et al. (2018) in Hawai. They established that use of biochar with vermicompost restored lost soil nutrients and restored weathered and nutrient depleted soils for use by plants. In cold regions where soils are stressed by freeze–thaw conditions, application of biochar has noticeable effects on water utilization rate by plants (Fu et al. 2021). Significance importance of biochar as soil conditioner is depicted in Fig. 3.

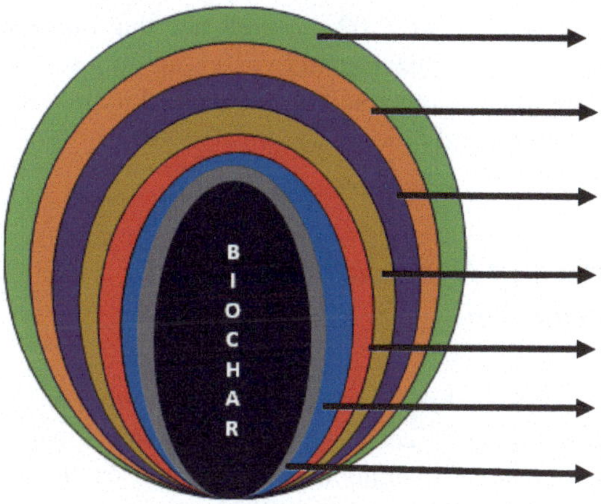

Fig. 3 Composite soil conditioning roles of biochar

6.2 Biochar as a Checker of Pollution of Groundwater Sources

Biochar checks pollution of ground water sources through sorption of pollutants (Priyanka Wood et al. 2024; Rawat et al. 2019; Zhao et al. 2024). Application of biochar in soil adsorbs inorganic fertilizers minerals and pesticides. As a result, their movement into adjacent or surrounding groundwater sources is controlled. This prevents degradation of quality of groundwater sources by agricultural activity borne chemicals. Poor management of wastewater also doubles as threat to quality of groundwater sources especially in urban and peri-urban settings where wastewater generation is alarmingly high. Thompson et al. (2016), Czekała et al. (2019), Sonsuphab et al. (2024), and Souza et al. (2021) confirmed that biochar detoxifies wastewater, rendering it harmless. Thus, biochar-treated wastewater does not pollute groundwater sources.

6.3 Biochar as an Enhancer of Microbial Abundance in Soil

Biochar increases availability of soil carbon. This is referred to as carbon farming (Edwards 2022; Institute for Carbon Removal Law and Policy 2018). Carbon is food for microorganisms (Bhattacharyya et al. 2022; Fang et al. 2024; Gopal et al. 2013; Nagappan 2021; Ruley et al. 2020a, b). Beyond stocking soil with carbon as a food and energy source for microbes, biochar improves soil water retention which in one way or the other creates conducive conditions for survival and growth of soil microorganisms (Malik et al. 2022; Song et al. 2021; Sousa et al. 2024).

A study by Warnock et al. (2007) and Sousa et al. (2024) established that biochar activates functioning and performance of mycorrhizal fungi by: conditioning soil to make it amenable and livable, exposing fungi to other soil microbes; and improving plant–fungus signaling interference and removal of toxic chemical elements.

6.4 Biochar as a Heavy Metal and Hydrocarbon Contaminant Remover from Soil

In any environment where nature has endowed the communities with oil resource, its exploration and consequent exploitation are greeted with applause. Specifically, proceeds from crude oil extraction transform livelihoods and national economic growth and development (Kaur and Roy 2021; Ruley et al. 2022; Su et al. 2024). On the other side of the coin, crude oil is a resource curse, leading to unspeakable havoc on environment. Notable examples include soil contamination with hydrocarbons and heavy metals from oil spills, illicit dumping of overburden, leakages during transportation, and produced water, among others (Ruley et al. 2020b, 2022). While several physical, chemical, and electrical methods are used to cleanse contaminated soils, several experimental studies have established that they cause secondary contamination. From the economic sense, these remediation strategies are prohibitive as colossal sums of money are required. This has attracted use of bioremediation strategies which are cheaper and ecofriendly. In bioremediation process, biochar is an agent (Aziz et al. 2020; Wang et al. 2024; Zahed et al. 2021; Zhang et al. 2019a, b). One of the ways in which biochar sustains bioremediation is by supporting growth of soil microorganisms and activation of their capacity to biodegrade organic pollutants and heavy metals (Guo et al. 2020; Itam 2020; Ouyang et al. 2021). Studies have shown that introduction of substantial amounts of biodegradable organic carbon by biochar improves potential of microorganisms to biodegrade hydrocarbons. The organic carbon serves as substrate for soil microbial communities (Guo et al. 2020; Zhang et al. 2019a, b; Saeed et al. 2022). Also, microbial community structure is improved by soil carbon, while it also enhances microbial activity through improved soil respiration rate, catalyzed microbial enzymatic activity, among others (Guo et al. 2020). Once such occurs, mineralization of organic pollutants is greatly increased. Minnikova et al. (2022) and Wang et al. (2019a, b) established that application of biochar in PHC-contaminated soil leads to abundance of enzymes in microorganisms such as catalase and dehydrogenases which amplify degradation potentials of microorganisms. Some studies have shown that organic amendments mixed with biochar lead to enormous growth of phytoremediation plant species and improve their capability to breakdown hydrocarbon contaminants, and one such example is discussed in Fig. 4 (Barati et al. 2018; Kudakwashe et al. 2022).

Besides enabling contaminant removers, biochar itself removes contaminants from soil through adsorption (Włodarczyk and Kwarciak-Kozłowska 2021; Wu et al.

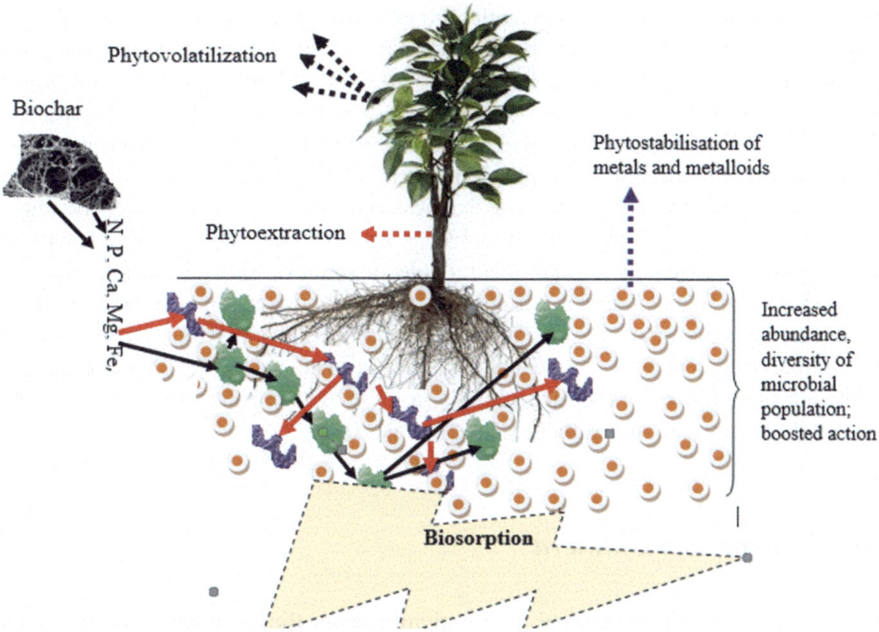

Fig. 4 Multidimensional ways in which biochar supports bioremediation

2022). This minimizes mobility of PAHs in soil. Equally, biochar is a carrier of metal-tolerant bacteria for degradation of heavy metals such as cadmium (Liu et al. 2020; Singh et al. 2020). Wu et al. (2019) and Dike et al. (2022) recommend co-application of biochar with bioremediation species for removal of contaminants from heavily polluted soils especially with recalcitrant hydrocarbon compounds. In agreement, Anbuganesan et al. (2024), Tawfik et al. (2024), Valizadeh et al. (2021), and Wu et al. (2024) noted that synergistic action leads to better results.

6.5 *Biochar and Sequestration of Carbon*

Biochar forms carbons sinks in soil (Mona et al. 2021; Tao et al. 2024). More evidence of this role is cited by Lombardi (2009) who regards biochar as coal made from biomass that becomes carbon sink when buried in soil. Lombardi (2009) further notes that production of biochar provides a better way of conserving environment as it prevents possibility of decaying plants releasing carbon dioxide into the atmosphere. Hornung et al. (2021) provide a concrete support for Lombardi (2009) and reported that biomass such stores about 650 billion tons of carbon worldwide, the same quantity produced by global anthropogenic emissions of carbon dioxide for about 65 years. Relatedly, Hornung et al. (2021) report that wood stores over 82% of carbon dioxide that would have been released into atmosphere by aerobic degradation of harvested shoots or dead remains. The emission of this carbon dioxide is checked by converting wood into biochar (Hornung et al. 2021; Meng et al. 2024; Villardon et al. 2024). As well, conversion of crop residues to biochar leads to sequestration of soil carbon (Majumder et al. 2019; Ruley 2024) which posts numerous environmental health benefits. For example, Yang et al. (2021) established that close to a ton of carbon dioxide equivalent was sequestered when one ton of crop residues was converted into biochar. Also, biochar as earlier noted induces the growth of microorganisms that play a fundamental role in sequestration of carbon. Indeed, Safraz et al. (2019) and Kapoor and Zdarta (2024) note that biochar favors growth of microorganisms that scavenge on carbon which controls its emission and consequent undesirable occurrences of climate change. Further to note is the fact that application of biochar in soil arrests other greenhouse gases such as nitrous oxide and prevents their emission into the atmosphere (Dawar et al. 2021).

6.6 Biochar as a Chimera

While most studies have painted a rosy picture about the invaluable contributions of biochar toward soil fertility restoration and environmental conservation, it goes without saying that in one way or the other, its use can have a deterrent effect. Six (2014) and Ghadirnezhad Shiade et al. (2024) enumerated the following drawbacks of biochar use.

1. When biochar has not been rightly applied in soil, it can lead to considerable decline in crop yields. Two scenarios explain decline in crop yields. Firstly, biochar can cause sorption of water and nutrients which reduces their availability for growing crops. As a result, crops cannot become luxuriant. Secondly, biochar can equally inhibit seed germination, which leads to reduction in crop yields.
2. In instances where pesticides and herbicides have been applied to crops, biochar can lead to their sorption, consequently reducing their efficacy.
3. Heavy metals, volatile organic compounds, dissolved organic carbon, and hydrocarbons such as PAHs can be introduced in soil by biochar, and these bring about soil pollution and turning of hitherto fertile soils into wasteland.
4. While it is orthodox that introduction of biochar causes a considerable reduction in emissions of greenhouse gases such as nitrous oxide and sulfur dioxide, it has equally been observed that even after biochar has been applied, the emissions of the two GHGs have at times been noted to increase.
5. The fine ash produced with biochar is emitted into atmosphere as dust leading to air pollution. This can cause airborne respiratory diseases when inhaled in substantial amounts.
6. As earlier noted, production of biochar is a little versatile as it entails use of biomass such as crop residues harvested shoot systems such as stems, leaves, and seed pods. In one way or the other, this practice interrupts the internal

nutrient recycling mechanisms and efficiencies of plants while also affecting microorganism residing in rhizosphere of harvested plants.

7. Soils with high clay or organic matter content may not necessarily benefit from addition of biochar. While it is common for soil CEC to increase with addition of biochar, the rates of increase are usually minimal in clay or organic matter-rich soils.
8. Biochar application affects the pH of soil. When applied to high alkaline soils, it may cause an increase in pH which might damage crops by altering morphological and biological activities.

7 Modification of Biochar for Efficiency

Several scientific discoveries have come up with novel techniques for modifying biochar so that its application becomes effectiveness in cleanup of both terrestrial and aquatic ecosystems. The modification processes (physical, chemical, or biological and co-application) are discussed below.

8 Physical Modification Processes

8.1 Ball-Milling

Also called solid-phase synthesis or solvent-free technique, this method is industry-based and normally used in the construction of biochar-based materials that are used in the treatment of environmental pollutants (Huang et al. 2020; Lamberti et al. 2024; Lyu et al. 2018). Through ball milling, the sorption capacity of biochar is strengthened manifold to the tune of over 200 times the original capacity. Usually, through this process, smaller nanoparticles of 10 microns or less are produced. This method is also used in improving physicochemical properties of biochar as well as its nanocomposites such as pore surfaces or even the micropores, hence enhancing the sorption capacity of the products. This accounts for why biochar nanocomposites produced through this process are excellent adsorbents of heavy metals, dyes, and a suite of other organic pollutants present in water (Chen et al. 2022; Zeng et al. 2024).

9 Chemical Modification Processes

Biochar can also be modified by chemical means. This process involves use of different chemicals to change surface properties of biochar such as surface properties to improve sorption. Chemical modification involves use of acids, alkaline, oxidizers, metal oxides and metal salts as well as other modifications (Lan et al. 2024; Nguyen et al. 2024). The processes are described below.

9.1 Acid Modification

This mechanism involves acid treatment of biochar which helps in reducing metals and other impurities present on surface of biochar. Besides purification, acid modification also increases acid functional groups on surface of biochar. In soil environment, acid-modified biochar is an excellent soil conditioner.

9.2 Alkali Modification

Alkali modifiers increase surface area of biochar, its porosity, and numbers of oxygen-containing functional groups. NaOH and KOH are used as reagents used in alkali modification of biochar. KOH specifically helps to increase surface area of pyrolysis prepared biochar. As well, KOH improves sorption capacity by boosting numbers of oxygen-containing functional groups. Like KOH, the use of NaOH increases in surface area of biochar and therefore and its sorption capacity.

9.3 Modification of Biochar with Metal Ions and Metal Oxides

Metal ions and oxides amplify performance of biochar. Biochar modified with metals ions and oxides has high sorption capacity for heavy metals and their oxides. Mainly, the metallized biochar increases exchange of ions and nitrogen-containing functional groups. The two ways of metallizing biochar are: through mixing metals or their oxides with raw materials for pyrolysis; and secondly by soaking biochar in metal ions or metal oxides under controlled conditions. The inoculation of biochar with metal ions and metal oxides transforms its original properties such as its magnetism, photocatalytic ability, and redox property.

10 Biological Modification of Biochar

This technique of modifying biochar entails deployment of living organisms (Lan et al. 2024; Tao et al. 2019; Yan et al. 2024). Processes involved in biological modification include grafting and functional group exchange. Under this technique, efficient microorganisms with excellence in adsorbing heavy metals from aqueous solutions are used (Muhammad et al. 2021). This technique is mainly used on biochar for treating wastewater (Diaz et al. 2024; Godwin et al. 2019). For example, alkaline biochar from rice husks and acidic biochar from sludge are modified with phosphate solubilizing bacteria to bioremediate Pb (II) from wastewater. In soil environment, inoculated biochar enhances bio-utilization of heavy metals such as Cd (II) and Cu (II) while also improving abundance of soil microbial communities. Consequently, this improves soil quality.

11 Co-Application of Biochar with Other Soil Amendments

In its original state, biochar can also be mixed or combined with other soil amendments (Rawat et al. 2019; Umair Hassan et al. 2024; Xia et al. 2024). The amendments in question can be manure, compost, or lime. These are mixed with biochar to improve on efficiency (Rawat et al. 2019; Rombel et al. 2021). As earlier noted, biochar has a high sorption capacity. Therefore, when it is mixed with other amendments, it can minimize possibilities of leaching of mineral nutrients which squares efficiency of other amendments such as organic manure. The resulting synergy improves soil physicochemical and biological properties than when either is used alone (Haider et al. 2024; Rawat et al. 2019). The following case examples provide illustrative evidence of synergistic action of co-application of biochar with other soil amendments.

11.1 Biochars with Coating

Degraded soils can best be restored through use of biochars with coating (Chakraborty and Das 2024). The coating materials can be polymeric and non-polymeric. This approach is mainly used when biochars exist in granule form. They are encapsulated in organic or inorganic materials which serve as diffusion barriers (Chen et al. 2018). In relative terms, the use of polymeric coats is discouraged because they are prohibitively expensive while also they are non-biodegradable (Zhang et al. 2019a, b). Besides polymeric coatings, amines, paraffin, synthetic wax, and mixture of amines and wax are used. Also, combinations of mineral oils make good candidates for coats (Naz and Sulaiman 2016). The viability applying biochars with coatings was proven by Ye et al. (2019). In their research, Ye et al. (2019) used biochar obtained from wheat straw by pyrolysis at 400 °C. Biochar was enriched with phosphorus and nitrogen and then soaked in potassium dihydrogen phosphate and potassium nitrate solution. This was followed by coating with bio-oil. The bio-oil coat led to slow release of phosphorus and nitrogen. The bio-oil coat adjusted kinetics of nutrient release and dissolution of nutrients by water to a considerable degree.

11.2 Mixing Biochars and Commercial Fertilizer

The nutrient richness of biochars is elevated by mixing them with inorganic fertilizers (Carey et al. 2015; Samoraj et al. 2022). NH_2SO_4, KCl, among others are key examples of inorganic fertilizers that can be mixed with less nutrient carrying biochars to boost their efficacy. Use of a combination of biochar and inorganic fertilizers is cost effective and enhances slow release of nutrients to soil. Thus, the introduction of fertilizers combined with biochar matrix is cost-effective way of releasing fertilizers into soil. Consequently, there is increase in yields of plants and in particular, crops.

11.3 Using a Mix of Inorganic Fertilizers with BC Coating

The effectiveness of biochars in soil can also be increased by coating inorganic fertilizers with biochars (Petrus et al. 2020). Compared with mixing of biochar with inorganic fertilizer, coating with biochar is slow and leads to a slower release of nutrients into soil. This modification strategy is far cost effective and is environmentally friendly compared to polymeric coatings (Azeem et al. 2014; Zhang et al. 2014). The usability of method is equally supplemented by availability of natural binders which increase efficiency of fertilizers. Natural binders in question include clay, starch, modified starch, lignin, modified lignin, or bentonite (Shi et al. 2020). The binders enhance bonding of inorganic fertilizer and biochar coating.

11.4 Mixing Biochar and Compost

The effect of biochar on soil ecosystems is magnified by application of a mixture of biochar and compost manure. This combination breeds synergy and promotes mutualism as weakness of one is compensated for by strength of other. The collective action of mixture greatly influences vigor and vitality of plants (crops) leading to increase in farmer productivity. Several studies (such as Abideen et al. 2020; Azeem et al. 2020; Goldan et al. 2019; Hannet et al. 2021; Radin et al. 2018; Rasool et al. 2021; Zahra et al. 2021) recommend mixing biochar with ready-made compost. However, co-composting biochar with compost raw material can also produce the mix (Awasthi et al. 2017; Bashir et al. 2020; Bello et al. 2021; Luo et al. 2017; Oldfield et al. 2018; Pandit et al. 2020; Wei et al. 2021).

11.5 Co-Composting Biochar with Organic Waste

Co-composting biochar with organic waste improves biochar efficacy in soil environment. Under this approach, biochar is added to compost raw materials such as sewage sludge, manure, agricultural residues, and other compostable materials before actual composting. The product from this process is referred to as COMBI, a short form for co-composted biochar. As in other combinations or mixes, COMBI posts superior advantages to the soil ecosystem over what biochar or compost can do alone (Antonangelo et al. 2021; Melo and Sanchez-Monedero 2024; Wang et al. 2019a, b). For better results, biochar should be co-composted with ready-made ingredients (Antonangelo et al. 2021; Mohamed et al. 2024). COMBIs hold many advantages such as ability to remediate heavy metal contaminated soils (Mounissamy et al. 2021), they improve soil fertility and structure (Luo et al. 2017) and play a key role in regulating kinetics of nutrient release from compost. This controls release of nutrients into soil (Iqbal et al. 2015). Further, it revitalizes nutrient available in biochar and can be priority solution to challenges of solid waste and sewage management that developing countries continue to grapple with. Therefore, at a future date, most environmentalists are optimistic that use of COMBIs will provide a sustainable solution to environmental management and conservation challenges while also providing a ray of hope that cheaper substitutes for mineral fertilizers in the future can be produced (Oldfield et al. 2018; Wang et al. 2019a, b).

12 Electro-Modified Biochar

Biochar can be modified by means of electrification to produce biochar with greater adsorption capacity (Gabhane et al. 2020; Rathee et al. 2024). Electro-modified biochar is purposed for removing certain contaminants from soil, water, and air. Under this process, biochar is mixed in iron, magnesium, or aluminum for a period of 2–12 h in the presence of electric current. This alters functional groups on the surface of pores of biochar leading to improvement in its adsorption capacity.

13 Conclusion

The innumerable ways cited in this chapter account for why biochar is widely recommended as soil fertility restoration agent and environmental clean-up materials. Biochar has composite roles for solving twin bedfellow challenges of food insecurity and climate change. The duo is feared to eliminate mankind, if their current and prospective effects are left unabated. However, much as it is widely believed that biochar is a "miraculous material", its application ought to consider the principles proposed by Guo et al. (2020). These are abbreviated as 3Rs for right source, right rate, and right placement. When borne in mind, soil health is persistently improved. Consequently, biochar becomes a panacea. This will vindicate the belief that biochar is a tripartite win–win solution; it causes improved crop yields, decreases global warming problem, and increases soil organic carbon. When the 3Rs are not closely followed, application of biochar might be chimeric.

References

Abd El-Mageed TA, Rady MM, Taha RS, Abd El Azeam S, Simpson CR, Semida WM (2020) Effects of integrated use of residual sulfur-enhanced biochar with effective microorganisms on soil properties, plant growth and short-term productivity of Capsicum annuum under salt stress. Sci Hortic 261:108930

Abideen Z, Koyro HW, Huchzermeyer B, Ansari R, Zulfiqar F, Gul BJ (2020) Ameliorating effects of biochar on photosynthetic efficiency and antioxidant defence of Phragmites karka under drought stress. Plant Biol 22(2):259–266

Adeniyi AG, Iwuozor KO, Emenike EC, Amoloye MA, Adeleke JA, Omonayin EO, Ezzat AO (2024) Leaf-based biochar: a review of thermochemical conversion techniques and properties. J Anal Appl Pyrolysis 106352

Agarwal H, Kashyap VH, Mishra A, Bordoloi S, Singh PK, Joshi NC (2022) Biochar-based fertilizers and their applications in plant growth promotion and protection. 3 Biotech 12(6):1–19

Agegnehu G, Srivastava AK (2017) Bird MI The role of biochar and biochar-compost in improving soil quality and crop performance: a review. Appl Soil Ecol 119:156–170

Aguado R, Escámez A, Jurado F, Vera D (2023) Experimental assessment of a pilot-scale gasification plant fueled with olive pomace pellets for combined power, heat and biochar production. Fuel 344:128127

Aloo BN, Mbega ER, Tumuhairwe JB, Makumba BA (2021) Advancement and practical applications of rhizobacterial biofertilizers for sustainable crop production in sub-Saharan Africa. Agric Food Secur 10(1):1–12

Anbuganesan V, Vishnupradeep R, Mehnaz N, Kumar A, Freitas H, Rajkumar M (2024) Synergistic effect of biochar and plant growth promoting bacteria improve the growth and phytostabilization potential of Sorghum bicolor in Cd and Zn contaminated soils. Rhizosphere 29:100844

Antonangelo JA, Sun X, Zhang H (2021) The roles of co-composted biochar (COMBI) in improving soil quality, crop productivity, and toxic metal amelioration. J Environ Manage 277:111443

Awasthi MK, Wang Q, Chen H, Wang M, Ren X, Zhao J, Li J, Guo D, Li DS, Awasthi SK, Sun X (2017) Evaluation of biochar amended biosolids co-composting to improve the nutrient transformation and its correlation as a function for the production of nutrient-rich compost. Bioresour Technol 237:156–166.

Azeem B, KuShaari K, Man ZB, Basit A, Thanh TH (2014) Review on materials & methods to produce controlled release coated urea fertilizer. J Control Release 181:11–21

Azeem M, Hale L, Montgomery J, Crowley D, McGiffen Jr ME (2020) Biochar and compost effects on soil microbial communities and nitrogen induced respiration in turfgrass soils. Plos one 15(11):e0242209

Aziz S, Ali MI, Farooq U, Jamal A, Liu FJ, He H, Huang Z (2020) Enhanced bioremediation of diesel range hydrocarbons in soil using biochar made from organic wastes. Environ Monit Assess 192(9):1–14

Azzi ES, Li H, Cederlund H, Karltun E, Sundberg C (2024) Modelling biochar long-term carbon storage in soil with harmonized analysis of decomposition data. Geoderma 441:116761

Bala R, Kalia A, Dhaliwal SS (2019) Evaluation of efficacy of ZnO nanoparticles as remedial zinc nanofertilizer for rice. J Soil Sci Plant Nutr 19(2):379–389

Bapat HD, Manahan SE (1998) Chemchar gasification of hazardous wastes and mixed wastes on a biochar matrix. In: Abstracts of papers of the American Chemical Society, vol 215. American Chemical Society, 1155 16th ST, NW, Washington, DC 20036 USA, pp U571–U571

Barati M, Bakhtiari F, Mowla D, Safarzadeh S (2018) Comparison of the effects of poultry manure and its biochar on barley growth in petroleum-contaminated soils. Int J Phytorem 20(2):98–103

Bashir A, Rizwan M, ur Rehman MZ, Zubair M, Riaz M, Qayyum MF, Alharby HF, Bamagoos AA, Ali S (2020) Application of co-composted farm manure and biochar increased the wheat growth and decreased cadmium accumulation in plants under different water regimes. Chemosphere 246:125809

Bekchanova M, Campion L, Bruns S, Kuppens T, Lehmann J, Jozefczak M, Malina R (2024) Biochar improves the nutrient cycle in sandy-textured soils and increases crop yield: a systematic review. Environ Evid 13(1):3

Bello A, Wang B, Zhao Y, Yang W, Ogundeji A, Deng L, Egbeagu UU, Yu S, Zhao L, Li D, Xu X (2021) Composted biochar affects structural dynamics, function and co-occurrence network patterns of fungi community. Sci Total Environ 775:145672

Berek AK, Hue NV, Radovich TJ, Ahmad AA (2018) Biochars improve nutrient phyto-availability of Hawai'i's highly weathered soils. Agronomy 8(10):203

Bhattacharyya SS, Ros GH, Furtak K, Iqbal HM, Parra-Saldívar R (2022) Soil carbon sequestration–an interplay between soil microbial community and soil organic matter dynamics. Sci Total Environ 152928

Bilias F, Kalderis D, Richardson C, Barbayiannis N, Gasparatos D (2023) Biochar application as a soil potassium management strategy: a review. Sci Total Environ 858:159782

Billa SF, Angwafo TE, Ngome AF (2019) Agro-environmental characterization of biochar issued from crop wastes in the humid forest zone of Cameroon. Int J Recycl Org Waste Agric 8(1):1–13

Bista P, Ghimire R, Machado S, Pritchett L (2019) Biochar effects on soil properties and wheat biomass vary with fertility management. Agronomy 9(10):623

Bolan N, Sarmah AK, Bordoloi S, Bolan S, Padhye LP, Van Zwieten L, Siddique KH (2023a) Soil acidification and the liming potential of biochar. Environ Pollut 317:120632

Bolan S, Hou D, Wang L, Hale L, Egamberdieva D, Tammeorg P, Bolan N (2023b) The potential of biochar as a microbial carrier for agricultural and environmental applications. Sci Total Environ 886:163968

Bonanomi G, Ippolito F, Cesarano G, Nanni B, Lombardi N, Rita A, Scala F (2017) Biochar as plant growth promoter: better off alone or mixed with organic amendments? Front Plant Sci 8:1570

Breulmann M, van Afferden M, Fühner C (2015) Biochar: bring on the sewage. Nature 518:483

Burrow BJ (2023) Biochar impacts on soil health, and plant growth. A case study of smallholder farms in Malawi (Master's thesis, Norwegian University of Life Sciences)

Campion L, Bekchanova M, Malina R, Kuppens T (2023) The costs and benefits of biochar production and use: a systematic review. J Clean Prod 408:137138

Carey DE, McNamara PJ, Zitomer DH (2015) Biochar from pyrolysis of biosolids for nutrient adsorption and turf grass cultivation. Water Environ Res 87(12):2098–2106

Chen S, Yang M, Ba C, Yu S, Jiang Y, Zou H, Zhang Y (2018) Preparation and characterization of slow-release fertilizer encapsulated by biochar-based waterborne copolymers. Sci Total Environ 615:431–437

Castillo D (2021) The Amazon Rainforest—500 Years of Exploitation in Brazil's Forests. Retrieved on 29 October 2022 from https://onetribeglobal.com/climate-action-news/the-amazon-rainforest-500-years-of-exploitation-in-brazils-forests/

Chakraborty V, Das P (2024) Investigation on efficiency of synthesized lanthanum oxide–coated biochar and graphene oxide–coated biochar on removal of fluoride: batch and fixed bed continuous reactor performance modelling. Biomass Convers Biorefinery 14(5):6507–6520

Chelsea Green Publishing (2022) How to make Biochar. Retrieved on 2 November 2022 from https://www.chelseagreen.com/2019/how-to-make-biochar/

Chen X, Wu W, Han L, Gu M, Li J, Chen M (2022) Carbon stability and mobility of ball milled lignin-and cellulose-rich biochar colloids. Sci Total Environ 802:149759

Chen X, Qian S, Ma Y, Zhu J, Shen S, Tang J, Zhang Z (2023) Efficient degradation of sulfamethoxazole in various waters with peroxymonosulfate activated by magnetic-modified sludge biochar: surface-bound radical mechanism. Environ Pollut 319:121010

Çığ F, Sönmez F, Nadeem MA, Sabagh AE (2021) Effect of biochar and pgpr on the growth and nutrients content of einkorn wheat (*Triticum monococcum* L.) and post-harvest soil properties. Agronomy 11(12):2418

Cyclic Carbon (2020) Biochar versus other forms of carbon. Retrieved on 12 October 2022 from https://greenmanchar.com.au/blogs/the-green-mans-blog/biochar-versus-other-forms-of-carbonhttps://greenmanchar.com.au/blogs/the-green-mans-blog/biochar-versus-other-forms-of-carbon

Czekała W, Jeżowska A, Chełkowski D (2019) The use of biochar for the production of organic fertilizers. J Ecol Eng 20(1)

Das SK, Ghosh GK (2020) Soil health management through low-cost biochar technology. In: Biochar applications in agriculture and environment management. Springer, Cham, pp 193–206

Dawar K, Fahad S, Jahangir MMR et al (2021) Biochar and urease inhibitor mitigate NH_3 and N_2O emissions and improve wheat yield in a urea fertilized alkaline soil. Sci Rep 11:17413. https://doi.org/10.1038/s41598-021-96771-0

Díaz B, Sommer-Márquez A, Ordoñez PE, Bastardo-González E, Ricaurte M, Navas-Cárdenas C (2024) Synthesis methods, properties, and modifications of biochar-based materials for wastewater treatment: a review. Resources 13(1):8

Dike CC, Hakeem IG, Rani A, Surapaneni A, Khudur L, Shah K, Ball AS (2022) The co-application of biochar with bioremediation for the removal of petroleum hydrocarbons from contaminated soil. Sci Total Environ 157753

de Figueiredo CC, Chagas JK, da Silva J, Paz-Ferreiro J (2019) Short-term effects of a sewage sludge biochar amendment on total and available heavy metal content of a tropical soil. Geoderma 344:31–39

Drewer JMZ, Abiven S, Hagemann N, Schmidt HP (2023) Permanence of soil applied biochar: conclusions from the natural pyrogenic carbon cycle validate carbon sink accounting. In: EGU general assembly conference abstracts, pp EGU–11054

Edwards T (2022) Carbon farming: applying biochar to increase soil carbon. https://www.agric.wa.gov.au/soil-carbon/carbon-farming-applying-biochar-increase-soil-carbon

El-Bassi L, Azzaz AA, Jellali S, Akrout H, Marks EA, Ghimbeu CM, Jeguirim M (2021) Application of olive mill waste-based biochars in agriculture: Impact on soil properties, enzymatic activities and tomato growth. Sci Total Environ. 755:142531

El Moussaoui H, Idardare Z, Bouqbis L (2024) Effect of integrated soil fertility management on water retention, productivity and physiological parameters of Alfalfa (*Medicago sativa*) without and under deficit irrigation. Sci Hortic 327:112816

Elmer WH, Pignatello JJ (2011) Effect of biochar amendments on mycorrhizal associations and Fusarium crown and root rot of asparagus in replant soils. Plant Dis 95(8):960–966

Elouafi I (2024) Why biodiversity matters in agriculture and food systems. Science 386(6718):eads8197

Ercan B, Alper K, Ucar S, Karagoz S (2023) Comparative studies of hydrochars and biochars produced from lignocellulosic biomass via hydrothermal carbonization, torrefaction and pyrolysis. J Energy Inst 109:101298

Ernsting, A., & Smolker, R. (2009). Biochar for climate change mitigation: fact or fiction. *Agrofuels and the Myth of the Marginal Lands*, 1–10.

Fang X, Zhang M, Zheng P, Wang H, Wang K, Lv J, Shi F (2024) Biochar-bacteria-plant combined potential for remediation of oil-contaminated soil. Front Microbiol 15:1343366

Feng W, Yang F, Cen R, Liu J, Qu Z, Miao Q, Chen H (2021) Effects of straw biochar application on soil temperature, available nitrogen and growth of corn. J Environ Manage 277:111331

Ferreira CSS, Keesstra S, Destouni G, Solomun MK, Kalantari Z (2024) Soil degradation in the Mediterranean region: drivers and future trends. In: Environmental sustainability in the mediterranean region: challenges and solutions. Springer International Publishing, Cham, pp 81–112

Gabhane JW, Bhange VP, Patil PD, Bankar ST, Kumar S (2020) Recent trends in biochar production methods and its application as a soil health conditioner: a review. SN Appl Sci 2(7):1–21

Gao ZW, Ding J, Ali B, Nawaz M, Hassan MU, Ali A, Sabagh AE (2024) Putting biochar in action: a black gold for efficient mitigation of salinity stress in plants. Review and future directions. ACS Omega 9(29):31237–31253

Ghadirnezhad Shiade SR, Fathi A, Minkina T, Wong MH, Rajput VD (2024) Biochar application in agroecosystems: a review of potential benefits and limitations. Environ Dev Sustain 26(8):19231–19255

Godbey H (2016) Biochar, a brief history. Retrieved on 3 November 2022 from https://www.permaculturenews.org/2016/11/18/biochar-brief-history/

Godwin PM, Pan Y, Xiao H, Afzal MT (2019) Progress in preparation and application of modified biochar for improving heavy metal ion removal from wastewater. J Bioresour Bioprod 4(1):31–42

Goldan E, Nedeff V, Sandu IG, Mosnegutu E, Panainte M (2019) Study of greenhouse use of biohazard wastewater and manure compost. Revista de Chimie 70(1)

Gopal M, Gupta A, Thomas GV (2013) Food for thought: do soil microbes need food too? Indeed, lest we don't need ours. Curr Sci 902–907

Guo M, Song W, Tian J (2020) Biochar-facilitated soil remediation: mechanisms and efficacy variations. Front Environ Sci 183

Gupta DK, Gupta CK, Dubey R, Fagodiya RK, Sharma G, Mohamed N, Shukla AK (2020) Role of biochar in carbon sequestration and greenhouse gas mitigation. In: Biochar applications in agriculture and environment management. Springer, Cham, pp 141–165

Haider FU, Khan I, Farooq M, Cai L, Li Y (2024) Co-application of biochar and plant growth regulators improves maize growth and decreases Cd accumulation in cadmium-contaminated soil. J Clean Prod 440:140515

Hannet G, Singh K, Fidelis C, Farrar MB, Muqaddas B, Bai SH (2021) Effects of biochar, compost, and biochar-compost on soil total nitrogen and available phosphorus concentrations in a corn field in Papua New Guinea. Environ Sci Pollut Res 28:27411–27419.

Hornung A, Stenzel F, Grunwald J (2021) Biochar-just a black matter is not enough. Biomass Convers Biorefinery 1–12

Hossain MZ, Bahar MM, Sarkar B, Donne SW, Ok YS, Palansooriya KN, Bolan N (2020) Biochar and its importance on nutrient dynamics in soil and plant. Biochar 2(4):379–420

Huang J, Zimmerman AR, Chen H, Gao B (2020) Ball milled biochar effectively removes sulfamethoxazole and sulfapyridine antibiotics from water and wastewater. Environ Pollut 258:113809

Ilango A, Lefebvre O (2016) Characterizing properties of biochar produced from simulated human feces and its potential applications. J Environ Qual 45(2):734–742

İlay R (2024) Biochar production from various low-cost marine wastes using different production methods: characterization of biochar and marine feedstock for agricultural purposes. Mar Pollut Bull 205:116623

Iqbal H, Garcia-Perez M, Flury M (2015) Effect of biochar on leaching of organic carbon, nitrogen, and phosphorus from compost in bioretention systems. Sci Total Environ 521:37–45

Institute for Carbon Removal Law and Policy (2018) Soil carbon and biochar. Retrieved on 6 November 2022 from https://www.american.edu/sis/centers/carbon-removal/upload/icrlp_fact_sheet_soil_carbon_biochar_181006.pdf

International Institute of Sustainable Development (IISD) (2020) World Population to Reach 9.9 billion by 2050. Retrieved on 2 November 2022 from https://sdg.iisd.org/news/world-population-to-reach-9-9-billion-by-2050/#:~:text=The%202020%20World%20Population%20Data%20Sheet%20indicates%20that%20world%20population,more%20than%2025%25%20from%202020

Itam DH (2020) The use of biochar for bioremediation of crude oil/hydrocarbon polluted soils (vol 1)–bioremediation of polluted soils, studies, results and possibilities. Hydrocarb Polluted Soils 1

Kapoor RT, Zdarta J (2024) Fabrication of engineered biochar for remediation of toxic contaminants in soil matrices and soil valorization. Chemosphere 142101

Kaur S, Roy A (2021) Bioremediation of heavy metals from wastewater using nanomaterials. Environ Dev Sustain 23(7):9617–9640

Khan S, Irshad S, Mehmood K, Hasnain Z, Nawaz M., Rais A, Ibrar D (2024) Biochar production and characteristics, its impacts on soil health, crop production, and yield enhancement: a review. Plants 13(2):166

King R (2013) Biochar: a brief history and developing future. Retrieved on 20 October 2022 from https://news.mongabay.com/2013/01/biochar-a-brief-history-and-developing-future/

Kostyniuk A, Likozar B (2024) Wet torrefaction of biomass waste into value-added liquid product (5-HMF) and high quality solid fuel (hydrochar) in a nitrogen atmosphere. Renew Energy 226:120450

Koulouri ME, Templeton MR, Fowler GD (2024) Enhancing the nitrogen and phosphorus content of faecal-derived biochar via adsorption and precipitation from human urine. J Environ Manage 352:119981

Kudakwashe MEKI, Qiang LIU, Shuai WU, Yanfei YUAN (2022) Plant-and microbe-assisted biochar amendment technology for petroleum hydrocarbon remediation in saline-sodic soils: a review. Pedosphere 32(1):211–221

Kumari R, Kumar V, Kumar A (2024) Soil and biochar: attributes and actions of biochar for reclamation of soil and mitigation of soil borne plant pathogens. J Soil Sci Plant Nutr 1–16

Lamberti E, Viscusi G, Kiani A, Boumezough Y, Acocella MR, Gorrasi G (2024) Efficiency of dye adsorption of modified biochar: a comparison between chemical modification and ball milling assisted treatment. Biomass Bioenerg 185:107247

Lan W, Zhao X, Wang Y, Jin X, Ji J, Cheng Z, Chen G (2024) Research progress of biochar modification technology and its application in environmental remediation. Biomass Bioenergy 184:107178

Li Y, Gupta R, Zhang Q, You S (2023) Review of biochar production via crop residue pyrolysis: development and perspectives. Biores Technol 369:128423

Li S, Skelly S (2023) Physicochemical properties and applications of biochars derived from municipal solid waste: a review. Environ Adv 100395.

Lian F, Xing B (2017) Black carbon (biochar) in water/soil environments: molecular structure, sorption, stability, and potential risk. Environ Sci Technol 51(23):13517–13532

Liang J, Li Y, Si B., Wang Y, Chen X, Wang X, Biswas A (2021) Optimizing biochar application to improve soil physical and hydraulic properties in saline-alkali soils. Sci Total Environ 771:144802

Liu Y, Paskevicius M, Wang H, Parkinson G, Veder JP, Hu X, Li CZ (2019) Role of O-containing functional groups in biochar during the catalytic steam reforming of tar using the biochar as a catalyst. Fuel 253:441–448

Liu Y, Tie B, Peng O, Luo H, Li D, Liu S, Du H (2020) Inoculation of Cd-contaminated paddy soil with biochar-supported microbial cell composite: a novel approach to reducing cadmium accumulation in rice grains. Chemosphere 247:125850

Liu D, Ding Z, Ali EF, Kheir A, Eissa MA, Ibrahim OH (2021) Biochar and compost enhance soil quality and growth of roselle (*Hibiscus sabdariffa* L.) under saline conditions. Sci Rep 11(1):1–11

Liu Q, Cui H, Yang W, Wang F, Liao H, Zhu Q, Lu P (2024) Soil conditioner improves soil properties, regulates microbial communities, and increases yield and quality of *Uncaria rhynchophylla*. Sci Rep 14(1):13398

Lombardi C (2009) Manmade biomass coal offers storage and fuel. Retrieved on 10 October 2022 from https://www.cnet.com/culture/manmade-biomass-coal-offers-storage-and-fuel/

Luo X, Liu G, Xia Y, Chen L, Jiang Z, Zheng H, Wang Z (2017) Use of biochar-compost to improve properties and productivity of the degraded coastal soil in the Yellow River Delta, China. J Soils Sediments 17:780–789

Lyu H, Gao B, He F, Zimmerman AR, Ding C, Huang H, Tang J (2018) Effects of ball milling on the physicochemical and sorptive properties of biochar: experimental observations and governing mechanisms. Environ Pollut 233:54–63

Mahawar N, Bamboriya JS, Rupesh T, Meena AK, Athnere S, Chiranjeeb K, Naik BSSS (2020) Legumes: a boon to keep soil alive and protect biodiversity. Int J Curr Microbiol App Sci (11):1103–1114

Mahmoud E, El-Beshbeshy T, El-Kader NA et al (2019) Impacts of biochar application on soil fertility, plant nutrients uptake and maize (*Zea mays* L.) yield in saline sodic soil. Arab J Geosci 12:719 (2019). Retrieved on 20 October 2022 from https://doi.org/10.1007/s12517-019-4937-4

Majumder S, Neogi S, Dutta T, Powel MA, Banik P (2019) The impact of biochar on soil carbon sequestration: meta-analytical approach to evaluating environmental and economic advantages. J Environ Manage 250:109466

Malik L, Sanaullah M, Mahmood F, Hussain S, Siddique MH, Anwar F, Shahzad T (2022) Unlocking the potential of co-applied biochar and plant growth-promoting rhizobacteria (PGPR) for sustainable agriculture under stress conditions. Chem Biol Technol Agric 9(1):1–29

Manikandan S, Vickram S, Subbaiya R, Karmegam N, Chang SW, Ravindran B, Awasthi MK (2023) Comprehensive review on recent production trends and applications of biochar for greener environment. Bioresour Technol 129725

Mansoor S, Kour N, Manhas S, Zahid S, Wani OA, Sharma V, Ahmad P (2021) Biochar as a tool for effective management of drought and heavy metal toxicity. Chemosphere 271:129458

Manuschevich D, Pfeiffer M, Perez-Quezada J (2023) Soil degradation and land cover change in Latin America. In: Routledge handbook of Latin America and the environment. Routledge, pp 23–35

Maroušek J, Minofar B, Maroušková A, Strunecký O, Gavurová B (2023) Environmental and economic advantages of production and application of digestate biochar. Environ Technol Innov 30:103109

Mašek O, Brown RC, Bakshi S (2024) Biochar production technology. In: Biochar for environmental management. Routledge, pp 85–110

Melo LCA, Sánchez-Monedero MÁ (2024) How biochar-based fertilizers and biochar compost affect nutrient cycling and crop productivity. Nutr Cycl Agroecosystems 1–4

Meng F, Rong G, Zhao R, Chen B, Xu X, Qiu H, Zhao L (2024) Incorporating biochar into fuels system of iron and steel industry: carbon emission reduction potential and economic analysis. Appl Energy 356:122377

Minnikova T, Kolesnikov S, Ruseva A, Kazeev K, Minkina T, Mandzhieva S, Sushkova S (2022) Influence of the biochar on petroleum hydrocarbon degradation intensity and ecological condition of Haplic Chernozem. Eurasian J Soil Sci 11(2):157–166

Mishra G, Danoglidis PA, Shah SP, Konsta-Gdoutos MS (2023) Carbon capture and storage potential of biochar-enriched cementitious systems. Cement Concr Compos 140:105078

Mohamed I, Farid IM, Siam HS, Abbas MH, Tolba M, Mahmoud SA, Bassouny MA (2024) A brief investigation on the prospective of co-composted biochar as a fertilizer for Zucchini plants cultivated in arid sandy soil. Open Agric 9(1):20220322

Mona S, Malyan SK, Saini N, Deepak B, Pugazhendhi A, Kumar SS (2021) Towards sustainable agriculture with carbon sequestration, and greenhouse gas mitigation using algal biochar. Chemosphere 275:129856

Mounissamy VC, Parihar RS, Dwivedi AK, Saha JK, Rajendiran S, Lakaria BL, Patra AK (2021) Effects of Co-composting of municipal solid waste and pigeon pea biochar on heavy metal mobility in soil and translocation to leafy vegetable spinach. Bull Environ Contam Toxicol 106:536–544

Muhammad H, Wei T, Cao G, Yu S, Ren X, Jia H, Li Y (2021) Study of soil microorganisms modified wheat straw and biochar for reducing cadmium leaching potential and bioavailability. Chemosphere 273:129644

Nagappan P (2021) Soil microbes and carbon emissions: the weather factor. https://newscenter.sdsu.edu/sdsu_newscenter/news_story.aspx?sid=78418

Naz MY, Sulaiman SA (2016) Slow-release coating remedy for nitrogen loss from conventional urea: a review. J Control Release, 225:109–120

Neogi S, Sharma V, Khan N, Chaurasia D, Ahmad A, Chauhan S, Bhargava PC (2021) Sustainable biochar: a facile strategy for soil and environmental restoration, energy generation, mitigation of global climate change and circular bioeconomy. Chemosphere 133474

Nerome M, Toyota K, Islam T, Nishijima T, Matsuoka T, Sato K, Yamaguchi Y (2005) Suppression of bacterial wilt of tomato by incorporation of municipal biowaste charcoal into soil. Retrieved on 19 October 2022 from https://worldveg.tind.io/record/1648/

Nguyen TB, Tran DPH, Nguyen VT, Vo TDH, Nguyen HV, Tra VT, Bui XT (2024) Utilization of modified biochar for removal of dyes from industrial effluent. In: Catalytic applications of biochar for environmental remediation: a green approach towards environment restoration, vol 1. American Chemical Society, pp 23–44

Nicholas HL, Devine A, Robertson I, Mabbett I (2024) Perspective chapter: agronomic properties of biochar from slow pyrolysis of human waste. In: Sustainable use of biochar-from basics to advances: from basics to advances, vol 109

Ogawa M (2009) Charcoal use in agriculture in Japan. In: First Asia Pacific biochar conference

Oldfield TL, Sikirica N, Mondini C, López G, Kuikman PJ, Holden NM (2018) Biochar, compost and biochar-compost blend as options to recover nutrients and sequester carbon. J Environ Manag 218:465–476

Oni BA, Oziegbe O, Olawole OO (2019) Significance of biochar application to the environment and economy. Ann Agric Sci 64(2):222–236

Ottani F, Morselli N, De Luca A, Puglia M, Pedrazzi S, Allesina G (2023) The conductivity dilemma: How biochar grain's chemical composition and morphology hinder the direct measurement of its electrical conductivity. Measurement 222:113662

Ouyang X, Yin H, Yu X, Guo Z, Zhu M, Lu G, Dang Z (2021) Enhanced bioremediation of 2, 3′, 4, 4′, 5-pentachlorodiphenyl by consortium GYB1 immobilized on sodium alginate-biochar. Sci Total Environ 788:147774

Pandit NR, Schmidt HP, Mulder J, Hale SE, Husson O, Cornelissen G (2020) Nutrient effect of various composting methods with and without biochar on soil fertility and maize growth. Arch Agron Soil Sci

Pandian K, Vijayakumar S, Mustaffa MRAF, Subramanian P, Chitraputhirapillai S (2024) Biochar–a sustainable soil conditioner for improving soil health, crop production and environment under changing climate: a review. Front Soil Sci 4:1376159

Pereira AKS, Cupertino GFM, da Silva ÁM, Oliveira TR, de Souza MP, Delatorre FM, Dias Júnior AF (2023) Perspectives and challenges of world charcoal production in technological, social, and climate change fields. In: Forest science: sustainable processes and wood products, pp 1–23

Petrus HT, Putera AD, Wangi IP, Ramadhian MA, Setiawan H, Prasetya A (2020) Characterization of nitrogen release in modified controlled-release-fertilizer using rice husk biochar. Int J Technol 11(4):774–783

Priyanka Wood IE, Al-Gailani A, Kolosz BW, Cheah KW, Vashisht D, Taylor MJ (2024) Cleaning up metal contamination after decades of energy production and manufacturing: reviewing the value in use of biochars for a sustainable future. Sustainability 16(20):8838

Radin R, Abu Bakar R, Ishak CF, Ahmad SH, Tsong LC (2018) Biochar-compost mixture as amendment for improvement of polybag-growing media and oil palm seedlings at main nursery stage. Int J Recycl Org Waste Agric 7:11–23

Rasool M, Akhter A, Soja G, Haider MS (2021) Role of biochar, compost and plant growth promoting rhizobacteria in the management of tomato early blight disease. Sci Rep 11(1):6092

Rathee J, Sharma A, Kaur M, Nandal K (2024) Biomass derived bio-chars: unlocking the potential for advanced energy storage. In: Materials for boosting energy storage: advances in sustainable energy technologies, vol 2. American Chemical Society, pp 153–175

Rawat J, Saxena J, Sanwal P (2019) Biochar: a sustainable approach for improving plant growth and soil properties. In: Biochar-an imperative amendment for soil and the environment, pp 1–17

Razzaghi F, Obour PB, Arthur E (2020) Does biochar improve soil water retention? A systematic review and meta-analysis. Geoderma 361:114055

Rombel A, Krasucka P, Oleszczuk P (2021) Sustainable biochar-based soil fertilizers and amendments as a new trend in biochar research. Sci Total Environ 151588

Roy A, Bharadvaja N (2021) Efficient removal of heavy metals from artificial wastewater using biochar. Environ Nanotechnol Monit Manag 16:100602

Ruley JA, Amoding A, Tumuhairwe JB, Basamba TA, Opolot E, Oryem-Origa H (2020a) Enhancing the phytoremediation of hydrocarbon-contaminated soils in the Sudd wetlands, South Sudan, using organic manure. Appl Environ Soil Sci

Ruley JA, Tumuhairwe JB, Amoding A, Westengen OT, Vinje H (2020b) Rhizobacteria communities of phytoremediation plant species in petroleum hydrocarbon contaminated soil of the Sudd ecosystem, South Sudan. Int J Microbiol

Ruley JA, Amoding A, Tumuhairwe JB, Basamba TA (2022) Rhizoremediation of petroleum hydrocarbon–contaminated soils: a systematic review of mutualism between phytoremediation species and soil living microorganisms. Phytoremediation 263–296

Ruley JA (2024) Crop residue management practices for sustaining soil health. In: Zero waste management technologies. Springer Nature Switzerland, Cham, pp 213–226

Saeed M, Ilyas N, Jayachandran K, Shabir S, Akhtar N, Shahzad A, Bano A (2022) Advances in Biochar and PGPR engineering system for hydrocarbon degradation: A promising strategy for environmental remediation. Environ Pollut 119282

Safarian S (2023) Performance analysis of sustainable technologies for biochar production: a comprehensive review. Energy Rep 9:4574–4593

Samoraj M, Mironiuk M, Witek-Krowiak A, Izydorczyk G, Skrzypczak D, Mikula K, Chojnacka K (2022) Biochar in environmental friendly fertilizers-Prospects of development products and technologies. Chemosphere 296:133975

Sanchez-Reinoso AD, Ávila-Pedraza EÁ, Restrepo-Díaz H (2020) Use of biochar in agriculture. Acta Biológica Colombiana 25(2):327–338

Sarfraz R, Hussain A, Sabir A, Ben Fekih I, Ditta A, Xing S (2019) Role of biochar and plant growth promoting rhizobacteria to enhance soil carbon sequestration-a review. Environ Monit Assess 191(4):1–13

Shetty R, Prakash NB (2020) Effect of different biochars on acid soil and growth parameters of rice plants under Aluminum toxicity. Sci Rep 10(1):1–10

Shi W, Ju Y, Bian R, Li L, Joseph S, Mitchell DR, Munroe P, Taherymoosavi S, Pan G (2020) Biochar bound urea boosts plant growth and reduces nitrogen leaching. Sci Total Environ 701:134424

Shoudho KN, Khan TH, Ara UR, Khan MR, Shawon ZBZ, Hoque ME (2024) Biochar in global carbon cycle: towards sustainable development goals. Curr Res Green Sustain Chem 100409

Silva LC, Corrêa RS, Wright JL, Bomfim B, Hendricks L, Gavin DG, Santos RV (2021) A new hypothesis for the origin of Amazonian Dark Earths. Nature Commun 12(1):1–11

Singh C, Tiwari S, Singh JS (2020) Biochar: a sustainable tool in soil pollutant bioremediation. In: Bharagava R, Saxena G (eds) Bioremediation of industrial waste for environmental safety. Springer, Singapore

Six J (2014) Biochar: is there a dark side? Retrieved on 5 November 2022 from https://ethz.ch/en/news-and-events/eth-news/news/2014/04/biochar-is-there-a-dark-side.html

Song L, Niu X, Zhang N, Li T (2021) Effect of biochar-immobilized *Sphingomonas sp.* PJ2 on bioremediation of PAHs and bacterial community composition in saline soil. Chemosphere 279:130427

Sonsuphab K, Toomsan W, Soontharo S, Supanchaiyamat N, Hunt AJ, Ngernyen Y, Siripattanakul-Ratpukdi S (2024) Integrated remediation and detoxification of triclocarban-contaminated water using waste-derived biochar-immobilized cells by long-term column experiments. Environ Pollut 357:124456

Sousa B, Soares C, Sousa F, Martins M, Mateus P, Rodrigues F, Fidalgo F (2024) Enhancing tomato plants' tolerance to combined heat and salt stress–the role of arbuscular mycorrhizae and biochar. Sci Total Environ 948:174860

Souza TRC, Kubeneck S, Warken A, Paliga L, Binotto D, Colla LM, Treichel H (2021) Synergistic approaches (use of biochar and microbes) in the bioremediation of industrial effluents. In: Biochar and its application in bioremediation. Springer, Singapore, pp 91–105

Spokas KA, Cantrell KB, Novak JM, Archer DW, Ippolito JA, Collins HP, Nichols KA (2012) Biochar: a synthesis of its agronomic impact beyond carbon sequestration. J Environ Qual 41(4):973–989

Stefanie S (2018) What is biochar? Retrieved on 1 November 2022 from https://regenerationinternational.org/2018/05/16/what-is-biochar/

Su ZJ, Zhang YM, Xu HW, Xu WM, Liu C, Rui S, Xiang P (2024) Preparation and applications of iron/biochar composites in remediation of heavy metal contaminated soils: current status and further perspectives. Environ Technol Innov 35:103671

Tan X, Zhang F, Wang H, Ho SH (2023) The magic of algae-based biochar: advantages, preparation, and applications. Bioengineered 14(1):2252157. https://doi.org/10.1080/21655979.2023.2252157

Tao Y, Feng W, He Z, Wang B, Yang F, Nafsun AI, Zhang Y (2024) Utilization of cotton byproduct-derived biochar: a review on soil remediation and carbon sequestration. Environ Sci Eur 36(1):79

Tao Q, Chen Y, Zhao J, Li B, Li Y, Tao S, Wang C (2019) Enhanced Cd removal from aqueous solution by biologically modified biochar derived from digestion residue of corn straw silage. Sci Total Environ 674:213–222

Tawfik A, Alhajeri NS, Nasr M (2024) Synergistic interaction of biochar, Phragmites australis, and rhizosphere microorganisms for enhanced 1, 4-dioxane biodegradation. J Water Process Eng 66:105964

Thompson KA, Shimabuku KK, Kearns JP, Knappe DR, Summers RS, Cook SM (2016) Environmental comparison of biochar and activated carbon for tertiary wastewater treatment. Environ Sci Technol 50(20):11253–11262

Tusar HM, Uddin MK, Mia S, Suhi AA, Wahid SBA, Kasim S, Anwar F (2023) Biochar-acid soil interactions—a review. Sustainability 15(18):13366

Umair Hassan M, Huang G, Munir R, Khan TA, Noor MA (2024) Biochar Co-compost: a promising soil amendment to restrain greenhouse gases and improve rice productivity and soil fertility. Agronomy 14(7):1583

United States Biochar Initiative (USBI) (2021) Biochar: then and now. Retrieved on 2 November 2022 from https://biochar-us.org/biochar-then-now

Valizadeh S, Lee SS, Baek K, Choi YJ, Jeon BH, Rhee GH, Park YK (2021) Bioremediation strategies with biochar for polychlorinated biphenyls (PCBs)-contaminated soils: a review. Environ Res 200:111757

Verma VC, Acharya S, Kumar R, Verma BC, Singh A, Tiwari VK (2022) Rhizobium as soil health engineer. In: Rhizosphere Engineering. Academic Press, pp 77–95

Villardon A, Alcazar-Ruiz A, Dorado F, Sanchez-Silva L (2024) Enhancing carbon dioxide uptake in biochar derived from husk biomasses: optimizing biomass particle size and steam activation conditions. J Environ Chem Eng 12(5):113352

Wang C, Gu L, Ge S, Liu X, Zhang X, Chen X (2019a) Remediation potential of immobilized bacterial consortium with biochar as carrier in pyrene-Cr (VI) co-contaminated soil. Environ Technol 40(18):2345–2353

Wang Y, Villamil MB, Davidson PC, Akdeniz N (2019b) A quantitative understanding of the role of co-composted biochar in plant growth using meta-analysis. Sci Total Environ 685:741–752

Wang F, Wang X, Song N (2021) Biochar and vermicompost improve the soil properties and the yield and quality of cucumber (*Cucumis sativus* L.) grown in plastic shed soil continuously cropped for different years. Agric Ecosyst Environ 315:107425

Wang L, Chen D, Zhu L (2023a) Biochar carbon sequestration potential rectification in soils: Synthesis effects of biochar on soil CO_2, CH_4 and N_2O emissions. Sci Total Environ 904:167047

Wang Y, Yu G, Xie S, Jiang R, Li C, Xing Z (2023b) Pyrolysis of food waste digestate residues for biochar: Pyrolytic properties, biochar characterization, and heavy metal behaviours. Fuel 353:129185

Wang X, Zhao R, Wu H, Jia X, Liu Y, Zhou G, Hu S (2024) Enhanced bioremediation of hexavalent chromium via Stenotrophomonas acidaminiphila 4–1 assisted with agricultural wastes-derived biochar. Biochem Eng J 208:109355

Warnock DD, Lehmann J, Kuyper TW, Rillig MC (2007) Mycorrhizal responses to biochar in soil–concepts and mechanisms. Plant Soil 300(1):9–20

Wei Y, Wang J, Chang R, Zhan Y, Wei D, Zhang L, Chen Q (2021) Composting with biochar or woody peat addition reduces phosphorus bioavailability. Sci Total Environ 764:142841

Włodarczyk R, Kwarciak-Kozłowska A (2021) Biochar for bioremediation of toxic metals. In: Handbook of assisted and amendment: enhanced sustainable remediation technology, pp 119–130

Wu B, Wang Z, Zhao Y, Gu Y, Wang Y, Yu J, Xu H (2019) The performance of biochar-microbe multiple biochemical material on bioremediation and soil micro-ecology in the cadmium aged soil. Sci Total Environ 686:719–728

Wu C, Zhi D, Yao B, Zhou Y, Yang Y, Zhou Y (2022) Immobilization of microbes on biochar for water and soil remediation: a review. Environ Res 212:113226

Wu J, Fu X, Zhao L, Lv J, Lv S, Shang J, Ma F (2024) Biochar as a partner of plants and beneficial microorganisms to assist in-situ bioremediation of heavy metal contaminated soil. Sci Total Environ 171442

Xia H, Wang J, Riaz M, Babar S, Li Y, Wang X, Jiang C (2024) Co-application of biochar and potassium fertilizer improves soil potassium availability and microbial utilization of organic carbon: a four-year study. J Clean Prod 469:143211

Xu L, Wu KC, Deng ZN, Huang CM, Verma KK, Pang T, Huang HR (2024) Biochar and its impact on soil profile and plant development. J Plant Interact 19(1):2401356

Yan L, Song X, Miao J, Ma Y, Zhao T, Yin M (2024) Removal of tetracycline from water by adsorption with biochar: a review. J Water Process Eng 60:105215

Yang Y, Qian X, Alamu SO, Brown K, Lee SW, Kang DH (2024) Qualities and quantities of poultry litter biochar characterization and investigation. Energies 17(12):2885

Yang R, Yang Q (2024) A review of emerged constructed wetlands based on biochar filler: wastewater purification and carbon sequestration/greenhouse gas reduction. Environ Eng Res 29(2)

Yang Q, Mašek O, Zhao L, Nan H, Yu S, Yin J, Cao X (2021) Country-level potential of carbon sequestration and environmental benefits by utilizing crop residues for biochar implementation. Appl Energy 282:116275

Yang X, Zhang S, Ju M, Liu L (2019) Preparation and modification of biochar materials and their application in soil remediation. Appl Sci 9(7):1365

Ye Z, Zhang L, Huang Q, Tan Z (2019) Development of a carbon-based slow-release fertilizer treated by bio-oil coating and study on its feedback effect on farmland application. J Cleaner Prod 239:118085

Yu H, Zou W, Chen J, Chen H, Yu Z, Huang J, Gao B (2019) Biochar amendment improves crop production in problem soils: a review. J Environ Manag 232:8–21

Yue Y, Jin X, Deng L (2023) Experimental study on properties of syngas, tar, and biochar derived from different gasification methods. Appl Sci 13(20):11490

Zahed MA, Salehi S, Madadi R, Hejabi F (2021) Biochar as a sustainable product for remediation of petroleum contaminated soil. Curr Res Green Sustain Chem 4:100055

Zahra MB, Fayyaz B, Aftab ZE, Haider MS (2021) Mitigation of degraded soils by using biochar and compost: a systematic review. J Soil Sci Plant Nutr 21(4):2718–2738

Zeng L, Ma J, Yang J, Yang J, Zeng X, Zhou Y (2024) Ball milling nano-sized biochar: bibliometrics, preparation, and environmental application. Environ Sci Pollut Res 1–16

Zhang S, Wei L, Trakal L, Wang S, Shaheen SM, Rinklebe J, Chen Q (2024) Pyrolytic and hydrothermal carbonization affect the transformation of phosphorus fractions in the biochar and hydrochar derived from organic materials: a meta-analysis study. Sci Total Environ 906:167418
Zhang B, Zhang L, Zhang X (2019a) Bioremediation of petroleum hydrocarbon-contaminated soil by petroleum-degrading bacteria immobilized on biochar. RSC Advances 9(60):35304–35311
Zhang M, Riaz M, Zhang L, El-Desouki Z, Jiang C (2019b) Biochar induces changes to basic soil properties and bacterial communities of different soils to varying degrees at 25 mm rainfall: more effective on acidic soils. Front Microbiol 10:1321
Zhao T, Ali A, Su J, Liu S, Yan H, Xu L (2024) Removal of sulfamethoxazole from water by biosurfactant-modified sludge biochar: Properties and mechanism. J Environ Chem Eng 12(6):114200
Zhang BG, Myers DE, Wallace GG, Brandt M, Choong PF (2014) Bioactive coatings for orthopaedic implants—recent trends in development of implant coatings. Int J Mol Sci 15(7):11878–11921
Zheng X, Xu W, Dong J, Yang T, Shangguan Z, Qu J, Tan X (2022) The effects of biochar and its applications in the microbial remediation of contaminated soil: a review. J Hazard Mater 129557
Zhu X, Luo Z, Zhang Q, He M, Tsang DC (2023) Valorization of slow pyrolysis vapor from biomass waste: comparative study on pyrolysis characteristics, evolved gas evaluation, and adsorption effects. Biores Technol 386:129543
Zinck JA, Montero PG (2023) Soil properties, formation, distribution, and classification. Psammic peinobiomes: nutrient-limited ecosystems of the upper Orinoco and Rio Negro basins. Springer International Publishing, Cham, pp 239–294

Chapter 2
Physical and Chemical Properties of Biochar

Ab Waheed Wani, C. Kiran, Sanjeev Kumar, Atul Khalangre, Meraj Ahmed, Monica Sharma, Harjinder Kaur, Princy Thakur, M. Radha Jogita, Anis A. Mirza, and Batulla Rashid

Abstract Biochar, a carbon-rich, porous material, is produced when organic biomass is pyrolysed. A sustainable solution for the management of illness and waste is the production of biochar. Due to the fact that it includes 50% of the original carbon, which is incredibly refractory, its synthesis aids in carbon sequestration by storing the carbon already present in plant biomass. When determining the basic properties of biochar and forecasting the numerous applications uses, both physical and chemical characterizations are required. Biochar is made from a variety of biomass with various chemical and physical properties, resulting in a wide range of compounds with various properties. The pH and electrical conductivity (EC) of the soil rise with the addition of biochar. The ash residue contains mostly silica, heavy metals, organic and inorganic nitrogen, and carbonates of alkaline and alkaline earth metals, suggesting that this may be the case. To assess the fundamental properties of biochar and determine its multiple application uses, physical and chemical characterizations are needed. Application of biochar improves soil fertility and quality by boosting soil pH, improving moisture retention, attracting more beneficial fungi and microorganisms, promoting cation exchange capacity, and preserving nutrients in the soil. This section of the book will go over the various physical and chemical properties of biochar as well as how they change depending on the feedstock.

A. W. Wani (✉) · S. Kumar · A. Khalangre · H. Kaur · M. Radha Jogita · A. A. Mirza
Department of Horticulture, Lovely Professional University, Jalandhar, Punjab, India
e-mail: waheedwani1@gmail.com

C. Kiran · M. Ahmed · P. Thakur
Department of Soil Science and Agricultural Chemistry, Lovely Professional University, Jalandhar, Punjab, India

M. Sharma
Department of Agronomy, Lovely Professional University, Jalandhar, Punjab, India

B. Rashid
Indian Institute of Integrative Medicine (IIIM) Srinagar, CSIR, Srinagar, India

S. Kumar
Department of Fruit Science, Dr. Yashwant Singh Parmar University of Horticulture and Forestry, Nauni, Solan, India

A. Roy et al. (eds.), *Biochar Production Engineering*,
https://doi.org/10.1007/978-3-031-92594-8_2

Keywords Biochar · Chemical and physical properties · pH and EC · Nutrients

1 Introduction

Plant biomass is pyrolysed in anaerobic settings to form biochar, an organic amendment rich in carbon and organic material with a pH that is suitable for usage in soil. Biochar is light, incredibly porous, finely grained, and all of these characteristics. Its colour is also black. In addition to the fundamental elements of carbon, hydrogen, sulphur, oxygen, and nitrogen, the ash fraction also contains minerals. Because of the spatial arrangement of carbon atoms in biochar, it is a highly stable high-carbon substance. Both a sustainable fuel and a soil amendment can be made from it (Campion et al. 2023; Koul et al. 2022). Biochar is biodegradable and rich in total and organic carbon. It also has the correct ratios of micro and macro elements, including copper, zinc, iron, potassium, sodium, magnesium, and calcium.

Porosity, pH, specific surface area, and surface functional group content are all highly elevated in biochar. It is very porous, with numerous longitudinal pores ranging in size from micro to macropores. Macropores produced from vascular bundles of raw biomass are crucial for enhancing soil quality because they can house symbiotic bacteria (Akpinar et al. 2023; Koul et al. 2022; Brar et al. 2024). According to Chaturvedi et al. (2024) and Nguyen et al. (2023), the condensed aromatic carbon structure of biochars can manifest itself in a variety of ways, including turbo static C (predominant at higher temperatures), graphite C, and amorphous C (dominant at low pyrolysis temperatures). Khan et al. (2024) found that the most striking chemical difference between biochar and other organic substances, such as lignin, is a substantially larger concentration of aromatic C and condensed aromatic structures in biochar.

2 Physical Properties of Biochar

2.1 Biochar Molecular Structure

The distribution of locally crystalline phases is influenced by the feedstock supply, and the carbonization process frequently enlarges any pre-existing micro-crystalline clusters (Tsolis and Barouchas 2023). Coal, charcoal, and coke are examples of common C-based materials that typically exhibit crystalline sections in their localised regions. Typically, these are graphite-based turbo stratified layers (Ahmad et al. 2023; Mou et al. 2023). However, X-ray diffraction investigations on biochar have revealed that it is largely an amorphous substance (Chaturvedi et al. 2024; Singh et al. 2022; Li et al. 2024). Elbarbary et al. (2023) discovered locally generated, highly conjugated aromatic compounds. In many ways, these structures mimic graphene-like structures. These tiny structures serve as trustworthy conductors despite their modest size (Maher and Torquato 2024). The defects (dislocations and vacancies) in the C lattice can be intercalated with metal and mineral compounds, and biochar that has been pyrolysed at temperatures lower than 500 °C has been said to have semiconductor properties (Ding et al. 2024). The majority of the biochar structure is made up of non-conductive pores, complicated aromatic-aliphatic chemical compounds, and minerals. As HTT grows, biochars exhibit higher crystallinity percentages. Layering in the turbostratum becomes more organised (Chaturvedi et al. 2024). In comparison with graphite, turbostratic regions with a wider planar spacing are frequently found in biochar (Mao et al. 2024; Anand et al. 2023). Although it is simple to identify the long-range two-dimensional order of these planes, it is impossible to detect the third direction of the crystallographic ordering (Mao et al. 2024). These substances are known as non-graphitic carbons.

The average HTT used to make biochar, however, is far lower than this and typically doesn't rise above 600 °C. Functional groups and oxygen atoms frequently collect along the edges of these organised C sheets (Ribeiro et al. 2024). When compared to the equilibrium interplanar distance of graphite, these species increase the interplanar distance of the C sheets by steric or electronic effects (Anand et al. 2023). Greater HTTs appear to facilitate the development of poly-aromatic crystallite clusters, as inferred from the drop in the H/C atomic ratio, according to (Emenike et al. 2024). Occasionally, graphite-like layers can be arranged crystallographically at the borders of these C layers (Ribeiro et al. 2024). Such arrangement may result in localised interstices and gaps inside hexagonal planes (Vasseghian et al. 2024). Tars, volatiles, or other products of decomposition can all obstruct these pores as well as any extra ones that could be present. As the HTT is increased, these chemicals volatilize, making the pore channels appear cleaner (Wood et al. 2024).

2.2 Atomic Proportions

Carbonization alters the chemical composition of fuel, primarily by dissociating functional groups. This is true for both artificial carbonizations used to manufacture charcoal and spontaneous organic and geochemical carbonization. The release of these hydrogen and oxygen-containing groups causes a decrease in the corresponding carbon-to-hydrogen ratios. As a result, changes in atomic ratios can reveal how the carbonization process evolves. The van-Krevelen diagram can be used to represent the evolution of the three basic fuel elements during carbonization: carbon, oxygen, and hydrogen (Anand et al. 2023). Until hard coal is formed, oxygen is emitted at nearly twice the rate of hydrogen during the natural carbonization process. The ensuing conversion into anthracite is distinguished by a decrease in the H/C ratio and a nearly constant, low oxygen concentration.

The greatest decline in atomic ratios occurs at temperatures ranging from 250 to 350 °C. A temperature drop of only about 100 °C is required to reduce the O/C ratio from 0.7 to 0.3. However, increasing the temperature from 0.3 to nearly zero requires an extra 700 °C. The same is true for H/C ratios; raising the temperature during treatment from 250 to 350 °C reduces the ratio from 1.5 to 0.8. The practically full release of hydrogen from biochar necessitates extremely high treatment temperatures. It is clear that the decrease in atomic ratios does not occur linearly with treatment temperature, but rather at different speeds.

2.3 *Composition of the Elements*

One of the primary purposes of biochar production is to change the chemical makeup of raw biomass, particularly to increase the carbon content. This is the result of the separation of the functional groups of oxygen and hydrogen. As a result, increasing the reaction temperature raises the concentration of carbon while decreases the concentrations of hydrogen and oxygen. Wood-based biochars represent the evolution of carbon and oxygen content. Untreated wood normally has slightly more than 50% carbon and slightly more than 40% oxygen (In terms of weight, devoid of dry ash).

High-temperature biochars may contain fewer than 5% oxygen and more than 95% carbon. During pyrolysis, the hydrogen concentration in wood decreases from 5 to 7% to $< 2\%$ or even $< 1\%$ at extremely high treatment temperatures. Regardless of the many species or process conditions studied, the carbon, oxygen, and hydrogen contents of woody biochars alter surprisingly uniformly with increasing temperature. Overall, there does not appear to be a substantial link between treatment temperature and nitrogen content. When data from the same feedstock and acquired under identical conditions are compared, it is discovered that nitrogen concentration increases somewhat with increasing temperature. This proportionate rise is due to the decrease in other components during devolatilization.

2.4 *Capability*

During carbonization, the thermal breakdown of the biomass structure results in the dissociation of functional groups and the release of oxygen and hydrogen. As a result, biochars with low H/C ratios (i.e. a higher level of carbonization) have fewer functional groups and more aromatic structures as compared to low-temperature chars. Because of their high thermal stability, aromatic structures are ideal for specialised applications such as soil amendment or metallurgical purposes, where biochar must remain stable over time. The feedstock (the chemical structures from which the chars originate) and process variables, most notably the treatment temperature, impact the number of aromatic structures formed during the carbonization process.

2.5 Distribution of Particle Sizes, Specific Gravity, and Dry Density

Because of its porous nature, biochar has a high value when used in agriculture. Its micropores may increase soil fertility by adsorbing plant nutrients and releasing them into the soil solution for plant absorption (Zhang et al. 2024). Macropores may aid in soil aeration and water infiltration, creating an environment conducive to microbial growth (Zhang et al. 2024; Razzak 2024). In addition to aiding soil water retention, biochar's porous nature provides a safe haven for fungus and bacteria from protist and microarthropod predation (Kumar et al. 2023). In charcoal, bacterial and fungal hyphae colonies have been identified (Gujre et al. 2021; Akpinar et al. 2023). Pore characteristics of biochars made with different feedstocks and temperatures. The overall impact of biochar application on soil environment is described in Fig. 1.

The naturally high porosity of biochar particles, (ii) soil particle rearrangement in interaction with biochar particles, (iii) better soil aggregation, and (iv) increased bioturbation can all influence the pore size distribution of the soil (e.g. earthworm activity). Biochar particles, due to their porosity, can even temporarily improve soil pore size distribution (Wei et al. 2023). ASTM D 422 and ASTM D 854 were utilised to define the particle size distribution and specific gravity of dry biochar samples, respectively. The Harvard small compaction test apparatus was used to evaluate the

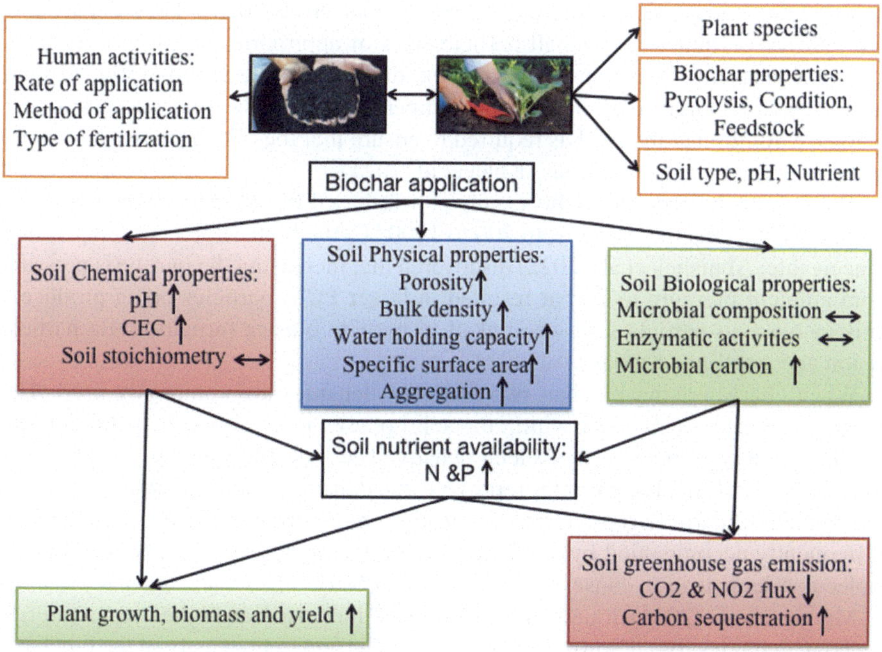

Fig. 1 Effect of biochar on soil physio-chemical, biological properties of soil

dry density utilising Wilson's suggested test protocol (Humboldt Mfg. Co. in 1970). The dried biochar sample was weighed before being placed in the empty Harvard small mould, and each layer was crushed with five strokes. Following the filling of the mould, biochar samples were balanced on top, and the weight of the biochar-filled mould was recorded. The dry densities of the biochars were estimated using the volume of the mould and the weight of the biochar compacted into it. The features of the original feedstock have a substantial influence on both the size distribution of biochar particles and the pore parameters. Shrinkage, attrition, and post-pyrolysis processing phases, on the other hand, will alter the initial feedstock size distribution. The pyrolysis and processing conditions determine the strength of these changes. The link between the particle size distribution of the starting material and the end biochar.

As the Highest Treatment Temperature (HTT) was increased, the particle size distribution of the resultant biochars exhibited a trend towards smaller particle sizes. This was due to changes in the mechanical properties of the biochar caused by the processing conditions. The heating rate has an effect on the particle size distribution as well. Smaller feedstock particles are required to achieve faster heating rates during pyrolysis due to heat and mass transmission constraints. For example, although continuous slow pyrolysis systems may handle feedstocks with diameters of several centimetres, fast pyrolysis techniques require very small feedstock particle sizes (sub-mm to mm). Traditional charcoal production technologies are capable of handling massive feedstocks such as forestry logs. To pyrolyse the biomass, several heating rates and particle sizes were used (Kumar et al. 2023). They determined that pyrolysis requires very small particle sizes of approximately 50 m to provide sufficient mass and heat transmission at the highest heating speeds (500 °C s^{-1}). Heat transfer to larger particles can be hampered by rapid heating rates. A longer residence duration at the HTT is required to ensure that the bigger particles respond appropriately in these conditions (Khan et al. 2024).

The feedstock particles (in this example, peat-based PCM) reduced by up to 20% when the HTT was increased from 200 to 1000 °C, the highest process temperature conceivable. Abhishek et al. (2022) discovered that increasing the pyrolysis pressure from ambient pressure to 20 bar resulted in larger PCM particles being produced. This increase in particle size was linked to particle cluster formation via particle fusion and swelling (Kumar et al. 2023).

When characterising biochar, two forms of densities are commonly used. The molecular density of biochar within the solid phases is measured by solid density, also known as real density or skeletal density. The bulk (or apparent) density of a larger volume of biochar particles represents the density of the solid phases, particle pores, and interparticle gaps. This relationship is often inverse: a rise in solid density is frequently accompanied by a reduction in bulk density. At HTTs below 700 °C, molecular densification causes an increase in porosity.

Myung et al. (2024) found that as porosity increased at higher temperatures, apparent densities rose at lower solid densities. The apparent density of the biochar's increased at very high HTT levels, while their porosity decreased due to the pores collapsing. Mohamed et al. (2023) discovered an inverse relationship between eucalyptus biochar solid and bulk density. Because graphite phases have better structural packing than the original feedstocks, the loss of volatiles from the unstructured fraction of the feedstock, along with an increase in the percentage of graphite-like crystalline phases, results in an increase in solid density.

2.6 *Testing of Hydraulic Characteristics and Leachability*

After being oven-dried for a full night at 60 °C, the biochar samples were tested for field capacity or water-holding ability (WHC). A known volume of deionised water was gently sprayed onto the biochar, and gravity was allowed to remove the water after roughly three hours. The ultimate moisture content of the biochar was calculated gravimetrically in accordance with ASTM D2216. WHC was computed by combining the sample's saturation moisture content and the relative proportion of water passing through the biochar sample after accounting for moisture absorbed by the filter paper. This method was previously used in a biochar characterisation investigation and was effective in determining the field capacity of biochar (Anand et al. 2023). Hydraulic conductivity was measured using the ASTM D2434 constant-head test methodology. Each biochar's hydraulic conductivity (k) was calculated by averaging the results from three distinct constant-head levels. Deionised water was used during the first pore volume to collect and evaluate the leachate for contaminants such as heavy metals and PAHs that could potentially leach.

2.7 *Surface Area Specific*

The surface area of the material is determined by both the temperature at which it was created and the type of feedstock used in its synthesis. Biochar has a surface area ranging from 1.4 to 500 m^2 g^1. At 300 °C, pine sawdust was pyrolysed, yielding a lower barrier of 1.4 m^2 g^1. Park et al. (2019) discovered that pyrolysis of Douglas fir wood at 600 °C yielded a maximum of 500 m^2 g^1.

The pyrolysis temperature affects biochar's surface area and porosity. This is presumably related to organic component breakdown and micropore formation (Anand et al. 2023). Degrading aliphatic alkyls and ester groups and exposing the aromatic lignin core may enhance surface area (Kumar et al. 2023). According to Myung et al. (2024), lignin does not transform into a hydrophobic PAH below 500 °C, whereas biochar does.

At temperatures exceeding 650 °C, biochar becomes more hydrophobic and thermally stable (Myung et al. 2024). However, it is unknown how the surface's hydrophobicity or hydrophilicity influences the sorption of aromatic compounds. The surface area of biochar expands as the temperature rises. Because pore-blocking chemicals are pushed aside or thermally broken as the pyrolysis temperature rises, the surface area exposed to the environment increases. By gradually decomposing organic components (such as cellulose and lignin) and forming vascular bundles or channel structure, pyrolysis can increase surface area and pore volume. As a result of the cellulose being destroyed during the pyrolysis process, some amorphous carbon structures are formed (Nguyen et al. 2023). Micropores have been discovered in amorphous carbon frameworks.

Biochar's pore size dispersion is proportional to its overall surface area. HTT creates more pores, increasing the surface area of the biochar. As the interplanar distances of aromatic C forms decrease, the surface area per volume increases with increasing ordering. Biochar has the ability to create pores of varying widths. However, the language used to describe these pores has caused considerable dispute in the literature. Notably, all pores < 0.2 m in diameter are referred to as micropores, whereas all pores greater than 10 m in diameter are referred to as macropores. A micropore is described in this chapter as a pore having a diameter of 2 nm or less, which is the most common diameter used by material scientists to estimate surface areas. These pores considerably increase the surface area of the biochars. As a result, biochar has excellent adsorptive properties and can absorb tiny molecules such as gases and liquids. Mesopores and macropores, with sizes ranging from 2 to 50 nm and greater than 50 nm, respectively.

Surface area has a substantial impact on soil features such as water, air, and nutrient cycling, as well as microbial activity. Sandy soils, e.g. are ineffective at retaining water and plant nutrients due to their small surface area. Despite having a low amount of water available to plants and poor aeration, clayey soils have a huge surface area and a high-water retention capacity. According to Brar et al. (2024), organic matter in soils can help with sandy and clayey soils that retain too much or too little water. Biochar has been proposed to affect the physical qualities of soil in a similar way. Microbial activity and populations may change as the surface area of biochars in soil increases. The various physical properties and implications of biochar for soil and environmental applications are depicted in Table 1.

2.8 Bulk Density

Bulk density is a measurement of the soil's health and level of compaction. It has an impact on crucial soil functions and processes, such as soil aeration, infiltration, water availability, plant nutrient availability, and soil microbial activity. The impact of temperature influences various properties of biochar depicted in Fig. 2.

Table 1 Overview of biochar's physical properties and their implications for soil and environmental applications

Property	Description	References
Molecular structure	Biochar primarily has an amorphous structure with localised crystalline phases resembling graphene-like structures. High-temperature treatment (HTT) increases crystallinity and organises turbostratic layers	Tsolis and Barouchas (2023), Mao et al. (2024) and Anand et al. (2023)
Atomic proportions	Carbonization reduces H/C and O/C atomic ratios, with rapid changes at 250–350 °C and slower declines thereafter. Higher HTT results in low H/C ratios, reflecting greater aromaticity	Anand et al. (2023) and Emenike et al. (2024)
Elemental composition	Carbon content increases with HTT, while oxygen and hydrogen content decrease. Wood-based biochars exhibit carbon content over 95% and oxygen below 5% at high temperatures	Kumar et al. (2023) and Zhang et al. (2024)
Capability	Aromatic structures increase with HTT, enhancing biochar's stability for applications like soil amendment. Functional groups dissociate with carbonization, leaving more aromatic structures	Kumar et al. (2023) and Ribeiro et al. (2024)
Particle size, density	Biochar has porous structures with micropores aiding nutrient adsorption and macropores enhancing soil aeration and water infiltration. These characteristics promote microbial growth	Gujre et al. (2021) and Akpinar et al. (2023)
Hydraulic characteristics	Field capacity and water-holding capacity (WHC) are determined gravimetrically. Pyrolysis increases porosity and improves WHC. Hydraulic conductivity and leachability are tested under ASTM standards	Anand et al. (2023) and Zhang et al. (2024)
Specific surface area	Surface area ranges from 1.4 to 500 m^2/g, increasing with HTT. Pore-blocking compounds volatilize, enhancing exposed surface area. Micropores (< 2 nm) contribute significantly to adsorption properties	Myung et al. (2024) and Nguyen et al. (2023)
Bulk density	Biochar's bulk density affects soil functions like aeration, infiltration, water, and nutrient availability. It is crucial for evaluating soil health and compaction levels	Brar et al. (2024)

3 Chemical Properties

3.1 Carbon Content and Ash Content

Pyrolysis temperature increases the carbon content of biochar. As seen by the high-carbon concentration, the biochars still retain some of the original organic plant remains, including cellulose (Yan et al. 2024). The ash content increased by 5.7–18.7% as the pyrolysis temperature was elevated found by Elbarbary et al. (2023). Nguyen et al. (2023) claim that the reason for the rise in ash content was the growing

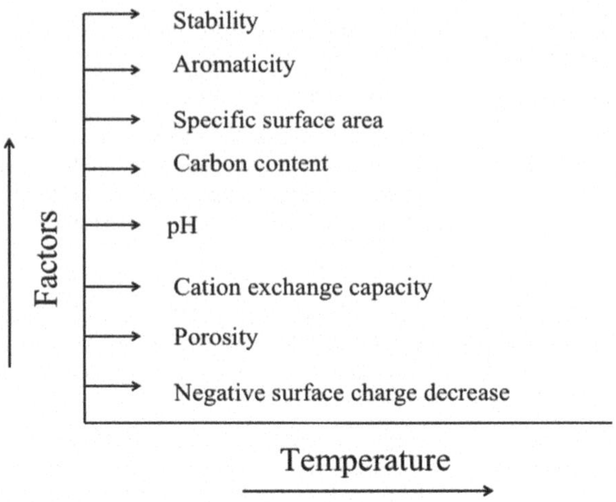

Fig. 2 Influences of temperature on the properties of biochar

concentration of inorganic elements and OM combustion by-products. Also, the rise in Mg, Ca, K, and P on biochars pyrolysed at high temperatures is caused by an increase in ash content ranging from 4.0 to 33.1%. Biochar still contains the minerals that result in ash following carbonization (Kumar et al. 2023). The carbon content of biochar increases between 62.2 and 92.4% as the pyrolysis temperature rises due to a higher degree of polymerization, which results in a more compacted carbon structure. The important chemical properties of biochar materials are presented in Table 2.

3.2 pH

The pH of biochars is influenced by the presence of inorganic alkalis and carbonate production. The pH is alkaline for the reasons listed above. It has been discovered that total base cation and carbonate concentrations rise with rising temperature, resulting in pH values between 6.5 and 10.8. Ash content and oxygen functional group increases caused by pyrolysis have been linked to greater pH levels with increasing temperature (Yan et al. 2024; Razzak 2024; Nguyen et al. 2023). Other factors to take into account include the disappearance of basic functional groups (–COOH) and the advent of acidic functional groups. However, the primary reason pH rises are due to alkali ions starting to dissociate from organic molecules as the pyrolysis temperature increases (Kumar et al. 2023). Alkali salts begin to dissociate from the organic matrix at temperatures above 300 °C, raising the pH of the end product. At temperatures between 200 and 300 °C, the breakdown of cellulose and hemicelluloses releases organic acids and phenolic chemicals, lowering the pH of the products. When all of the alkali salts have been liberated from the pyrolytic framework, the pH stabilises at roughly 600 °C.

Table 2 Overview of the chemical properties of biochar, highlighting variations in carbon content, pH, and cation exchange capacity based on pyrolysis conditions and feedstock types

Property	Key observations	References
Carbon content and ash content	- Carbon content increases with pyrolysis temperature, ranging between 62.2% and 92.4% - Ash content increases by 5.7–18.7% due to combustion of organic matter and concentration of inorganic elements	Yan et al. (2024), Elbarbary et al. (2023), Nguyen et al. (2023) and Kumar et al. (2023)
pH	- Influenced by alkalis and carbonate production - Ranges between 6.5 and 10.8, increasing with temperature - Alkalinity varies with feedstock and pyrolysis conditions	Yan et al. (2024), Razzak (2024), Li et al. (2024) and Kumar et al. (2023)
Cation exchange capacity (CEC)	• Varies with feedstock type: - Chicken manure: 81.4 cmol/kg - Pig manure: 32.7 cmol/kg - Sugarcane bagasse: 122.0 cmol/kg - Non-wood biochars have higher CEC than wood-derived biochars	Abhishek et al. (2022), Yan et al. (2024) and Al-Rumaihi et al. (2022)

Biochars have an alkaline pH that ranges from 7.1 to 10.5. Li et al. (2024) found that when heated to 300 °C, biochar made from canola straw was acidic and alkaline biochar made from corn straw, peanuts, and soybeans (9.4, 8.6, and 7.7, respectively) (around 6.5). As an example of how biomass type can affect pH differences, biochar made from wood has an average pH that is 2 pH units lower than the values for other biomasses synthesised under similar pyrolysis conditions. The amount of lignin, hemicellulose, or cellulose in a biochar as well as the presence of oxygen compounds are likely to have an impact on its pH (Yan et al. 2024). The –COO– and –O–groups as well as the carbonate concentration, according to Li et al. (2024), were in charge of the biochars' alkaline properties.

3.3 CEC

The feedstock used to make biochar has an impact on its CEC. At 500 °C, pig manure-derived biochar had a lower CEC (32.7 cmol/kg) than chicken dung-derived biochar (81.4 cmol/kg) (Abhishek et al. 2022). The CEC of biochars made from paper mill waste, on the other hand, varied from 9.0 to 18.0 cmol/kg. Biochar made from sugar cane bagasse has a significantly lower CEC (122.0 cmol/kg). This discrepancy was discovered because biomass with a high ash content can create biochar with a higher CEC (Yan et al. 2024). A possible explanation for why manure-derived biochar has a greater CEC than woody biochar is because alkali and alkali metals in biomass enhance the synthesis of O-containing surface functional groups. According to Al-Rumaihi et al. (2022), the biochars made from orange pomace (29.9 cmol/kg) and chicken litter (48.4 cmol/kg) had the lowest CEC at 500 °C. The biochar made from algae (57.5 cmol/kg) and poultry litter (48.4 cmol/kg) had the highest CEC. Comparing non-wood biochars to wood biochars, non-wood biochars showed higher CEC and surface acidity.

4 Conclusion

Plant biomass is pyrolysed in anaerobic settings to form biochar, an organic amendment rich in carbon and organic material with a pH that is suitable for usage in soil. Biochar is light, incredibly porous, finely grained, and all of these characteristics. Its colour is also black. In addition to the fundamental elements of carbon, hydrogen, sulphur, oxygen, and nitrogen, the ash fraction also presents in the biochar. The preparation is done by half burning of the biomass. This is done to increase the carbon content in the biomass. In this chapter, we have discussed the various physical and chemical properties of the biochar. The Highest Treatment Temperature (HTT) is the deciding factor for the physical as well as the chemical composition of the biochar. By studying many papers on the biochar, it can be concluded that the biochar is the best soil amendment material as it is the best naturally produced material which have desired chemical and physical properties for soil amendment.

References

Abhishek K, Shrivastava A, Vimal V, Gupta AK, Bhujbal SK, Biswas JK et al (2022) Biochar application for greenhouse gas mitigation, contaminants immobilization and soil fertility enhancement: a state-of-the-art review. Sci Total Environ 853:158562

Ahmad S, Siddiqi WA, Ahmad S (2023) Sustainable nanocomposite porous absorbent and membrane sieves: definition, classification, history, properties, synthesis, applications, and future prospects. J Environ Chem Eng 11(2):109367

Akpinar D, Tian J, Shepherd E, Imhoff PT (2023) Impact of wood-derived biochar on the hydrologic performance of bioretention media: effects on aggregation, root growth, and water retention. J Environ Manage 339:117864

Al-Rumaihi A, Shahbaz M, Mckay G, Mackey H, Al-Ansari T (2022) A review of pyrolysis technologies and feedstock: a blending approach for plastic and biomass towards optimum biochar yield. Renew Sustain Energy Rev 167:112715

Anand A, Gautam S, Ram LC (2023) Feedstock and pyrolysis conditions affect suitability of biochar for various sustainable energy and environmental applications. J Anal Appl Pyrol 170:105881

Brar B, Saharan BS, Seth CS, Kamboj A, Bala K, Rajput VD et al (2024) Nanobiochar: soil and plant interactions and their implications for sustainable agriculture. Biocata Agric Biotechnol 57:103077

Campion L, Bekchanova M, Malina R, Kuppens T (2023) The costs and benefits of biochar production and use: a systematic review. J Clean Prod 408:137138

Chaturvedi K, Singhwane A, Dhangar M, Mili M, Gorhae N, Naik A et al (2024) Bamboo for producing charcoal and biochar for versatile applications. Biomass Convers Biorefinery 14(14):15159–15185

Ding Z, Huang R, Ouyang Z (2024) Scaling laws behind penetrative turbulence: history and perspectives. Adv Atmos Sci 41(10):1881–1900

Elbarbary AM, Sharaf El-Deen SEA, Abu Elgoud EM, Sharaf El-Deen GE (2023) Radiation fabrication of hybrid activated carbon and functionalized terpolymer hydrogel for sorption of Eu (III) and Sm (III) ions. Radiochim Acta 111(6):439–457

Emenike EC, Iwuozor KO, Ighalo JO, Bamigbola JO, Omonayin EO, Ojo HT et al (2024) Advancing the circular economy through the thermochemical conversion of waste to biochar: a review on sawdust waste-derived fuel. Biofuels 15(4):433–447

Gujre N, Soni A, Rangan L, Tsang DC, Mitra S (2021) Sustainable improvement of soil health utilizing biochar and arbuscular mycorrhizal fungi: a review. Environ Pollut 268:115549

Khan S, Irshad S, Mehmood K, Hasnain Z, Nawaz M, Rais A et al (2024) Biochar production and characteristics, its impacts on soil health, crop production, and yield enhancement: a review. Plants 13(2):166

Koul B, Yakoob M, Shah MP (2022) Agricultural waste management strategies for environmental sustainability. Environ Res 206:112285

Kumar A, Bhattacharya T, Shaikh WA, Roy A, Chakraborty S, Vithanage M, Biswas JK (2023) Multifaceted applications of biochar in environmental management: a bibliometric profile. Biochar 5(1):11

Li C, Liang D, Tian Y, Liu S, He W, Li Z et al (2024) Sorting out the latest advances in separators and pilot-scale microbial electrochemical systems for wastewater treatment: concomitant development, practical application, and future perspective. Environ Sci Technol 58(22):9471–9486

Maher CE, Torquato S (2024) Local order metrics for many-particle systems across length scales. Phys Rev Res 6(3):033262

Mao Q, Huang JP, Mu XR, Yang JD, Zhang YJ (2024) Accurate simulations of pure-viscoacoustic wave propagation in tilted transversely isotropic media. Pet Sci 21(2):866–884

Mohamed BA, Ruan R, Bilal M, Khan NA, Awasthi MK, Amer MA et al (2023) Co-pyrolysis of sewage sludge and biomass for stabilizing heavy metals and reducing biochar toxicity: a review. Environ Chem Lett 21(2):1231–1250

Mou Y, Wu X, Qin C, Chen J, Zhao Y, Jiang L et al (2023) Linkage microenvironment of azoles-related covalent organic frameworks precisely regulates photocatalytic generation of hydrogen peroxide. Angew Chem Int Ed 62(36):e202309480

Myung E, Kim H, Choi N, Cho K (2024) The biochar derived from *Spirulina platensis* for the adsorption of Pb and Zn and enhancing the soil physicochemical properties. Chemosphere 364:143203

Nguyen TB, Sherpa K, Bui XT, Nguyen VT, Chen CW, Dong CD (2023) Biochar for soil remediation: a comprehensive review of current research on pollutant removal. Environ Pollut 337:122571

Park YK, Jung JS, Jae J, Hong SB, Watanabe A, Kim YM (2019) Catalytic fast pyrolysis of wood plastic composite over microporous zeolites. Chem Eng J 377:119742

Razzak SA (2024) Municipal solid and plastic waste derived high-performance biochar production: a comprehensive review. J Anal Appl Pyrol 181:106622

Ribeiro TS, de Araújo Sobrinho I, Gonçalves MA, da Silva Lima V, Figueira BAM, da Rocha Filho GN, da Conceição LRV (2024) Green synthesis of biodiesel from magnetic basic biochar

derived from *Amazonian murici* residual biomass: optimization, kinetic, thermodynamic, and environmental studies. J Environ Chem Eng 12(6):114725
Singh H, Northup BK, Rice CW, Prasad PV (2022) Biochar applications influence soil physical and chemical properties, microbial diversity, and crop productivity: a meta-analysis. Biochar 4(1):8
Tsolis V, Barouchas P (2023) Biochar as soil amendment: the effect of biochar on soil properties using VIS-NIR diffuse reflectance spectroscopy, biochar aging and soil microbiology—a review. Land 12(8):1580
Vasseghian Y, Nadagouda MM, Aminabhavi TM (2024) Biochar-enhanced bioremediation of eutrophic waters impacted by algal blooms. J Environ Manage 367:122044
Wei B, Peng Y, Lin L, Zhang D, Ma L, Jiang L et al (2023) Drivers of biochar-mediated improvement of soil water retention capacity based on soil texture: a meta-analysis. Geoderma 437:116591
Wood R, Mašek O, Erastova V (2024) Developing a molecular-level understanding of biochar materials using public characterization data. Cell Rep Phys Sci 5(7):102036
Yan B, Jia X, Li J, Li Z, Che Y, Zhou Z et al (2024) In-situ elimination of biomass gasification tar based on the understanding of tar formation process: a review. J Energy Inst 112:101477
Zhang Y, Wang B, Hassan M, Zhang X (2024) Biochar coupled with multiple technologies for the removal of nitrogen and phosphorus from water: a review. J Environ Manage 370:122407

Chapter 3
Agricultural Wastes for Production of Biochar

Ravichandra Rangappa, Jyothi C Abbar, and M. Nagamadhu

Abstract Although agriculture activity is declining since 1970 to till date, it has been considered as one of the major contributor to the GDP for many of the African and Asian countries. In these countries, managing the waste and weed biomass produced by the agricultural activity has become one of the major concern. Crop Residue Open Burning has been a usual practice among farmers of Indian and Southeast Asian countries considering it as one of the quickest and least expensive methods of clearing land for the next crop. Due to the lack of proper waste management systems, Crop Residue Open Burning has been producing a large quantity of particulate matter emissions and air pollution. Lignocellulosic biomass (LB) is an abundant and renewable source available in the agricultural waste and weed biomass employed to produce the carbon rich biochar via pyrolysis process. The characterization of biochar has shown enhanced physicochemical properties than the hydrochars in terms of pH value, lower polarity and aromaticity. Through microbiological techniques as well as by enhancing the physical qualities of the soil, such as pH and base cation retention in the rooting zone, biochar may potentially increase crop production. Many studies have proven that one time application of biochar for agriculture soil can benefit to several cropping seasons for at least four years by improving soil fertility and productivity especially in humid tropical regions practicing industrial and subsistence agriculture. Thus, biochar is considered to be a beneficial tool for the management of waste of agriculture and weed biomass. The processes and equipment available for the continuous production of biochar from agricultural biomass are seeking more attention among the researchers these days. This chapter summarizes various feedstocks, processes, their parameters and the efficiency for the production of biochar.

Keywords Biochar · Pyrolysis · Agriculture waste · Lignocellulosic biomass · Crop residue open burning

R. Ravichandra · J. C. Abbar (✉) · M. Nagamadhu
BMS Institute of Technology and Management, Affiliated to Visvesvaraya Technological University, Belagavi, Bengaluru, Karnataka 560119, India
e-mail: jyothiabbar@gmail.com

R. Ravichandra
Atria Institute of Technology, Affiliated to Visvesvaraya Technological University, Belagavi, Bengaluru, Karnataka 560024, India

A. Roy et al. (eds.), *Biochar Production Engineering*,
https://doi.org/10.1007/978-3-031-92594-8_3

1 Introduction

The UN estimates that the population of world will increase to roughly 8.5 billion in 2030 and 9.7 billion by 2050 (UN 2016). Agriculture will have to generate 50% more food in order to keep up with the population's rising demand. In addition, for living, cotton, bamboo, and other wood requirements also will need to increase their production. Consequently, agriculture is now the largest employer globally and one of the most significant and substantial economic sectors for many nations (FAO 2017). Produce from agriculture is now used for more than just feeding the world's population; it is also used to feed animals and create biofuel for industrial purposes. In terms of electric energy conversion, Biomass gasifier and gas turbine with 35% efficiency produces 60 g/kWh in comparison with coal steam cycle with 33% efficiency produces 1000 g/kWh (Hazell and Pachauri 2006) thereby bioenergy has been considered as promising energy source. Thus, agriculture is also regarded as one of the highest producer of biomass production (Duque-Acevedo et al. 2020). According to the global snapshot of solid waste management to 2050, the total global waste generation would increase from 2.01 billion of tonnes (2016) to 3.4 billion of tonnes by 2050 (Awogbemi and Kallon 2023b; SilpaKaza et al. 2018). In the current scenario, out of total global waste, wood, food, and green being the major contributor with 46% by composition (Fig. 1).

This chapter provides a thorough explanation about the various agricultural wastes, information on biochar and its composition, properties, synthesis, and various

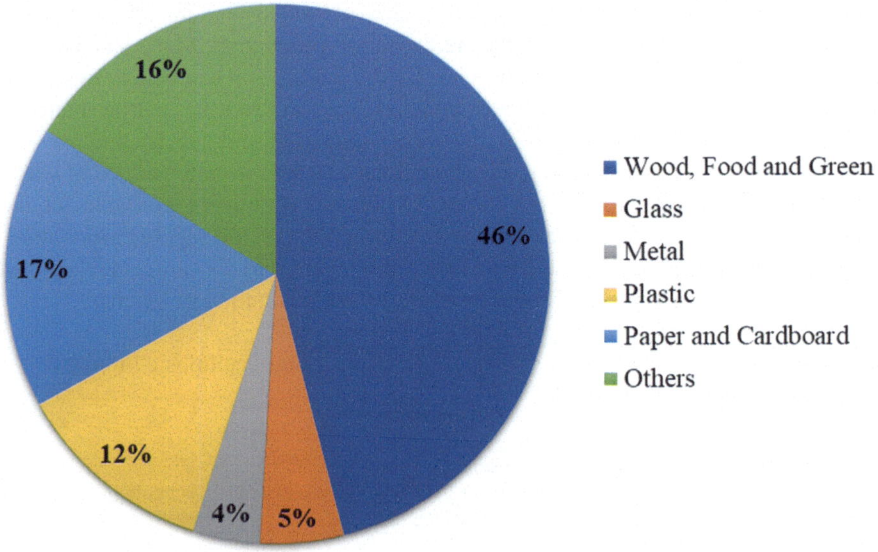

Fig. 1 Composition of global solid waste

factors that influence the synthesis. It also covers the importance of biochar production, as well as its effectiveness, applications in various fields, challenges and recommendations for overcoming the challenges.

2 Agricultural Waste as Biomass

Biomass is an inexhaustible carbon resource where its constituents are derived from animal and plants. Additional categories for biomass include virgin biomass and biomass waste. Typically, biomass waste produced from wood, leaves, peels, husks, trunk, frond, and nuts either from forestry or agricultural residues, carcass, and excreta from animals and humans is a topic of interest as they contribute to the environmental pollution. Although it is difficult for accurate estimation of food/agriculture waste per year, it has been evident that a total of 1.3 billion tonnes of food has been annually wasted which is meant for human consumption. It constitutes the one-third of the eatable part of the food. It has also created a negative impact on the environment (FAO 2017).

In most popular nations for agriculture produce, farmers generally adopt Crop Residue Open Burning (CROB) for easy and faster method to clear the land for next crop which has been a major contributing factor for environmental pollution. India is being the highest producer of agriculture waste in the Asia, producing 500 million tons/year of agricultural waste. Out of this, 92 million tons is burner every year. Similarly, farmers from three of Southeast Asian countries being listed in the World's Top 10 exporter of rise are practicing CROB as a convenient way to clean the agricultural land for more than decades. Although, CROB enriches the soil, heat produced due to burning can damage the soil structure and the microorganisms in the soil (Kim Oanh et al. 2018). CROB not only contributes to global environmental change it also disrupts rainfall patterns. According to a study on the state of the air in Asia, burning biomass is a major cause of air pollution, especially in India, China, Indonesia, and the nations of Mekong River Basin Sub-Region (Andini et al. 2018). Some of the major agricultural corps producing enormous waste which are potential feedstock for the production of biochar. Figure 2 illustrates various agricultural wastes used in the production of bio-gas, bio-oil, and bio-char through a pyrolysis reactor. Each of them are discussed in detail as below.

2.1 Wheat and Rice Husk

Wheat and rice are considered to be the highly consumed food commodities. Combining the both, current consumption is about 1250 Mt and projected consumption is expected to be 1400 Mt by 2030 (OECD-FAO 2021). As per FAO 2021 collectively wheat and paddy are 16% of the total agricultural produce, where China being first and India is second largest producer of wheat and rice (FAO 2021). In China,

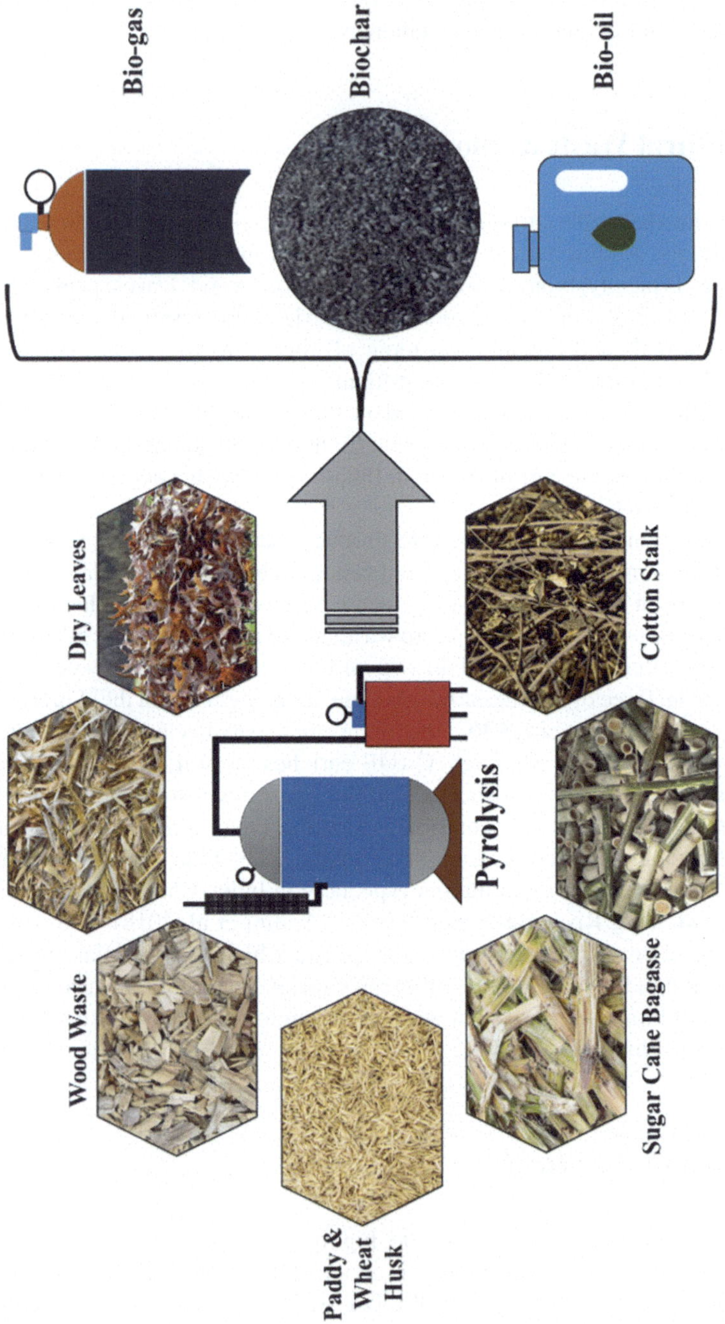

Fig. 2 Schematic representation of various agricultural wastes used as feedstock for pyrolysis

it is reported that 62% of the wheat and paddy straw were burnt to clear the land to cultivate the second crop (Ren et al. 2019). Similarly, under Indian context, 75% of paddy remains are burnt. However, 48% of wheat remains are burnt as another 45% is used as animal fodder (Singh et al. 2018). Eventually, burning wheat and paddy straw largely effects on environmental pollution. There exist huge amount of waste from wheat and rice that could be converted to useful by-product without affecting the environment.

2.2 *Sugarcane Bagasse*

As per OECD-FAO (2021), 250 Mt of sugar is consumed every year (OECD-FAO 2021) where major portion of the sugar is obtained from sugar cane (86%) and rest is by sugar beet (14%). Apart from edible sugar, both sugar cane and sugar beet are used for the production of ethanol one of the prominent biofuel. It is projected to see the production of 1960 Mt of sugar cane by 2030. Brazil being the highest producer with 38% followed by India being second with 27% total world production of sugar cane. Generally, before harvesting, to make the job easy dried leaves of sugar cane are burnt. Other than that, after the extraction of juice out of sugar cane, the bagasse would results in huge amount of agricultural bio-waste. Burning of bagasse would cause major impact on the environment.

2.3 *Maize/Corn Straw*

According to FAO, maize is considered to be the agricultural product with highest demand with its consumption as animal feed, food, fuel, and other applications. Current global consumption of maize is 1180 Mt and expected to cross 1300 Mt by 2030 (OECD-FAO 2021). Out of which nearly 700 Mt is used for animal feed, which will increase further to 800 Mt by 2030. China is the largest consumer of maize followed by United States and European Union. While production, China is the largest producer of maize which takes the 43% of global production share followed by United States with 43% and in the third place Brazil with 9% production share. China being highest in burning maize biomass followed by India and African countries. It can also be noticed that maize is the highest contributor among the Crop Residue Burning with 48% share (Cassou 2018). In order to control the crop burning and combat GHG emission an alternative measures to be taken to handle the biomass produced by maize/corn.

2.4 Cotton Stalk

Cotton is one among the natural fiber having huge industrial and domestic applications. According to Cotton and Wool Outlook 2022, projection of cotton production and mill use for 2022 is approximately 120 bales (1 bale = 480 pounds). India is the largest producer of cotton taking 25% of world market share followed by China with 22% and USA in third position with 15% of total global market share (OECD-FAO 2021). According to U.S. Department of Agriculture, United State alone is producing 16.5 bales of cotton during 2022 by cultivating across 12.2 million acres of land (Meyer and Dew 2022).

For many years, cotton crop yield has a major reduction due to a pest, Pink Ball Worm (PBW). To eradicate the ill effect of the pest, huge effort were put through insecticide, genetically modified cotton crop and by introducing sterile PBW. Due to which the sulk produced by cotton plants were with chemicals and sterile PBW which need to be disposed with care. Without a strong pest management strategy, farmers from many countries are finding open burning of crop residue is the easiest way to get rid of sterile PBW. According to some researchers, one of the greatest technological solutions for this issue is to process the cotton stalks waste through pyrolysis (Afif et al. 2012).

2.5 Other Agricultural and Wood Waste

Wood waste produced by the wood and furniture industries, fallen tree branches, truck, and other parts. In recent past, bamboo is considered as one of the major material and various products have been introduced made by bamboo. Especially, in Africa and India government authorities have provided special support for the development of bamboo-based products to mitigate the unemployment issues. Waste produced by bamboo can be potential bio-waste. Dried leaves from trees in the gardens and forests are also considered to be other bio-waste. Collectively, all these items can contribute to a huge quantity of biomass.

To mitigate the environmental emission, instead of burning the agricultural waste, reusing the same for useful conversion to energy source is most preferable. Direct combustion of agriculture waste has been one of the ancient method of reusing the waste to produce domestic heating requirements. Similarly, anaerobic digestion is also used to produce methane gas which can be further used as fuel. However, pyrolysis a process of heating the agriculture waste at 400–600 °C in the absence of oxygen to prepare char is most efficient way to convert the waste to biochar. Pyrolysis process is explored in details in this chapter to understand composition, properties, and factor affecting the synthesis of biochar and their applications.

Table 1 Physical and chemical properties of biochar

Physiochemical properties
Particle size
Area of surface
Functional groups
Distribution of pore size
Capacity to cation exchange
Carbon speciation and structure
Hydrophilicity/Hydrophobicity
Ash content

3 Biochar—Composition and Properties

No matter the process temperature or the feedstock, biochar which is rich in carbon, a product of pyrolysis that has all non-combustible components of the feedstock (ash), is always produced during pyrolysis and gasification. When using a slow pyrolysis method, biochar yield is increased, however under high-temperature quick pyrolysis conditions, biochar yield is decreased. Slow pyrolysis uses temperatures between 300 and 700 °C, heating rates between 1 and 100 °C/s, and pyrolysis timeframes between minutes to days (Qambrani et al. 2017; Woolley and Hallowell 2018).

The chemical composition and physical characteristics of biochar can vary greatly since it can be made in a variety of ways utilizing a variety of feedstocks. When analyzing biochar made from a single kind of feedstock, the amount and caliber of the biochar produced are affected by the pyrolysis temperature, time, and oxygen exposure. Nevertheless, there are developments in the production of biochar. Physicochemical properties listed in Table 1 are imbibed into the biochar products based on the parameter process and the feedstock type employed in the process. Aside from that, by adjusting the pyrolytic conditions and procedures, the yield of biogases, bio-oil, and biochar can be specifically targeted (Foong et al. 2020; Roy et al. 2021).

4 Conventional Pyrolysis

Biochar is obtained when biomass is heated to a temperature above 350 °C without any oxygen present. Pyrolysis is the name for the thermochemical procedure that turns biomass into biochar. Biochar is prepared using a variety of organic resources, including plant leftovers, wood, manure, animal carcasses, food wastes, biosolids, biowaste from industries, and wastes from municipal (Purakayastha et al. 2019). The pyrolysis procedure is said to produce during the heat reaction, the gases that are non-condensable and bio-oil that is combustible, while also preserving 1/3rd of the biomass carbon as the resultant char material. Because of the selective organic molecules removal and the structural changes made to the initial carbon framework, the pyrolysis product, biochar, has a specific area of surface and porosity rather than the initial biomass and is more resistant to biological degradation (Bolan et al. 2022).

As depicted in Fig. 3, the pyrolysis of biomass occurs in three steps. Since biomass contains some moisture, removing the moisture must come first. Then, lignocellulosic materials must be depolymerized and fragmented, which removes CO_2 and bound water. Finally, re-polymerization and re-condensation, step that leads to cracking and charring to form the biochar. The physiochemical properties of biochar are significantly influenced by process variables utilized during the pyrolysis process, viz. residence time, temperature, heating rate, pressure, and method (Table 2) (Foong et al. 2020).

Singh et al. (2010) in their study, they have synthesized eleven variants of biochars using five feedstocks as listed in Table 3. Slow pyrolysis was adopted with process temperature 400 and 550 °C with and without steam activation. Heating rate of 5–10 °C/min and residence time 40 min was used during the process and nitrogen was added during cooling-off period to create an environment that is inert (Singh et al. 2010).

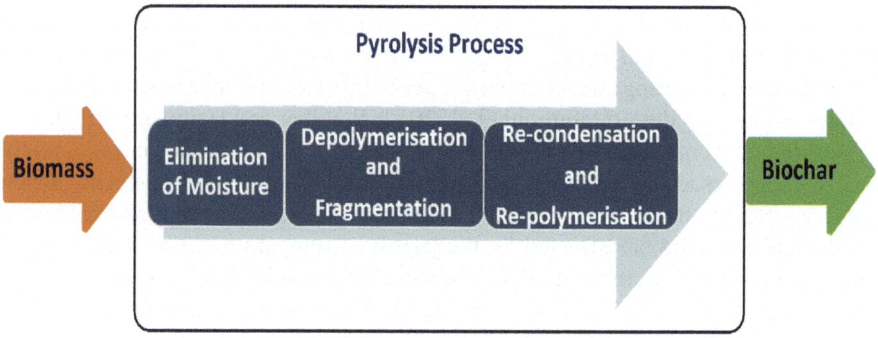

Fig. 3 Stages involved in pyrolysis process

Table 2 Pyrolysis process parameters and rate of pyrolysis (Foong et al. 2020)

Process parameters	Methods of pyrolysis		
	Slow	Fast	Flash
Heating rate(°C/s)	0.1–10	10–200	> 1000
Residence time	> 5 min to several hours	10–25 min	< 1 min
Biochar yield	25–50%	15–25%	5–15%

Table 3 Biochar pyrolysis feedstock and process parameters (Singh et al. 2010)

Feedstocks	Carbon (C) and nutrient (N) content	Process parameters	
		Temperature (°C)	Steam treatment
Eucalyptus saligna wood	47.9% C, ~ 0.03% N	400	Activated/ Non-activated
		550	Activated/ Non-activated
Eucalyptus saligna leaves	50.1% C, 1.33% N	400	Activated
		550	Activated
Poultry litter	39.3% C, 6.09% N	400	Non-activated
		550	Activated
Cow manure	20% C, 1.96% N	400	Non-activated
		550	Activated
Paper sludge	33.5% C, 0.20% N	400	Activated

5 Advanced Pyrolytic Methods

To enhance the yield of pyrolysis, oxygen-containing functional groups, surface area, and the porous architecture of the products of biochar, various advanced pyrolytic methods need to be adopted upon conventional method. These techniques are achieved with adoption of various modifications on the conventional technique. Advanced pyrolysis methods involves (i) Microwave-assisted pyrolysis, (ii) Steam-assisted pyrolysis, (iii) Hydro/wet pyrolysis, (iv) NH_3 ambiance pyrolysis, (v) Co-pyrolysis, and (vi) Catalytic pyrolysis (Lee et al. 2020; Wang et al. 2020). Hybrid pyrolysis can be achieved by combining two or more technologies discussed above for desired biochar output. Only hydro-pyrolysis and microwave aided pyrolysis can handle fresh biomass without pre-drying. These methods additionally enhance the nano-architecture of pores and create distinctive nano-carbon structures as a result of the uniform heating given to the feedstock and the self-extrusion and microscale explosion of volatile organic matter at the time of the thermal reaction. Many modern pyrolytic techniques achieve the requisite biochar characteristics while operating at low temperatures, making them substantially more energy- and cost-efficient than traditional techniques (Kambo and Dutta 2015; Ponnusamy et al. 2020).

5.1 Microwave-Assisted Pyrolysis

Microwave is an electromagnetic radiation which is used for dielectric heating method at a typical frequencies of 2.45 GHz. With no direct physical contact in between the heat source and heated object, electromagnetic energy can be converted into heat by microwave radiation with wavelengths between 0.001 and

1 m. Microware is capable to produce relatively rapid heating rates. It is essential for the feedstock to efficiently absorb microwaves during microwave-assisted pyrolysis procedures in order to produce polarization effects. Given that water can rotate and align into dipoles, causing collision and friction that contribute for the heating mechanism, it is thought that a high water content helps the absorption of microwaves (Kong et al. 2019; Li et al. 2016). It is also clear that, in comparison with typical slow pyrolysis, microwave-assisted pyrolysis can increase the degree of carbonization of biochar at low pyrolysis temperature, which has an impact on the material's physical and chemical properties (Mašek et al. 2013). Under microwave irradiation, feedstock was heated uniformly both internally and externally, and microscale explosions produced the characteristic morphology. In order to help with environmental mitigation, biochar produced by microwave-assisted pyrolysis has employed for the organic pollutants adsorption, heavy metals, nitrate, or phosphorus.

5.2 Co-Pyrolysis

Co-pyrolysis is a useful method for changing the biochar characteristics and producing composite products. Additionally, because of the better immobilization of intrinsic metals within the biochar matrix, it can be a useful strategy to lower the environmental concerns that are connected with biochars produced from feedstocks that are rich in metals. Metal mobility in the biochar might be greatly reduced after pyrolysis with unpolluted biomass (such as waste tea, walnut shells, bamboo, and hazelnut shells), lowering the hazards of metal leaching.

5.3 Steam-Assisted Pyrolysis

Steam-assisted pyrolysis has significant impact on the biochar physiochemical properties. Many researchers have observed the enhanced specific surface area and pore volume due to removal of tar and other by-products due to addition of steam during the pyrolysis process. This helps the promotion of adsorption ability of the biochar. In general, conventionally produced biochar can absorb metal and organic contaminants from water. A steam-assisted pyrolysis biochar can absorb sulfur dioxide from fuel gas. This is achieved due to decreased oxygen-containing functional group on the biochar surfaces during steam activation, which increases aromaticity and intensity of delocalized π bonds. Researchers have regarded the steam-assisted pyrolysis as one of the easiest and relatively energy-efficient technique to produce biochar in comparison with other methods aimed at increasing surface area.

5.4 *Wet Pyrolysis*

Wet pyrolysis, which simultaneously converts biomass into biochar and modifies that biochar, is a more energy-efficient and economical method of pyrolysis than conventional methods. Wet pyrolysis is achieved by mixing the biomass with phosphoric acid with 85% weight ratio. After stirring the mixture, it will be soaked before send for autoclave with 200 °C for 5 h. Wet pyrolysis helps in introducing hydroxyl and carboxyl onto the surface of biochar. This will increase the oxygen-containing functional groups. Biochar produced by wet pyrolysis are used in adsorption of lead and cadmium with higher capacity. This technique is getting more popularity and becoming a new mania, as there is no necessity for creating an oxygen-limited environment, thus makes process of pyrolysis more simple.

5.5 *Ammonia Ambiance Pyrolysis*

Biomass oxygen-containing species including esters, ketones, aldehydes, and furans may react with ammonia (NH_3) during pyrolysis processes to produce nitrogen (N)-bearing heterocyclic moieties like pyrrole, piperidine, pyridine, and indole. Metal adsorption can be accomplished with N-doped biochar. The surface complexation of Cr(VI) with N-bearing groups and subsequent Cr(III) production post-catalytic reduction of Cr(VI) by magnetic biochar allowed the resultant biochar to successfully adsorb Cr (VI) (adsorption capacity 142.86 mg g^{-1}). N-doped biochar may additionally aid in the catalytic breakdown of methylene blue. Mian et al. (2019) discovered that NH_3 ambiance pyrolysis may considerably lower the bandgap energy of TiO_2/Fe_3O_4/biochar composite, allowing for the excitation of electrons from the valence band to the conduction band by proportionately low-energy visible light. The mechanics underlying this process were not entirely known, though. Future studies are advised to look more closely at the function of N-containing groups in the catalytic elimination of pollutants and to estimate the potential of N-doped biochar as a metal immobilizing agent.

6 Factors Affecting Biochar During Synthesis

There are various factors that influence the synthesis of biochar like feedstock type, particle size, temperature of pyrolysis, and heating rate. The details are as follows:

6.1 Feedstock Type

Type of feedstock used in pyrolysis has a high impact on biochar properties, its elemental composition and the proportions of organic and inorganic compounds. Agriculture and plant feedstocks largely contains cellulose, lignin, hemicelluloses, and few inorganic compounds, whereas, feedstock from animal manure are with high nutrient content. Thus, biochars produced by animal manure are rich in minerals compared to wood feedstocks. Under wood and agriculture, feedstock with ligno-cellulose degrades at approx. 120 °C, hemicellulose degrades at 200–260 °C, and cellulose degrades at 240–350 °C, whereas lignin degrades at 280–350 °C. Thus, the pyrolysis temperature largely depends on type of feedstocks used in the process (Qambrani et al. 2017).

In terms of quality of biochar to be produced, both feedstock composition and pyrolysis parameters have large influence on both physical and chemical properties that have control over porosity, surface area, pH, cation exchange capacity, extended carbon, and adsorption ability. As the pyrolysis temperature rose, volatile compounds; it helps in increasing the area of surface and content of ash for the biochar and to decrease the surface functional groups (Guo and Rockstraw 2007).

6.2 Feedstock Particle Size

Feedstock particle size used in the pyrolysis has direct control over the transfer of heat and mass and the degree of secondary reactions within the particles. As a result, it has an important role in regulating the output of pyrolysis products. The particle size of input material greatly depend on the type of feedstock used in pyrolysis. It is common practice to use smaller particle size in case the fast pyrolysis is preferred. Because smaller particles have tendency to heat up uniformly, and helps to release more volatile matter, eventually leads to increased bio-oil yield and gas-product. It has been noticed that for particle size more than 1.8 mm has higher temperature gradients within them, resulting a lower core temperature. Such process increases the yield in liquid and biochar, subsequently decreasing in bio-oil and gas-product yields.

In an experiment conducted by Yorgun and Yildiz on pyrolysis process, they have considered wood with varied particles sizes ranging from 0.224 mm to 0.425 mm. They have set the process temperature at 500 °C with a heating rate of 223.15 °C min^{-1} to get the biochar yield up to 28% (highest). They have noticed that, beyond the said range, particle size has no significant impact on biochar conversion and product yields (Yorgun and Yildiz 2015). Other similar researchers have similar kind of results on pyrolysis product yield for differently sized biomass particles (Demiral and Şensöz 2006; Septien et al. 2012; Yorgun and Şimşek 2003; Yorgun and Yildiz 2015).

6.3 Pyrolysis Temperature

Process temperature is an important parameter under pyrolysis process which controls the characteristics, physiochemical properties of biochar. As the temperature rises, ash content, surface area, and pH of biochars made from biomass rise, while the production of biochar diminishes. Further analyses on biochar have revealed that the lower temperature promotes for higher evolution of CO_2 and water. Whereas, higher temperatures results in the liberation of CO, H_2, and CH_4. At lower process temperature, aliphatic and cellulose-type structures generated in the biochar due to which, highly diversified organic characters are imbibed in to the biochar material. These organic characters present in biochar are good ingredients to promote the mineralization by bacterial and fungi action that are important in aggregate formation and nutrient turnover. Biochar produced by temperatures in the range of 400–700 °C contain substantial amount of C in poly-condensed aromatic structures. However, they have very few ion exchange functional groups because of decarboxylation and dehydration (Qambrani et al. 2017).

6.4 Heating Rate

Biomass pyrolysis yields three variant products, biochar, bio-oil, and bio-gases with varied proportions. Quantity and quality of the products produced can be controlled via wide range by controlling the parameters of pyrolysis. Heating rates are second important factor affecting the pyrolysis process, where biochar structure as well as the yields of bio-oil and syngas are highly influenced. As illustrated in Fig. 4, for higher biochar production, slower heating rates and lower processing temperatures are suitable. Whereas, higher heating rates, optimum processing temperatures, and short residence time supports for the bio-oils production. Long residence times, low heating rates, and high pyrolysis temperatures encourage the production of gaseous products from biomass (Qambrani et al. 2017).

It is clear from the pyrolysis of sawdust that lower heating rates of 20 °C s^{-1} permitted the sawdust's natural porosity to transfer to the biochar without any significant morphological changes and that higher heating rates of 500 °C s^{-1} disrupted the cell structure of the sawdust via devolatilization (Xu and Chen 2013).

7 Importance of Biochar Production

Biochar is a multifunctional material which is emerged out of bio-waste. Bio-absorbent properties of the biochar is also effectively utilized in eliminating pollutants such as phenols, chromium (VI), cadmium (II), dyes, and others present in the wastewater (Awogbemi and Kallon 2023b). However, researchers have noticed that

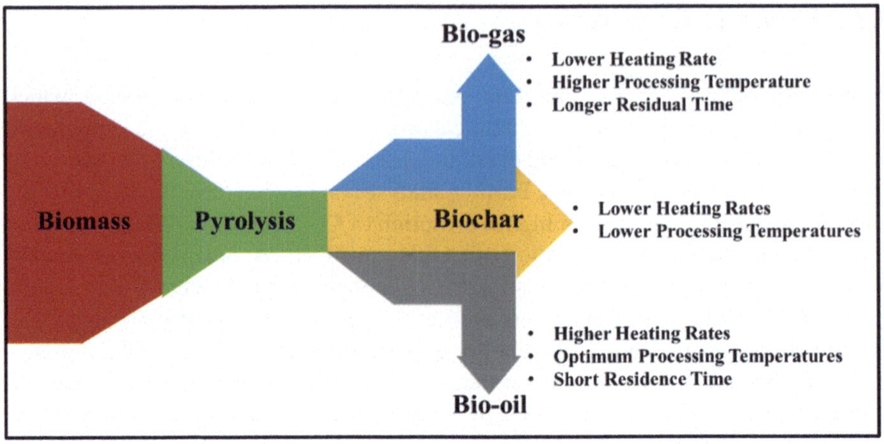

Fig. 4 Three variant products derived from pyrolysis process

eliminating pharmaceuticals from the water or wastewater is highly challenging. Conventional methods like chlorination, activated sludge, phytoremediation, flocculation are ineffective in removing the pharmaceutical contaminants. Advanced oxidation techniques may seem to be promising yet, its demand for high energy and low efficiency in mineralization have made it not viable. Thus, biochar-based adsorption technique is more effective yet economical to operate (Hama Aziz et al. 2024). As supply and demand for bio-fuel is increasing day-by-day, to energize the bio-fuel production, biochar has been used as catalyst in the bio-fuel production process (Awogbemi and Kallon 2023a).

Presently, advancement in the technology has made viable to produce biochar in nanoscale making it nano-biochar which has enhanced physical and biological properties that could assist biochar to enhance its ability to absorb pollutants from wastewater as well as potential element in the production of nano-carbon composites and electroactive material in energy storage devices (Sani et al. 2023) such as supercapacitors with metal-oxide and biochar composites (Khedulkar et al. 2024). A bibliometric review on current and future prospects of biochar has noticed substantial potential for biochar as an active material in the energy storage devices in the near future (Ma et al. 2024).

Apart from, Fig. 5 illustrates the major applications of biochar in various other fields. It has been used as a mitigating agent for greenhouse gases, and as a soil amendment to enrich agriculture soil, it can also be used as a supplement in animal feed and has potential application in medicine.

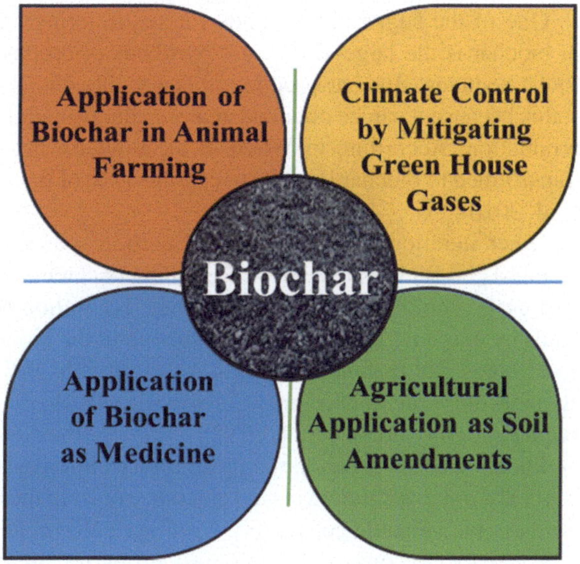

Fig. 5 Importance of biochar in various fields

7.1 Mitigate the Greenhouse Gas Emission

Disposing of biomass in huge quantity would leads to the decomposition as time goes, where the process of decomposition releases huge amount of greenhouse gases, like CO_2 or CH_4 to the atmosphere. The capacity of biochar to store carbon is based on the stability of carbon that prevents the carbon from near term decomposing into lighter hydrocarbons and releasing greenhouse gases into the environment. The bigger role biochar can play in carbon sequestration schemes, the longer it can hold its structure (better carbon stability) and keep its carbon from re-entering the atmosphere. According to several studies and proponents of biochar, the production of biochar might reduce US greenhouse gas emissions annually by about 12%, or around 1.8 $PgCO_2$-C (Woolf et al. 2010). Carbon rich, biochar is also instrumental in regulate the carbon-to-nitrogen ratio during the compost matrix, thereby control the formation of greenhouse gases. Researchers have found the biochar one of the most efficient additive for the composting process. It can also enrich the quality of composting and adsorption of pollutants (Noor et al. 2024).

The half-life or mean residence time, of the carbon in the biochar is frequently employed to measure the biochar's carbon stability. The half-life is the period of time needed for half of the carbon that was initially available in the biochar to decompose, and the mean residence time is the typical amount of time the carbon stays in the soil until it is released back into the atmosphere. But it is known that the pyrolysis conditions, the type of feedstock, and the soil in which the biochar is added all have an impact on the biochar half-life or the mean residence time (Fang et al. 2014; Singh et al. 2012).

One of the suggested methods for sequestering carbon through the production of biochar is the large-scale mass pyrolysis of crop residues, animal manures, and forest residues. Biochar production is a viable choice regardless of socioeconomic status because it can be obtained at any scale, from domestic production to industrial production. According to Woolf et al. annually 5 trillion lbs of biomass will be transformed to biochar in order to capture 12% of the world's CO_2 emissions (Woolf et al. 2010).

Other iterations of biochar-based carbon sequestration methods involve more emphasis on fuel displacement. Researchers calculated that 25% of the US's fossil fuel usage could be offset by harvesting 1.1 billion tonnes of biomass from croplands by maximizing bioenergy production in the biochar production process. It was ascertained that this method might offset roughly 10% of the yearly US CO_2 emissions through the displacement of fossil fuels, the biochar production, and carbon sequestration (Galinato et al. 2011).

For the successful biochar sequestration techniques, the biomass used for pyrolysis cannot be obtained either via land clearance or from the food sources which eventually destroys the valuable ecosystem. Even agricultural lands utilized for production of food cannot be converted to biomass crops for the production of biochar as this would have a detrimental effect on food security and encourage land clearing in other locations (Woolf et al. 2010). Thus, involving agricultural waste as feedstock for biochar plants would solve both waste management issue and greenhouse gas emission issue. Additionally, the production of biochar would require the use of contemporary pyrolysis technology with minimal emissions and significant biochar yield, as the emissions from ineffective biochar units can negate any advantages.

It is crucial to do thorough study on the carbon stability of biochar as well as how the pyrolysis conditions and feedstock may be employed to maximize the carbon stability of biochar before establishing any biochar production plants into operation. Before evaluating the potential for carbon sequestration from biochar, the carbon stability of the material must be adjusted because the sequestration of carbon by biochar depends on the material's capacity to prevent the release of carbon into the atmosphere (Woolf et al. 2010).

7.2 Application of Biochar in Agriculture

Biochar has a notable effect on soil quality and fertility. It can enhance the soil's ability to retain water and nutrients. Biochar works effectively as a soil amendment instead of a fertilizer.

Various research groups carried out remediation studies to enrich the soil used in paddy fields. A comparative study to investigate highly potential remediation agents among peat, biochar, and lime was part of the study. Reduction of cadmium level in rice grains and enrichment of organic matter in soil were considered the major factors in the study. It was noticed that remediation has resulted in a reduction of the cadmium level from 0.41 mg/kg to 0.08 mg/kg. Although peat and lime had a slightly better influence on cadmium reduction, biochar not only reduced cadmium but also enriched organic matters without having any major impact on the pH value of the soil (Osman et al. 2024).

Researchers have noticed that using biochar as fertilizer would reduce the yield of crops. However, biochar as soil amendment would develop the physical properties like porosity, water holding capacity, and decrease in bulk density which is most essential for agriculture. It also enhances the chemical properties by improving the pH value, nutrients, cation exchange capability in the soil. Additionally, it has been claimed that applying biochar to soils can help lower emissions of greenhouse gases such carbon dioxide, methane, and nitrous oxide (Osman et al. 2022).

Agricultural soil of tropical lands with low altitude and high rainfall are with low fertility due to instant mineralization of organic matter and the presence of secondary minerals. To mitigate the issue, farmers are advised to use organic and inorganic fertilizers to improve the fertility. The major concern in using inorganic fertilizers is their low accessibility to resource-poor farmers (Garrity 2004). Whereas, organic fertilizers are capable of improving nutrient since they can quickly mineralize under tropical conditions. However, they could last only for few cycle of crop. In contrast, biochar a biomass-derived black carbon (C) is highly stable and capable of mineralize in soil.

Major et al. (2010) carried a field experiment at Matazul farm in the Llanos Orientales, non-flooded savannas of Colombia. Nine experimental plots were created with a measurement of 4 m by 5 m. It was made sure that the land chosen is not used for agriculture or cultivation in the past. The lands are capable of growing two crops in a year. Biochar was introduced to the soil with three variants, 0 t/ht, 8 t/ht, and 20 t/ht. Maize and soybean were seeded using hand tools for the experimental purpose. Harvest Index (HI) was used to measure agricultural yield and soil properties were also tested to know the growth pattern across the years from 2003 to 2006. Results were favoring for the amendment of soil with biochar, yield of the crop were improved annually for first 4 years after applying the biochar (Major et al. 2010).

A comparable field experiment with five crop seasons spread over three years and low fertility soil was reported from Central China. The biochar soil amendment was tested in the field experiment with or without nitrogen fertilizer (urea) for crop yield. Each treatment was carried out three times on a single 4 m $\times$ 5 m area. Following the harvest of wheat at 0, 10, and 40 t/ha in June 2010, a single soil amendment with wheat biochar was applied. By hand, biochar material was applied to the soil's surface in an equal layer to a depth of around 15 cm. Throughout the five crop seasons of this study, no additional biochar amendments were used. Nitrogen fertilizer (urea) was sprayed twice in a single crop season, 60% as a foundation fertilizer prior to sowing and another 40% as a dressing at the vegetative stage. Throughout the experiment's five seasons, alternative crops like maize and wheat were grown. In dry conditions, irrigation was done to wet the soil after maize seeds were sown and when wheat plants turned green. Over the treatment plots, all other crop management techniques remained constant. For all of the agricultural seasons, soil sample, GHGs emission monitoring, and crop yield measurement were conducted (Liu et al. 2014).

Over the course of five crop seasons following a single amendment, biochar amendment significantly and consistently increased SOC and K availability and decreased N_2O emissions. In a single crop season, there were neither simultaneous changes in soil fertility and crop productivity nor synergistic changes in soil improvement and greenhouse gas emission reduction. In order to achieve a low carbon and sustainable agriculture and combat the rapidly worsening effects of climate change, biochar soil amendment could therefore be seen as a viable, dependable, and affordable option (Liu et al. 2014).

7.3 Application of Biochar as Medicine

As of 2014, no research had been done to assess the use of biochar in food and medicine. Although biochar continues to be a significant component of numerous herbal remedies and dietary supplements, its efficacy in these contexts has not been scientifically established. Although more research is required to support this use, using biochar as a low-grade activated carbon could have future applications in medicine, particularly in rural and low-income areas where activated carbon is not easily accessible. Because of its high adsorption capacity, activated carbon is frequently used in medicine to eliminate various narcotics and toxins from the gastrointestinal tract. Studies have shown that this method is superior to stomach pumping in terms of drug/toxin clearance. Despite being most effective at adsorbing toxins and drugs that are in the stomach at the time of consumption, activated carbon can still aid in the elimination of toxins even after systematic adsorption. To make a significant impact, activated carbon should be taken as quickly as possible after intoxication. Alcohol, cyanide, and certain heavy metals, including lithium and iron, cannot be adsorbed by activated carbon (Neuvonen and Olkkola 1988).

7.4 Application of Biochar in Animal Farming

The effects of co-feeding biochar with animal diets on poultry, cattle, pigs, and fish have recently been the subject of research by a number of researchers (Abakari et al. 2020; Al-Azzawi et al. 2021; Goiri 2021; Kalus et al. 2020). Cofeeding with biochar led to a 3.43% rise in milk output, a 2.63–6.32% increase in protein-fat content, and a 30% decrease in enteric methanogens in Australian dairy cattle (Al-Azzawi et al. 2021). In addition, laying hens' daily feed intake increased when co-fed with 2.5% biochar, and their laying efficiency and thickness were both improved by 10% and 6%, respectively (Kalus et al. 2020). Similar findings were made by Goiri et al. who found that co-feeding broilers with biochar at a concentration of 30 g kg^{-1} diet boosted daily weight increase, average weight of body, and decreased feed conversion ratios compared to non-co-fed groups (Goiri 2021). It has also been demonstrated to promote the survival and development of aquatic fish.

8 Challenges and Recommendations

Recent innovations in the use of agricultural waste for biochar production have been transformative, addressing both environmental challenges and economic feasibility. The article published in Biochar Today (Aug 2024), mentioned about the innovations in the production of biochar from agricultural waste is totally focusing on making the process more efficient, sustainable, with few developments in this area like the tractor-mounted systems and other mobile biochar production units are enabling the on-site conversion of residues like wheat straw and rice husks, lowering transportation costs, and promoting adoption. Modern pyrolysis procedures, such as hydrothermal and microwave-assisted techniques, are increasing the energy efficiency and broadening the application of wet wastes. Furthermore, production is optimized through automation using real-time sensors and artificial intelligence (AI) and the integration of renewable energy sources, such as employing process syngas.

Although agricultural waste has the potential to be used to produce biochar, a number of obstacles prevent its widespread use. The quality of biochar and process efficiency are impacted by the chemical makeup and moisture content of agricultural leftovers. Feedstock input standardization is still challenging. Implementing pyrolysis technology, especially sophisticated techniques like microwave-assisted pyrolysis, can be costly. For small farmers, the expense of energy, transportation, and scaling up continues to be a major obstacle. Pollutants include volatile organic compounds (VOCs) and particulate matter can be released during the manufacture of biochar if improperly handled. Pyrolysis systems must be kept hygienic and effective. Additionally, it can be difficult and expensive to collect and transport agricultural waste from farms to facilities that produce biochar, particularly in rural or isolated places.

To overcome challenges in biochar production from agricultural waste, standardizing feedstock quality, implementing awareness, establish supportive policies like subsidies and carbon credits, adopting cost-effective and energy-efficient pyrolysis technologies are important. Even though, classical pyrolysis is a well-known method for producing pyrolytic products. But sophisticated pyrolysis is thought to be very recent. Nevertheless, the advancement on such advanced or non-conventional pyrolysis technologies has indeed been improved steadily during the previous 5 years. There is a tonne of room for research in the area of optimization because the pyrolysis techniques' process parameters are not optimized for their intended outputs (Lee et al. 2020). Biochar synthesis and production must initially be recognized as a carbon sequestration process by a carbon market in order to become an economically viable choice for carbon sequestration. The Kyoto Protocol, developed by the United Nations Framework on Climate Change, currently does not acknowledge biochar as an acceptable method for sequestering carbon (UNFCCC). It is doubtful that biochar will be a commercially viable option for agricultural soil amendment or climate change mitigation unless national or global carbon markets adopt it (Galinato et al. 2011).

9 Conclusions

The utilization of agricultural waste for biochar production presents a promising avenue for sustainable agriculture, waste management, and climate action. Biochar's primary uses include improving soil fertility, enhancing water retention, and acting as a long-term carbon sink to mitigate climate change. It also serves as an effective amendment in mitigating greenhouse gases, animal farming, as a medicine and in agricultural prospects. Emerging applications, such as nutrient-enriched biochar fertilizers highlight its versatility. However, challenges like feedstock variability, high costs, and limited awareness persist. Innovations such as portable pyrolysis units, renewable energy integration, and advanced biochar customization are addressing these issues. To unlock its full potential, policy support, research investments, and market development are crucial. By leveraging its wide-ranging benefits and overcoming barriers, biochar can revolutionize waste management, enhance agricultural productivity, and contribute significantly to climate resilience.

References

Abakari G, Luo G, Meng H, Yang Z, Owusu-Afriyie G, Kombat EO, Alhassan EH (2020) The use of biochar in the production of tilapia (*Oreochromis niloticus*) in a biofloc technology system - BFT. Aquacult Eng 91(September):102123. https://doi.org/10.1016/j.aquaeng.2020.102123

Afif RA, Pfeifer C, Pröll T (2012) Bioenergy recovery from cotton stalk. In: Intech. https://doi.org/10.1039/C7RA00172J. https://www.intechopen.com/books/advanced-biometric-technologies/liveness-detection-in-biometrics. https://doi.org/10.1016/j.colsurfa.2011.12.014

Al-Azzawi M, Bowtell L, Hancock K, Preston S (2021) Addition of activated carbon into a cattle diet to mitigate GHG emissions and improve production. Sustain (Switz) 13(15):1–17. https://doi.org/10.3390/su13158254

Andini A, Bonnet S, Rousset P, Hasanudin U (2018) Impact of open burning of crop residues on air pollution and climate change in Indonesia. Curr Sci 115(12):2259–2266. https://doi.org/10.18520/cs/v115/i12/2259-2266

Awogbemi O, Von Kallon DV (2023a) Application of biochar derived from crops residues for biofuel production. Fuel Commun 15(November 2022):100088. https://doi.org/10.1016/j.jfueco.2023.100088

Awogbemi O, Von Kallon DV (2023b) Progress in agricultural waste derived biochar as adsorbents for wastewater treatment. Appl Surf Sci Adv 18(December):100518. https://doi.org/10.1016/j.apsadv.2023.100518

Bolan N, Hoang SA, Beiyuan J, Gupta S, Hou D, Karakoti A, Joseph S, Jung S, Kim KH, Kirkham MB, Kua HW, Kumar M, Kwon EE, Ok YS, Perera V, Rinklebe J, Shaheen SM, Sarkar B, Sarmah AK, Van Zwieten L (2022) Multifunctional applications of biochar beyond carbon storage. Int Mater Rev 67(2):150–200. https://doi.org/10.1080/09506608.2021.1922047

Cassou E (2018) Agricultural pollution. Field Burning. World Bank Group, 1–7. http://documents.worldbank.org/curated/en/989351521207797690/pdf/124342-repl-WB-Knowledge-Burning.pdf%0Ahttps://openknowledge.worldbank.org/bitstream/handle/10986/29504/124342-BRI-p153343-PUBLIC-march-22-9-pm-WB-Knowledge-Burning.pdf?sequence=1&isAllowed=

Demiral I, Şensöz S (2006) Fixed-bed pyrolysis of hazelnut (*Corylus avellana* L.) bagasse: influence of pyrolysis parameters on product yields. Energy Sources Part A Recover Util Environ Eff 28(12):1149–1158. https://doi.org/10.1080/009083190966126

Duque-Acevedo M, Belmonte-Ureña LJ, Cortés-García FJ, Camacho-Ferre F (2020) Agricultural waste: review of the evolution, approaches and perspectives on alternative uses. Glob Ecol Conserv 22. https://doi.org/10.1016/j.gecco.2020.e00902

Fang Y, Singh B, Singh BP, Krull E (2014) Biochar carbon stability in four contrasting soils. Eur J Soil Sci 65(1):60–71. https://doi.org/10.1111/ejss.12094

FAO (2017) The future of food and agriculture: trends and challenges. In: The future of food and agriculture: trends and challenges, vol 4, Issue 4. www.fao.org/publications. www.fao.org/3/a-i6583e.pdf. http://siteresources.worldbank.org/INTARD/825826-1111044795683/20424536/Ag_ed_Africa.pdf. www.fao.org/cfs. http://www.jstor.org/stable/4356839. https://ediss.uni-goettingen.de/bitstream/han

FAO (2021) World food and agriculture—statistical yearbook 2021. In: World food and agriculture—statistical yearbook 2021. https://doi.org/10.4060/cb4477en

Foong SY, Liew RK, Yang Y, Cheng YW, Yek PNY, Wan Mahari WA, Lee XY, Han CS, Vo DVN, Van Le Q, Aghbashlo M, Tabatabaei M, Sonne C, Peng W, Lam SS (2020) Valorization of biomass waste to engineered activated biochar by microwave pyrolysis: progress, challenges, and future directions. Chem Eng J 389:124401. https://doi.org/10.1016/j.cej.2020.124401

Galinato SP, Yoder JK, Granatstein D (2011) The economic value of biochar in crop production and carbon sequestration. Energy Policy 39(10):6344–6350. https://doi.org/10.1016/j.enpol.2011.07.035

Garrity DP (2004) Agroforestry and the achievement of the Millennium Development Goals. 61(July):5–17. https://doi.org/10.1007/978-94-017-2424-1_1

Goiri I (2021) Assessing the potential use of a feed additive based on biochar on broilers feeding upon productive performance, pH of digestive organs, cecum fermentation and bacterial community. Anim Feed Sci Technol 279(1):115039

Guo Y, Rockstraw DA (2007) Activated carbons prepared from rice hull by one-step phosphoric acid activation. Microporous Mesoporous Mater 100(1–3):12–19. https://doi.org/10.1016/j.micromeso.2006.10.006

Hama Aziz KH, Mustafa FS, Hassan MA, Omer KM, Hama S (2024) Biochar as green adsorbents for pharmaceutical pollution in aquatic environments: a review. Desalination 583(April):117725. https://doi.org/10.1016/j.desal.2024.117725

Hazell P, Pachauri RK (2006) Bioenergy and agriculture: promises and challenges. Bioenergy Agric Promises Chall. https://doi.org/10.2499/2020focus14

Kambo HS, Dutta A (2015) A comparative review of biochar and hydrochar in terms of production, physico-chemical properties and applications. Renew Sustain Energy Rev 45:359–378. https://doi.org/10.1016/j.rser.2015.01.050

Kalus K, Konkol D, Korczyński M, Koziel JA, Opaliński S (2020) Effect of biochar diet supplementation on chicken broilers performance, NH_3 and odor emissions and meat consumer acceptance. Anim 10(9):1–13. https://doi.org/10.3390/ani10091539

Khedulkar AP, Dang VD, Thamilselvan A, Doong R, Pandit B (2024) Sustainable high-energy supercapacitors: metal oxide-agricultural waste biochar composites paving the way for a greener future. J Energy Storage 77(December 2023), 109723. https://doi.org/10.1016/j.est.2023.109723

Kim Oanh NT, Permadi DA, Hopke PK, Smith KR, Dong NP, Dang AN (2018) Annual emissions of air toxics emitted from crop residue open burning in Southeast Asia over the period of 2010–2015. In: Atmospheric environment, vol 187. https://doi.org/10.1016/j.atmosenv.2018.05.061

Kong SH, Lam SS, Yek PNY, Liew RK, Ma NL, Osman MS, Wong CC (2019) Self-purging microwave pyrolysis: an innovative approach to convert oil palm shell into carbon-rich biochar for methylene blue adsorption. J Chem Technol Biotechnol 94(5):1397–1405. https://doi.org/10.1002/jctb.5884

Lee XJ, Ong HC, Gan YY, Chen WH, Mahlia TMI (2020) State of art review on conventional and advanced pyrolysis of macroalgae and microalgae for biochar, bio-oil and bio-syngas production. Energy Convers Manage 210(March):112707. https://doi.org/10.1016/j.enconman.2020.112707

Li J, Dai J, Liu G, Zhang H, Gao Z, Fu J, He Y, Huang Y (2016) Biochar from microwave pyrolysis of biomass: a review. Biomass Bioenerg 94:228–244. https://doi.org/10.1016/j.biombioe.2016.09.010

Liu X, Ye Y, Liu Y, Zhang A, Zhang X, Li L, Pan G, Kibue GW, Zheng J, Zheng J (2014) Sustainable biochar effects for low carbon crop production: a 5-crop season field experiment on a low fertility soil from Central China. Agric Syst 129:22–29. https://doi.org/10.1016/j.agsy.2014.05.008

Ma J, Zheng L, Yu F (2024) Current status and future prospects of biochar application in electrochemical energy storage devices: a bibliometric review. Desalination 581(April):117597. https://doi.org/10.1016/j.desal.2024.117597

Major J, Rondon M, Molina D, Riha SJ, Lehmann J (2010) Maize yield and nutrition during 4 years after biochar application to a Colombian savanna oxisol. Plant Soil 333(1):117–128. https://doi.org/10.1007/s11104-010-0327-0

Mašek O, Budarin V, Gronnow M, Crombie K, Brownsort P, Fitzpatrick E, Hurst P (2013) Microwave and slow pyrolysis biochar—comparison of physical and functional properties. J Anal Appl Pyrol 100:41–48. https://doi.org/10.1016/j.jaap.2012.11.015

Meyer LA, Dew T (2022) Cotton and wool outlook : May 2022 global cotton mill use projected above production for third consecutive year in 2022/23. May, 1–8

Mian MM, Liu G, Yousaf B, Fu B, Ahmed R, Abbas Q, Munir MAM, Ruijia L (2019) One-step synthesis of N-doped metal/biochar composite using NH_3-ambiance pyrolysis for efficient degradation and mineralization of Methylene Blue. J Env Sci (China). 78(April):29–41. https://doi.org/10.1016/j.jes.2018.06.014

Neuvonen PJ, Olkkola KT (1988) Toxicology management review role of single and repeated doses intoxications 58:33–58

Noor RS, Shah AN, Tahir MB, Umair M, Nawaz M, Ali A, Ercisli S, Abdelsalam NR, Ali HM, Yang SH, Ullah S, Assiri MA (2024) Recent trends and advances in additive-mediated composting technology for agricultural waste resources: a comprehensive review. ACS Omega 9(8):8632–8653. https://doi.org/10.1021/acsomega.3c06516

OECD-FAO (2021) OECD-FAO agricultural outlook 2021–2030. In: OECD-FAO Agricultural Outlook 2021–2030. 10.1787/agr-outl-data-%0A10.1787/agr-outl-data-%0Ahttp://www.fao.org/documents/card/en/c/cb5332en

Osman AI, Farghali M, Rashwan AK (2024) Life cycle assessment of biochar as a green sorbent for soil remediation. Curr Opin Green Sustain Chem 46:100882. https://doi.org/10.1016/j.cogsc.2024.100882

Osman AI, Fawzy S, Farghali M, El-Azazy M, Elgarahy AM, Fahim RA, Maksoud MIAA, Ajlan AA, Yousry M, Saleem Y, Rooney DW (2022) Biochar for agronomy, animal farming, anaerobic digestion, composting, water treatment, soil remediation, construction, energy storage, and carbon sequestration: a review. Environ Chem Lett 20(4). Springer International Publishing. https://doi.org/10.1007/s10311-022-01424-x

Ponnusamy VK, Nagappan S, Bhosale RR, Lay CH, Duc Nguyen D, Pugazhendhi A, Chang SW, Kumar G (2020) Review on sustainable production of biochar through hydrothermal liquefaction: physico-chemical properties and applications. Biores Technol 310:123414. https://doi.org/10.1016/j.biortech.2020.123414

Purakayastha TJ, Bera T, Bhaduri D, Sarkar B, Mandal S, Wade P, Kumari S, Biswas S, Menon M, Pathak H, Tsang DCW (2019) A review on biochar modulated soil condition improvements and nutrient dynamics concerning crop yields: pathways to climate change mitigation and global food security. Chemosphere 227:345–365. https://doi.org/10.1016/j.chemosphere.2019.03.170

Qambrani NA, Rahman MM, Won S, Shim S, Ra C (2017) Biochar properties and eco-friendly applications for climate change mitigation, waste management, and wastewater treatment: a

review. Renew Sustain Energy Rev 79(February):255–273. https://doi.org/10.1016/j.rser.2017.05.057
Ren J, Yu P, Xu X (2019) Straw utilization in China-status and recommendations. Sustainability (Switzerland) 11(6):1–17. https://doi.org/10.3390/su11061762
Roy PS, Garnier G, Allais F, Saito K (2021) Strategic approach towards plastic waste valorization: challenges and promising chemical upcycling possibilities. ChemSusChem 14(19):4007–4027. https://doi.org/10.1002/cssc.202100904
Sani MNH, Amin M, Siddique AB, Nasif SO, Ghaley BB, Ge L, Wang F, Yong JWH (2023) Waste-derived nanobiochar: a new avenue towards sustainable agriculture, environment, and circular bioeconomy. Sci Total Environ
Septien S, Valin S, Dupont C, Peyrot M, Salvador S (2012) Effect of particle size and temperature on woody biomass fast pyrolysis at high temperature (1000–1400 °C). Fuel 97:202–210. https://doi.org/10.1016/j.fuel.2012.01.049
SilpaKaza, Yao L, Bhada-Tata P, Van Woerden F (2018) What a Waste 2.0. The World Bank with. https://openknowledge.worldbank.org/bitstream/handle/10986/30317/9781464813290.pdf
Singh B, Singh BP, Cowie AL (2010) Characterisation and evaluation of biochars for their application as a soil amendment. Aust J Soil Res 48(1):516–525. https://doi.org/10.1071/SR10058
Singh BP, Cowie AL, Smernik RJ (2012) Biochar carbon stability in a clayey soil as a function of feedstock and pyrolysis temperature. Environ Sci Technol 46(21):11770–11778. https://doi.org/10.1021/es302545b
Singh J, Singhal N, Singhal S, Sharma M, Agarwal S, Arora S (2018) Environmental implications of rice and wheat stubble burning in north-western States of India. In: Advances of health and environment safety, springer transaction in civil and environmental engineering, January, 317–325. https://doi.org/10.1007/978-981-10-7122-5
UN (2016) World Population Prospects. In: World Population Prospects. https://doi.org/10.18356/cd7acf62-en
Wang L, Ok YS, Tsang DCW, Alessi DS, Rinklebe J, Wang H, Mašek O, Hou R, O'Connor D, Hou D (2020) New trends in biochar pyrolysis and modification strategies: feedstock, pyrolysis conditions, sustainability concerns and implications for soil amendment. Soil Use Manag 36(3):358–386. https://doi.org/10.1111/sum.12592
Woolf D, Amonette JE, Street-Perrott FA, Lehmann J, Joseph S (2010) Sustainable biochar to mitigate global climate change. Nat Commun 1(5). https://doi.org/10.1038/ncomms1053
Woolley S, Hallowell B (2018) Biochar: an overview www.BiomassControls.com. February. www.BiomassControls.com
Xu Y, Chen B (2013) Investigation of thermodynamic parameters in the pyrolysis conversion of biomass and manure to biochars using thermogravimetric analysis. Biores Technol 146:485–493. https://doi.org/10.1016/j.biortech.2013.07.086
Yorgun S, Şimşek YE (2003) Fixed-bed pyrolysis of miscanthus × giganteus: product yields and bio-oil characterization. Energy Sources 25(8):779–790. https://doi.org/10.1080/00908310390207828
Yorgun S, Yildiz D (2015) Slow pyrolysis of paulownia wood: effects of pyrolysis parameters on product yields and bio-oil characterization. J Anal Appl Pyrol 114:68–78. https://doi.org/10.1016/j.jaap.2015.05.003

Chapter 4
Production of Biochar from Sewage Sludge and Its Environmental Implications

Umair Riaz, Ahmad Waqas, Shazia Iqbal, Humera Aziz, Abid Hussain, Tanveer-ul-Haq, Ghulam Murtaza, and Sana Farooq

Abstract Wastewater is now becoming an emerging issue in the world with the increase in population. Treatment of wastewater results in increased sewage sludge production which consists of over 80% moisture content, making it a big problem for treatment. Only China is currently producing over 13 million tons of municipal sewage sludge (MSS) every year to treat municipal wastewater. The MSS and industrial sewage sludge (ISS) are hazardous as its potentially toxic elements like heavy metal(oids)s, organic pollutants like polycyclic aromatic hydrocarbons (PAHs), Persistent organic pollutants (POPs), sometimes known as "forever chemicals" and pathogens including viral, bacterial, protozoan, fungal, and helminth pathogens. Currently used technologies for reducing the volume of sewage sludge, e.g., land filling, composting, etc., are not promising because of their side effects on plant, soil, human as well as on environment. The sewage sludge usage as a raw material for biochar production is gaining popularity in the scientific community as a better treatment strategy that benefits both the soil and plants. The sludge-derived biochar (SDBC) reduced the risks associated with its direct application to soil. This biochar decreases the heavy metals leaching to the ground water and additionally enhances the soil's physical and chemical properties. Nutrients present in sewage sludge-derived biochar provide better growth conditions for plants. This approach

U. Riaz (✉) · A. Hussain · Tanveer-ul-Haq
Department of Soil and Environmental Sciences, MNS University of Agriculture Multan, Multan, Pakistan
e-mail: umair.riaz@mnsuam.edu.pk

A. Waqas
Soil Chemistry Lab, Central Cotton Research Institute, Multan, Pakistan

S. Iqbal
Department of Soil and Environmental Science, University of Sargodha, Sargodha, Pakistan

H. Aziz · S. Farooq
Department of Environmental Sciences, Government College University, Faisalabad, Pakistan

G. Murtaza
Institute of Soil and Environmental Sciences, University of Agriculture, Faisalabad, Pakistan

A. Roy et al. (eds.), *Biochar Production Engineering*,
https://doi.org/10.1007/978-3-031-92594-8_4

results in a reduced volume of sewage sludge, providing benefits to soil and plants and turning waste into beneficial products, and could be an eco-friendly, low-cost strategy for sewage sludge management. This chapter explored the possibilities of sewage sludge-derived biochar and their environmental implications.

Keywords Biochar · Sewage sludge · Industrial pollution · Environmental issues · Management strategies

1 Introduction

The population of the world is increasing and concentrating in cities, claims a recent Uni study. The current global population of 7.6 billion people is projected to increase to 8.6 billion in 2030, 9.8 billion in 2050, and 11.2 billion in 2100, according to a recent United Nations study (Lin et al. 2018). Cities' growing populations generate billions of tons of trash each year, including sludge. These wastes can sometimes be collected or not and treated or not. With few and partial exceptions, information on waste streams are rare and dispersed in the literature, and thorough studies and analyses at the international level are missing. Nevertheless, efforts are from organizations such as the Global Water Intelligence (nd FAO/IWMI through AQUASTAT, UN-Habitat (2008), and no prominent measures are being taken that's why new and updated assessments are required. The wastewater treatment plant produces MSS. China, as one of the leading contributors to waste generation, produces about 13 million tons (dry weight) of MSS per year, with an expected 10% annual growth rate during the last ten years (Havukainen et al. 2017). For some years, the huge volume of MSS has been recognized as "a future waste concern" and has drawn increasing attention. MSS is suitable for composting because it frequently contains high levels of organic and mineral nutrients like phosphorus, nitrogen, potassium, and trace elements (Tian et al. 2013; Uchimiya et al. 2011). However, sludge contains high concentrations of contaminants such as heavy metals, pathogens, and organic matter that can interfere with soil uptake (Yuan et al. 2015). At least 20% of the entire amount of sludge in China is improperly disposed of, leading to severe sanitary, environmental, and social problems). Wastewater treatment methods produce a by-product called sewage sludge (SS), which frequently includes a variety of potentially harmful substances such as pathogenic bacteria, metal(loids), parasite ova, and refractory organics. China produces between 32 and 52 million tons of sewage sludge annually, with an 80% water content. This amount might rise by 10% annually (Yang et al. 2015). The Water Pollution Control Action Plan adopted in 2015 stipulates that the safe treatment rate of urban sewage sludge should reach 90% by the end of 2020, but currently, about 70% of China's sewage treatment plant's sewage sludge is not fully treated (Feng et al. 2015). The three most popular conventional disposal methods are incineration, sanitary landfills, and composting for agricultural use.

The technology's use and advancement are constrained by the possibility of dioxin and other potentially dangerous gases being released during the incineration of sewage sludge, as well as the need for further treatment of the leftover ash (Cieślik et al. 2015) and higher energy levels (Capodaglio and Olsson 2020). A growing number of nations have outlawed the disposal of untreated sewage sludge in landfills and composting facilities due to worries about leachates, odors, potentially harmful substances contamination, and the spreading of possibly dangerous microbes in the surroundings (Wang et al. 2019). Numerous researches demonstrated that during the composting of sewage sludge, various organic compounds are released, including volatile compounds of organic sulfur, volatile fatty acids, alcohols, aldehydes, terpenes, ketones, and aromatics (Zhang et al. 2016a; Nie et al. 2019). The primary precursors that speed ozone loss in the convective boundary layer are volatile organic chemicals, which result in environmental problems such as photochemical smog, global warming, and the production of airborne ozone (Nie et al. 2019). The enormous and ongoing growth in the volume of sewage sludge necessitates the creation of alternative processes for volume reduction, added value, and risk-free treatment. A method that can produce valuable bioenergy, pyrolyze potentially harmful organic matter, and eradicate pathogens is the pyrolysis of sewage sludge (Racek et al. 2020; Paz-Ferreiro et al. 2018). A solid carbonaceous by-product of pyrolysis known as "sewage sludge biochar," or SSB, can be used to make bioenergy, as a substrate for landscape plants, or as an adsorbent for metallic (loid) or organic impurities (Vardon et al. 2012). However, the use of this technology is hampered by their high ash content, enrichment with potentially harmful elements, high moisture content, and inorganic constituents, particularly the metal(loids), which make the process energy-intensive and degrade the quality of their biochar (Zhang et al. 2020). A carbon-rich substance known as biochar is created when organic matter thermally decomposes at low temperatures relatively with a finite amount of oxygen (Lehmann and Joseph 2009). It can be simply described as a fine-grained, black colored, incredibly porous, stable, and charcoal-like carbon (Laird 2008). Biochar, in contrast to other solid biomass pyrolysis products, is primarily used for long-term carbon storage as opposed to producing feedstock for fuel or other processing industries (Mašek et al. 2013). Wood and biomass plants, agricultural by-products such as wheat straw and hazelnuts, different wastes (including kitchen garbage), food, and sewage sludge may all be pyrolyzed to produce biochar (Hossain et al. 2011). Biochar is a solid by-product of pyrolysis. Biochar has been demonstrated in extensive studies to significantly enhance soil chemical, physical, and biological qualities (Uchimiya et al. 2011; Gaskin et al. 2008). The functional groups in biochar not only improve cation exchange capacity and soil nutrient uptake efficiency (CEC) (Prapagdee and Tawinteung 2017), but also lower pollutant bioavailability and mobility in soils (Bielská et al. 2018). Due to biochar's low density and porous structure, aeration, soil water retention, as well as permeability may all be improved (Obia et al. 2016). Significant liming effects from biochar ash can lessen heavy metal toxicity and soil acidity (Al-wabel et al. 2015). Numerous cereals, roots, tubers, and fibers benefit from the addition of biochar in tropical, subtropical, and even temperate climates (Biederman and Harpole 2013). Additionally, adding biochar encourages carbon sequestration, boosts soil carbon storage, and lowers greenhouse gas emissions (Wu et al. 2017). Biochar has a variety of effects on soil, depending on factors like plant species, application rate, biochar properties, and soil type (Ding et al. 2016). Both biochar type and addition ratio may have a significant impact on the effects of biochar amendment on soil and plants (Luo et al. 2014). The feedstock source, as well as pyrolysis variables such as temperature, residence time, and activation treatment, can all have a substantial impact on biochar characteristics (Cao and Harris 2010). Because pyrolysis can offset some of the drawbacks of direct application to soil, sewage sludge (SL) makes for an intriguing feedstock for the production of biochar. For instance, using SL biochar can reduce the amount of nickel and zinc that leaches out of raw SL (Méndez et al. 2012b). The use of SL biochar can benefit soil in ways like: B. Increases the soil's ability to hold water (Méndez et al. 2012a). Furthermore, Paz-Ferreiro et al. (2012) reported that adding biochar (8 wt%) from SL to soils supplemented with only raw sewage led to an increase in enzyme activity (Mendez et al. 2013). Electrical conductivity, phosphorus content, and nitrogen content of soils have all been shown to significantly improve with the addition of sewage sludge-derived biochar (SDBC) (Hossain et al. 2010). In addition, biochar has recently been demonstrated to be a successful way to remove organic and heavy metal contaminants from contaminated soil and water from tainted water and soil (Cao and Harris 2010; Cao et al. 2009).

In agricultural applications, soil amendment is frequently used to enhance the nutrient and water retention capacity, surface drainage, physical qualities of soils, such as air permeability. Biochar made from sewage sludge has recently attracted interest due to its potential to enhance soil and offer additional benefits for the treatment of sewage sludge (Beesley et al. 2011). A wide range of organic and inorganic molecules in the environment interact with biochar in specific ways that are controlled by these variations in physicochemical qualities. For instance, biochars made from manures, seaweed, and crop wastes (such as rye, maize), are often less and finer robust. Such biochars has high nutrient concentrations, making microbial populations in the environment more easily able to degrade them (Sohi and Naghdabadi 2009). Lower surface acidity and cation exchange capacity (CEC), comparative to low temperature biochars, high-temperature biochars have a reduced ability to retain nutritional cations in soil (Beesleyet al. 2011). Manure-based biochars may mineralize and release nutrients from the soil more quickly in comparison with plant-based biochars that are rich in compressed aromatic compounds (Novak et al. 2009; Singh et al. 2010). Additionally, it was shown that soils in tropical regions, supplemented with range of burned constituents had higher concentrations of extractable phosphorus (Lehmann et al. 2003). The various effects of sewage sludge biochar on soil and plants are presented in Table 1.

Table 1 Effect of sewage sludge-derived biochar on plant and soil

Treatment	Studied plant/ medium	Effects on soil and plants	References
Sewage sludge biochar produces at 400 and 600 °C)	Soil incubation	CO_2 emissions decreased by 11 and 32%	Méndez et al. (2013)
Composted sewage sludge biochar produces at 450 and 700	E5CK-T muffle furnace	Highest N concentration, yield, and ash content, high Ni, Cu, Cr, Zn, and lowest DOC concentrations	Titova and Baltrėnaitė (2021)
Feedstocks of wastewater sludge-derived biochar at 300–600	Fixed bed reactor	The largest reduction in DTPA-extractable metals by biochar (300 °C). Biochar of 400 °C tends to attain a more uniform surface charge	Lu et al. (2013)
Hydrothermal pre-treatment of sewage sludge biochar at 180 °C	UASB reactor	Heavy metals were immobilized well Water content reduced from 85 to 33 wt%	Li et al. (2018a)
Biochar at 200 to 700 °C	Turf grass in pots	There were notable increases in the amount of soil organic C fractions, accessible K, P, as well as increase in turf grass dry biomass and total heavy metal levels	Tian et al. (2019)
Sewage sludge + bamboo sawdust and rice husk at 400 and 700 °C	Bench-scale chamber-type atmospheric furnace	Compared to co-pyrolysis with sawdust, sewage sludge + husk generated increased immobilization of metal but appeared to have lower biochar stability	Zhang et al. (2020)
Sewage biochar at a different rate in heavy metal (Pb, Cd, Ni) contaminated soils	Incubation	Heavy metal concentration (Fe, Pb, Cu, Zn, Cr, As) decreased	Tsang et al. (2016)
Sewage sludge biochar at different rates	Rice in Field	The biochar increased the biomass and yield of rice plants. The biochar raised the pH of the soils. The amount of organic matter was greatly increased by applying biochar at a rate of 3 t/ha Less Cd bioaccumulation was seen in rice grains, plant roots, and husk	Zhang et al. (2016b)
Biochar derived from sewage sludge at different temperatures and with soybean straw, peanut shells and rice straw	Rice in the growth room	Sewage sludge enhanced the number of seedlings. Rice plants' bioaccumulation of Cd and As significantly decreased, although it rose in the soil	Li et al. (2018b)

(continued)

Table 1 (continued)

Treatment	Studied plant/ medium	Effects on soil and plants	References
Biochar is derived from sewage sludge	Rice in a greenhouse in Hg contaminated soil	Methylmercury (MeHg) content and pH of the soil were both raised by biochar Hg levels not affected in either soil or plant. When rice plants were treated with sewage sludge biochar, the Hg and MeHg concentrations in the plants were reduced	Zhang et al. (2019)
Biochar derived from sewage sludge, soybeans, and rice straw	Turnip in greenhouse	The turnip obtained its maximum levels when 2% of sewage sludge biochar was applied As, bioaccumulation of heavy metals greatly reduced. Soil levels of As, Cd, Cu, Pb, and Zn dropped	Khan et al. (2015)
Biochar derived from sewage sludge at different application rates	Cucumber in greenhouse	In soil, biochar decreased the availability of hydrocarbons, particularly when 10% biochar was applied Increase in sewage sludge application rate, the quantities of Cd, Cu, Pb, and Zn also rose. As, Pb, Cu, and Zn levels were decreased whereas Cd levels were raised by the biochar In the cucumber, the biochar considerably decreased As, Cu, Pb, and Zn while increasing Cd	Waqas et al. (2014)
Sewage sludge biochar	Cherry tomato in greenhouse	There was no discernible variation among the plant heights The tomatoes treated with biochar exhibited decreased quantities of trace/minor elements overall, with the exception of Sb and Sr	Hossain et al. (2015)
Different rates of sewage sludge and ammonium nitrate	Maize in field	Heavy metal concentration (Pb, Cu and Zn) in soil rose. P, N, C, K, and Na contents did not differ appreciably Plant Zn and Cu uptake was decreased when sewage sludge biochar utilized for treatment method. The highest results in the treatment with sewage sludge + AN were observed for Zn and Cu	Gwenzi et al. (2016)

(continued)

Table 1 (continued)

Treatment	Studied plant/ medium	Effects on soil and plants	References
Biochar (sewage sludge) at different temperatures and rates	Maize in the greenhouse	Sewage sludge biochar increased total N with the increase in dose and increased total P reported at 600 °C. The plant growth was hindered by the biochar generated at 550–700 °C	Gonzaga et al. (2017)
Biochar (sewage sludge and wheat chaff)	Sunflower in field	The sewage sludge biochar decreased C, maximized N values, decreased EC, and increased pH	de la Rosa et al. (2018)

2 Sewage Sludge Generation and Disposal

Sewage sludge is a residual material obtained as a result of municipal wastewater treatment. It is usually 1–2% of the total volume of treated wastewater (Guangyin and Youcai 2017). Sewage sludge can be grouped as primary and secondary sewage sludge. The primary sewage sludge is formed due to the gravitational settlement of suspended organic material in the wastewater and the secondary sewage sludge formation includes the role of microorganisms (Lamastra et al. 2018). The type and source of the wastewater determine the sewage sludge's chemical composition. Municipal wastewater is a multiphase medium made up of a wide range of different elements, such as microbial aggregates of different strains, organic and inorganic particles, polymeric compounds, and water (Guangyin and Youcai 2017). Wastewater treatment plants in USA is estimated to annually produce 6.5 million metric tons of dry solids (NEBRA 2007). On the other hand, China and Japan produce 3 and 2 million metric tons of solid waste annually, respectively (Ministry of Land, Infrastructure, Transport, and Tourism, Japan 2014). As the developing countries are also focusing on installation of wastewater treat plants, sewage sludge production is increasing at a high rate. The economic development status of a country also plays a crucial role in deciding the reuse and disposal rate of sewage sludge. Most developed countries are reusing their sewage sludge into producing building materials, agriculture, bioenergy generation, etc. However, the sewage sludge reuse in developing countries is very limited.

In recent times, the improper sewage sludge disposal has become a global challenge as it leads to environmental degradation of the ecosystem through soil and water pollution and also resulting into wastage of valuable resources (Praspaliauskas and Pedišius 2017). The safe disposal of sewage sludge is a difficult and costly maneuver. The common sewage sludge disposal routes are field application as an amendment, landfill, and incineration (Babatunde and Zhao 2007). The field-applied process of sewage sludge removal is affordable and enriches the soil with organic matter and nutrients. However, the risk is also increased by the presence of pathogens, organic contaminants, and heavy metals (Lamastra et al. 2018). The high moisture percentage of sewage sludge also makes landfilling of sewage sludge as a difficult

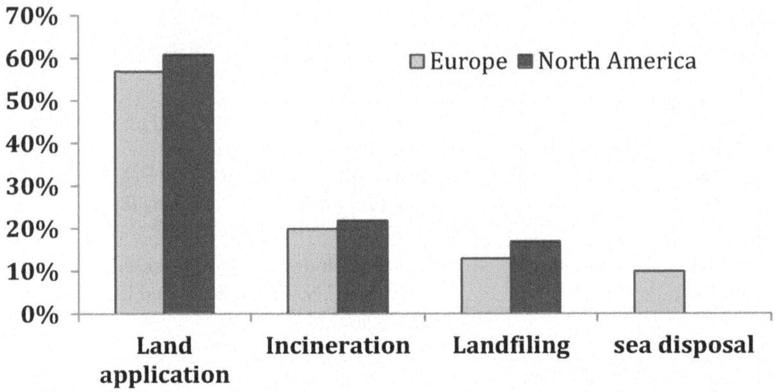

Fig. 1 Comparison of sewage sludge disposal in Europe and North America

disposal approach as it carries risk of sliding upon use of operation machinery and the sewage sludge required certain conditioning with solidifying agents before landfilling (Guangyin and Youcai 2017). The incineration is another approach for sewage sludge disposal as it reduces sludge volume to almost 90% (Hjelmar 1996). The ash content of sludge used as industry raw material to develop different by-products. However, this disposal method also added much cost to the total expenditure of wastewater treatment operation. The sewage sludge disposal strategies performed in Europe and North America are depicted in Fig. 1.

3 Sewage Sludge Derived Biochar (SDBC)

The land application of sewage sludge is a cost-effective method for sewage sludge disposal. However, the presence of heavy metals and other environmentally harmful object makes its use limited in agriculture. Therefore, use of biochar derived through sewage sludge carbonization for land application is an effective approach to safely dispose sewage sludge. The biochar is produced through pyrolytic conversion (heating at 300–700 °C in oxygen absence) of sewage sludge (Pathak et al. 2009). Due to its unique capabilities to make safe reuse of sewage sludge as an amendment in agriculture, improve soil structure and carbon sequestration, biochar is getting attention of scientist and policy makers worldwide (Steiner et al. 2008). Due to its porous surface and higher functional groups at its surface, biochar has large surface area, very high porosity, and extraordinary cation exchange capacity (Ahmad et al. 2014). Owing to these properties biochar has can retain high amount of nutrients, water in the soil. It also contains labile plant nutrients and organic carbon, which help to improve the soil structure and plant growth (Song and Guo 2012). Table 2 outlined the primary benefits of using SDBC in agriculture.

Table 2 Main advantages of SDBC

Sr.	Advantages	Explanation	References
1	Decrease heavy metals leaching	Biochar production significantly decrease the mobility and bioavailability of hazardous heavy metals through binding it within its structure, so protecting water quality and making soil suitable for plant growth	Velli et al. (2021)
2	Pathogen elimination	High temperature used in production of biochar kill pathogens present in sewage sludge reducing the risk of harmful pathogens in soil and water	Qian et al. (2019)
3	Carbon sequestration	Porous structure of biochar may efficiently capture and trapped CO_2 or help in fighting against climate change	Ghorbani et al. (2022)
4	Enhance plant growth	By delaying the release of micronutrients and slowing mineral formation, SDBC increases the favorable conditions for plant growth in soil compared to sewage sludge compost	Jakubus and Graczyk (2020)
5	Soil amendment	Biochar improves soil structure by increasing its nutrient uptake capacity, bulk density, electrical conductivity, high surface area, ion exchange capacity and lower pollutant load make it good option for soil amendment	Janus et al. (2015)

In addition, sewage sludge conversion to biochar is an excellent way to bring about enormous quantities of waste with significant environmental benefits which includes carbon sequestration, heavy metals immobilization, soil water retention, greenhouse gas reduction, etc. (Racek et al. 2020). Although, the presences of organic matter as well as large concentrations and numbers of essential nutrients in sewage sludge can improve soil and plant condition, its metal concentration is the main constraint for its direct land application. Typical sewage sludge is usually containing lead, nickel, chromium, copper, and other harmful metals which possess a great risk to the environment. These metals are soluble and highly mobile in the sewage sludge. However, many studies have showed that all these metals are immobile and stable in biochar made from sewage sludge. The pyrolysis reduces their release potential and makes them unable to get into the soil solution (Agrafioti et al. 2013). The sewage

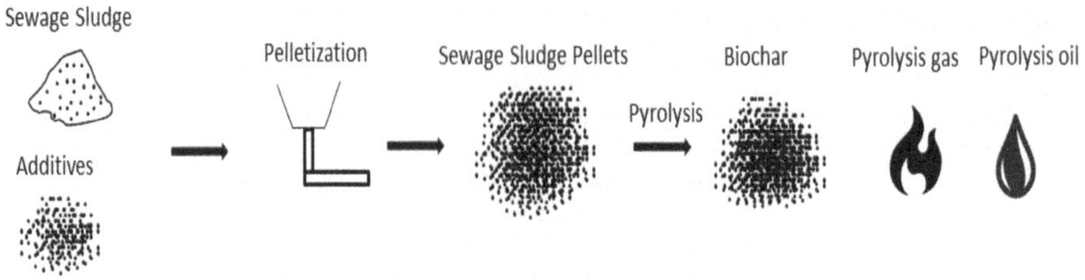

Fig. 2 Steps involved in biochar production through sewage sludge

sludge is pretreated before the pyrolysis process. The three steps pre-treatment of sewage sludge consist of dewatering, drying and pelletization of sewage sludge. The reduction of water content in sewage sludge helps to reduce the transportation and treatment cost of sewage sludge. It is also necessary to dewater and dry sludge before pyrolysis process (Liu et al. 2018). During pelletizing, additives could be added with the sewage sludge to enhance the quality of biochar and to meet regulatory standards of the respective country (Ševčík et al. 2018). During pyrolysis, the organic components of the sewage sludge are heated in O_2 absence at a certain time on a fixed temperature to carry out carbonization of sewage sludge and converting it to charcoal, producing oil and gas as the as end product in the process (Abnisa et al. 2011). The factors which most regulate the carbonization are the peek temperature, residence time and heating temperature (Raček et al. 2018). The temperature of pyrolysis is typically between 300 and 700 °C. A deeper level of sewage sludge is transformed quickly and at a faster rate thanks to the higher temperature, which also speeds up the carbonization process (Chen et al. 2017). Determination of residence time for solid is also important to achieve the complete pyrolysis of sewage sludge and it depends upon pyrolysis temperature and heating rate (Guo et al. 2019). In case of incomplete pyrolysis, uncarbonized carbon could be present and noticeable in biochar. The schematic representation of different steps involved in the process of biochar production from sewage sludge are depicted in Fig. 2.

4 Biochar Elemental Structure

As discussed earlier, biochar composition is mainly dependent on the pyrolysis type, process temperature, feedstock composition, residence time, and additives. The chemical composition of biochar is very hard to check due to large amount of inorganic component derived from heterogeneous sewage sludge. The sewage sludge biochar is comprises of organic portion which mainly contains organic carbon and a inorganic portion which contains heavy metal, bicarbonates and nutrients like Ca, Mg, K, and P (Racek et al. 2020). Biochar is resistant to microbial decay because of its high aromaticity (Mukome and Parikh 2015). Additionally, biochar has high levels of organic carbon and low levels of inorganic carbon (Xu et al. 2012). The

H:C, O:C, and C:N ratios are helpful indicators of biochar's aromacity, stability, and N mineralization (Racek et al. 2020).

The ash content of biochar is its inorganic component derived from sewage sludge. This mainly includes Ca, Mg, K, P, Fe, Na, and Zn, etc. (Özçimen and Ersoy-Meriçboyu 2010). Biochar also contains large concentration of essential macronutrients such as P an N, as they are also in abundance in wastewater and sewage sludge. Therefore, biochar made from sewage sludge could be used as fertilizer if use appropriately (Thangarajan et al. 2015). In addition, biochar contains higher surface area which makes it able to provide habitat to microorganisms, soil aggregation, water retention, and reduce metal contamination. The relationship between the surface area of biochar and pyrolysis temperature is direct and surface area can be determined by sorption of N_2 or CO_2 on biochar surface and calculated by surface area per gram of sample (Özçimen et al. 2010). In addition, biochar is generally alkaline in nature and could be used in acidic soil as an amendment to raise its pH, which could increase its base saturation, concentration of Ca, Mg, and improve soil microbial activity (Van Zwieten et al. 2010). The EC and CEC of biochar, which is mainly responsible for exchange of Ca, Mg, Na, and K, is mainly depend upon sewage sludge and pyrolysis process (Racek et al. 2020). Comparison of SDBC on soil quality and plant growth are describe in Fig. 3.

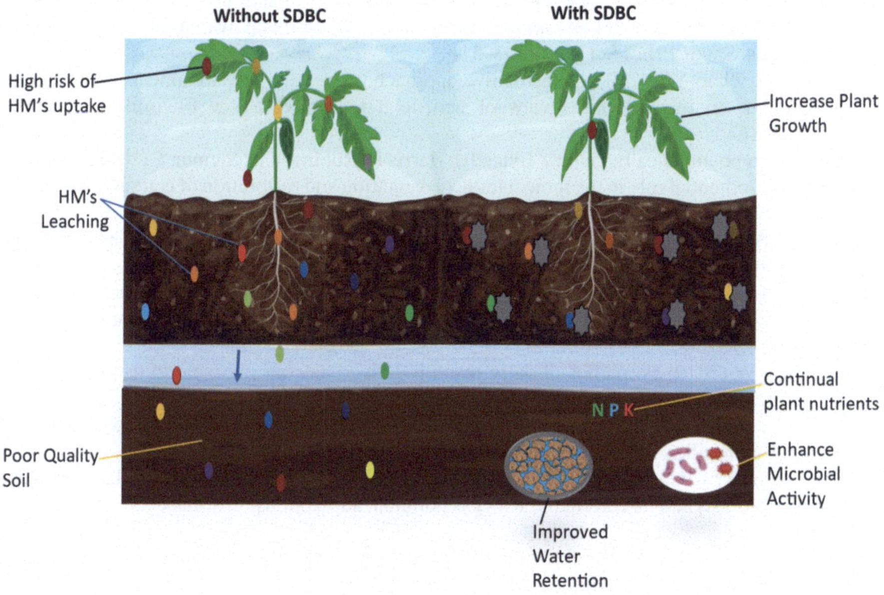

Fig. 3 Comparative effects of SDBC on soil with and without the addition of SDBC

5 Environmental Implications and Future Prospective

Among all treatment methods discussed previously in this chapter regarding sewage sludge, the use of sewage sludge usage as a feedstock for biochar can be a good, economically effective and environment friendly strategy to deal with huge waste. Through this method, nutrients (N, P, K, S, Zn, Cu, Fe, Mn, B), present in sewage sludge can be used as a fertilizer/amendment for crop production but the application rates or recommended doses should be defined through extensive field trials on different soil–plant–environment system.

References

Abnisa F, Daud WW, Husin WNW, Sahu JN (2011) Utilization possibilities of palm shell as a source of biomass energy in Malaysia by producing bio-oil in pyrolysis process. Biomass Bioenerg 35(5):1863–1872

Agrafioti E, Bouras G, Kalderis D, Diamadopoulos E (2013) Biochar production by sewage sludge pyrolysis. J Anal Appl Pyrol 101:72–78

Ahmad M, Rajapaksha AU, Lim JE, Zhang M, Bolan N, Mohan D, Ok YS (2014) Biochar as a sorbent for contaminant management in soil and water: a review. Chemosphere 99:19–33

Al-Wabel MI, Usman AR, El-Naggar AH, Aly AA, Ibrahim HM, Elmaghraby S, Al-Omran A (2015) Conocarpus biochar as a soil amendment for reducing heavy metal availability and uptake by maize plants. Saudi J Biol Sci 22(4):503–511

Babatunde AO, Zhao YQ (2007) Constructive approaches toward water treatment works sludge management: an international review of beneficial reuses. Crit Rev Environ Sci Technol 37(2):129–164

Beesley L, Moreno-Jiménez E, Gomez-Eyles JL, Harris E, Robinson B, Sizmur T (2011) A review of biochars' potential role in the remediation, revegetation and restoration of contaminated soils. Environ Pollut 159(12):3269–3282

Biederman LA, Harpole WS (2013) Biochar and its effects on plant productivity and nutrient cycling: a meta-analysis. GCB Bioenergy 5(2):202–214

Bielská L, Škulcová L, Neuwirthová N, Cornelissen G, Hale SE (2018) Sorption, bioavailability and ecotoxic effects of hydrophobic organic compounds in biochar amended soils. Sci Total Environ 624:78–86

Cao X, Harris W (2010) Properties of dairy-manure-derived biochar pertinent to its potential use in remediation. Biores Technol 101(14):5222–5228

Cao X, Ma L, Gao B, Harris W (2009) Dairy-manure derived biochar effectively sorbs lead and atrazine. Environ Sci Technol 43(9):3285–3291

Chen Y, Zhang X, Chen W, Yang H, Chen H (2017) The structure evolution of biochar from biomass pyrolysis and its correlation with gas pollutant adsorption performance. Biores Technol 246:101–109

Cieślik BM, Namieśnik J, Konieczka P (2015) Review of sewage sludge management: standards, regulations and analytical methods. J Clean Prod 90:1–15

de la Rosa JM, Rosado M, Paneque M, Miller AZ, Knicker H (2018) Effects of aging under field conditions on biochar structure and composition: implications for biochar stability in soils. Sci Total Environ 613:969–976

Ding Y, Liu Y, Liu S, Li Z, Tan X, Huang X, Zeng G, Zhou L, Zheng B (2016) Biochar to improve soil fertility. A review. Agron Sustain Dev 36(2):1–18

Feng L, Luo J, Chen Y (2015) Dilemma of sewage sludge treatment and disposal in China. Environ Sci Technol 49:4781–4782
Gaskin JW, Steiner C, Harris K, Das KC, Bibens B (2008) Effect of low-temperature pyrolysis conditions on biochar for agricultural use. Trans ASABE 51(6):2061–2069
Ghorbani M, Konvalina P, Walkiewicz A, Neugschwandtner RW, Kopecký M, Zamanian K et al (2022) Feasibility of biochar derived from sewage sludge to promote sustainable agriculture and mitigate GHG emissions—a review. Int J Environ Res Public Health 19(19):12983
Gonzaga MIS, Mackowiak CL, Comerford NB, da Veiga Moline EF, Shirley JP, Guimaraes DV (2017) Pyrolysis methods impact biosolids-derived biochar composition, maize growth and nutrition. Soil Tillage Res 165:59–65
Guangyin Z, Youcai Z (2017) Chapter one. Sewage sludge generation and characteristics. In: Pollution control and resource recovery, pp 1–11
Guo M, Xiao P, Li H (2019) Valorization of agricultural byproducts through conversion to biochar and bio-oil. In: Byproducts from agriculture and fisheries: adding value for food, feed, pharma, and fuels, pp 501–522
Gwenzi W, Muzava M, Mapanda F, Tauro TP (2016) Comparative short-term effects of sewage sludge and its biochar on soil properties, maize growth and uptake of nutrients on a tropical clay soil in Zimbabwe. J Integr Agric 15(6):1395–1406
Havukainen J, Zhan M, Dong J, Liikanen M, Deviatkin I, Li X, Horttanainen M (2017) Environmental impact assessment of municipal solid waste management incorporating mechanical treatment of waste and incineration in Hangzhou, China. J Clean Prod 141:453–461
Hossain MK, Strezov V, Chan KY, Nelson PF (2010) Agronomic properties of wastewater sludge biochar and bioavailability of metals in production of cherry tomato (*Lycopersicon esculentum*). Chemosphere 78(9):1167–1171
Hossain MK, Strezov V, Chan KY, Ziolkowski A, Nelson PF (2011) Influence of pyrolysis temperature on production and nutrient properties of wastewater sludge biochar. J Environ Manage 92(1):223–228
Hossain MK, Strezov V, Nelson PF (2015) Comparative assessment of the effect of wastewater sludge biochar on growth, yield and metal bioaccumulation of cherry tomato. Pedosphere 25(5):680–685
Jakubus M, Graczyk M (2020) Microelement variability in plants as an effect of sewage sludge compost application assessed by different statistical methods. Agronomy 10(5):642
Janus A, Pelfrêne A, Heymans S, Deboffe C, Douay F, Waterlot C (2015) Elaboration, characteristics and advantages of biochars for the management of contaminated soils with a specific overview on Miscanthus biochars. J Environ Manage 162:275–289
Khan S, Waqas M, Ding F, Shamshad I, Arp HPH, Li G (2015) The influence of various biochars on the bioaccessibility and bioaccumulation of PAHs and potentially toxic elements to turnips (*Brassica rapa* L.). J Hazard Mater 300:243–253
Laird DA (2008) The charcoal vision: a win–win–win scenario for simultaneously producing bioenergy, permanently sequestering carbon, while improving soil and water quality. Agron J 100(1):178–181
Lamastra L, Suciu NA, Trevisan M (2018) Sewage sludge for sustainable agriculture: contaminants' contents and potential use as fertilizer. Chem Biol Technol Agric 5(1):1–6
Lehmann J, Pereira da Silva J, Steiner C, Nehls T, Zech W, Glaser B (2003) Nutrient availability and leaching in an archaeological Anthrosol and a Ferralsol of the Central Amazon basin: fertilizer, manure and charcoal amendments. Plant Soil 249(2):343–357
Li C, Wang X, Zhang G, Li J, Li Z, Yu G, Wang Y (2018a) A process combining hydrothermal pretreatment, anaerobic digestion and pyrolysis for sewage sludge dewatering and co-production of biogas and biochar: pilot-scale verification. Biores Technol 254:187–193
Li G, Khan S, Ibrahim M, Sun TR, Tang JF, Cotner JB, Xu YY (2018b) Biochars induced modification of dissolved organic matter (DOM) in soil and its impact on mobility and bioaccumulation of arsenic and cadmium. J Hazard Mater 348:100–108

Lin L, Xu F, Ge X, Li Y (2018) Improving the sustainability of organic waste management practices in the food-energy-water nexus: a comparative review of anaerobic digestion and composting. Renew Sustain Energy Rev 89:151–167

Liu T, Liu Z, Zheng Q, Lang Q, Xia Y, Peng N, Gai C (2018) Effect of hydrothermal carbonization on migration and environmental risk of heavy metals in sewage sludge during pyrolysis. Biores Technol 247:282–290

Lu H, Zhang W, Wang S, Zhuang L, Yang Y, Qiu R (2013) Characterization of sewage sludge-derived biochars from different feedstocks and pyrolysis temperatures. J Anal Appl Pyrol 102:137–143

Luo Y, Jiao YJ, Zhao XR, Li GT, Zhao LX, Meng HB (2014) Improvement to maize growth caused by biochars derived from six feedstocks prepared at three different temperatures. J Integr Agric 13(3):533–540

Mašek O, Brownsort P, Cross A, Sohi S (2013) Influence of production conditions on the yield and environmental stability of biochar. Fuel 103:151–155

Méndez AM, Barriga S, Guerrero F, Guerrero GG (2012) The effect of paper mill waste and sewage sludge amendments on acid soil properties. Soil Sci 177(7):451–457

Méndez A, Gomez A, Paz-Ferreiro J, Gasco G (2012) Effects of sewage sludge biochar on plant metal availability after application to a Mediterranean soil. Chemosphere 89(11):1354–1359

Méndez A, Tarquis AM, Saa-Requejo A, Guerrero F, Gascó G (2013) Influence of pyrolysis temperature on composted sewage sludge biochar priming effect in a loamy soil. Chemosphere 93(4):668–676

Mukome FN, Parikh SJ (2015) Chemical, physical, and surface characterization of biochar. CRC Press, Boca Raton, FL, pp 67–98

NE Biosolids Residuals Association (2007) A national biosolids regulation, quality, end use and disposal survey. Tamworth, NH

Nie E, Zheng G, Gao D, Chen T, Yang J, Wang Y, Wang X (2019) Emission characteristics of VOCs and potential ozone formation from a full-scale sewage sludge composting plant. Sci Total Environ 659:664–672

Novak JM, Busscher WJ, Laird DL, Ahmedna M, Watts DW, Niandou MA (2009) Impact of biochar amendment on fertility of a southeastern coastal plain soil. Soil Sci 174(2):105–112

Obia A, Mulder J, Martinsen V, Cornelissen G, Børresen T (2016) In situ effects of biochar on aggregation, water retention and porosity in light-textured tropical soils. Soil Tillage Res 155:35–44

Özçimen D, Ersoy-Meriçboyu A (2010) Characterization of biochar and bio-oil samples obtained from carbonization of various biomass materials. Renew Energy 35(6):1319–1324

Pathak A, Dastidar MG, Sreekrishnan TR (2009) Bioleaching of heavy metals from sewage sludge: a review. J Environ Manage 90(8):2343–2353

Paz-Ferreiro J, Gasco G, Gutiérrez B, Mendez A (2012) Soil biochemical activities and the geometric mean of enzyme activities after application of sewage sludge and sewage sludge biochar to soil. Biol Fertil Soils 48(5):511–517

Paz-Ferreiro J, Nieto A, Méndez A, Askeland MPJ, Gascó G (2018) Biochar from biosolids pyrolysis: a review. Int J Environ Res Public Health 15(5):956

Prapagdee S, Tawinteung N (2017) Effects of biochar on enhanced nutrient use efficiency of green bean, Vigna radiata L. Environ Sci Pollut Res 24(10):9460–9467

Praspaliauskas M, Pedišius N (2017) A review of sludge characteristics in Lithuania's wastewater treatment plants and perspectives of its usage in thermal processes. Renew Sustain Energy Rev 67:899–907

Qian T, Yang Q, Jun DCF, Dong F, Zhou Y (2019) Transformation of phosphorus in sewage sludge biochar mediated by a phosphate-solubilizing microorganism. Chem Eng J 359:1573–1580

Racek J, Sevcik J, Chorazy T, Kucerik J, Hlavinek P (2020) Biochar–recovery material from pyrolysis of sewage sludge: a review. Waste Biomass Valor 11(7):3677–3709

Raček J, Chorazy T, Novák V, Čáslavský J, Hlavínek P (2018) Energy efficiency of dry sewage sludge before and after low-temperature microwave pyrolysis. In: Naxos 2018 6th International Conference on Sustainable Solid Waste Management. Athens, Greece: National Technical University of Athens, pp 1–11

Ševčík J, Raček J, Hluštík P, Hlavínek P, Dvořák K (2018) Microwave pyrolysis full-scale application on sewage sludge. Desalin Water Treat 112:161–170

Singh B, Singh BP, Cowie AL (2010) Characterisation and evaluation of biochars for their application as a soil amendment. Soil Res 48(7):516–525

Sohi AN, Naghdabadi R (2009) Stability of single-walled carbon nanopeapods under combined axial compressive load and external pressure. Phys E 41(3):513–517

Song W, Guo M (2012) Quality variations of poultry litter biochar generated at different pyrolysis temperatures. J Anal Appl Pyrol 94:138–145

Steiner C, Glaser B, Geraldes Teixeira W, Lehmann J, Blum WE, Zech W (2008) Nitrogen retention and plant uptake on a highly weathered central Amazonian Ferralsol amended with compost and charcoal. J Plant Nutr Soil Sci 171(6):893–899

Thangarajan R, Bolan N, Mandal S, Kunhikrishnan A, Choppala G, Karunanithi R, Qi F (2015) Biochar for inorganic contaminant management in soil. In: Biochar production, characterization, and applications. CRC Press, Taylor& Francis, Boca Raton, pp 100–138

Tian Y, Zhang J, Zuo W, Chen L, Cui Y, Tan T (2013) Nitrogen conversion in relation to NH3 and HCN during microwave pyrolysis of sewage sludge. Environ Sci Technol 47(7):3498–3505

Tian Y, Cui L, Lin Q, Li G, Zhao X (2019) The sewage sludge biochar at low pyrolysis temperature had better improvement in urban soil and turf grass. Agronomy 9(3):156

Titova J, Baltrėnaitė E (2021) Physical and chemical properties of biochar produced from sewage sludge compost and plants biomass, fertilized with that compost, important for soil improvement. Waste Biomass Valor 12(7):3781–3800

Tsang DC, Zhou F, Zhang W, Qiu R (2016) Stabilization of cationic and anionic metal species in contaminated soils using sludge-derived biochar. Chemosphere 149:263–271

Uchimiya M, Wartelle LH, Klasson KT, Fortier CA, Lima IM (2011) Influence of pyrolysis temperature on biochar property and function as a heavy metal sorbent in soil. J Agric Food Chem 59(6):2501–2510

UNEP, WHO, UN-Habitat (2014) Monitoring sustainable development goal on water task team on earth observations, novel data collection and data integration. Report on the first meeting, Tokyo: 30–31 May 2014

UN-Habitat (2008) Global Atlas of excreta, wastewater sludge and biosolids management: moving forward the sustainable and welcome uses of a global. United Nations Human Settlements Programme, Nairobi. http://esa.un.org/iys/docs/san_lib_docs/habitat2008.pdf. Accessed 30 September 2022

Van Zwieten L, Kimber S, Morris S, Chan KY, Downie A, Rust J et al (2010) Effects of biochar from slow pyrolysis of papermill waste on agronomic performance and soil fertility. Plant Soil 327(1):235–246

Vardon DR, Sharma BK, Blazina GV, Rajagopalan K, Strathmann TJ (2012) Thermochemical conversion of raw and defatted algal biomass via hydrothermal liquefaction and slow pyrolysis. Biores Technol 109:178–187

Velli P, Manolikaki I, Diamadopoulos E (2021) Effect of biochar produced from sewage sludge on tomato (*Solanum lycopersicum* L.) growth, soil chemical properties and heavy metal concentrations. J Environ Manag 297:113325

Wang Z, Xie L, Liu K, Wang J, Zhu H, Song Q, Shu X (2019) Co-pyrolysis of sewage sludge and cotton stalks. Waste Manage 89:430–438

Waqas M, Khan S, Qing H, Reid BJ, Chao C (2014) The effects of sewage sludge and sewage sludge biochar on PAHs and potentially toxic element bioaccumulation in Cucumis sativa L. Chemosphere 105:53–61

Wu H, Lai C, Zeng G, Liang J, Chen J, Xu J, Dai J, Li X, Liu J, Chen M, Lu L (2017) The interactions of composting and biochar and their implications for soil amendment and pollution remediation: a review. Crit Rev Biotechnol 37(6):754–764

Xu G, Lv Y, Sun J, Shao H, Wei L (2012) Recent advances in biochar applications in agricultural soils: benefits and environmental implications. Clean Soil Air Water 40(10):1093–1098

Yang G, Zhang G, Wang H (2015) Current state of sludge production, management, treatment and disposal in China. Water Res 78:60–73

Yuan H, Lu T, Huang H, Zhao D, Kobayashi N, Chen Y (2015) Influence of pyrolysis temperature on physical and chemical properties of biochar made from sewage sludge. J Anal Appl Pyrol 112:284–289

Zhang Y, Chen T, Liao Y, Reid BJ, Chi H, Hou Y, Cai C (2016a) Modest amendment of sewage sludge biochar to reduce the accumulation of cadmium into rice (*Oryza sativa* L.): a field study. Environ Pollut 216:819–825

Zhang Z, Zhou Y, Zhang J, Xia S, Hermanowicz SW (2016b) Effects of short-time aerobic digestion on extracellular polymeric substances and sludge features of waste activated sludge. Chem Eng J 299:177–183

Zhang J, Wu S, Xu Z, Wang M, Man YB, Christie P, Liang P, Shan S, Wong MH (2019) The role of sewage sludge biochar in methylmercury formation and accumulation in rice. Chemosphere 218:527–533

Zhang J, Jin J, Wang M, Naidu R, Liu Y, Man YB, Liang X, Wong MH, Christie P, Zhang Y, Song C (2020) Co-pyrolysis of sewage sludge and rice husk/bamboo sawdust for biochar with high aromaticity and low metal mobility. Environ Res 191:110034

Chapter 5
Production of Biochar from Algal Biomass and Their Applications

Shilpa Borehalli Mayegowda, M. Yashwant, Gitartha Sarma, Nagachandra Reddy, and N. G. Manjula

Abstract Algal biomass represents one of the most promising targets as a source of biofuel production, bioenergy, as sustainable feedstock and other valuable products. Algal biomass with its physicochemical properties and potential use has gained immense interest presently for converting it into biochar. Algal biochar synthesis has been done using conventional methods like pyrolysis, slow, and microwave-assisted methods along with modern ones like hydrothermal carbonization and torrefaction. Algal biochar has been variedly derived from the macroalga from fresh, brackish, and marine waters. This algal biochar has been extracted for maximum nutrient contents and capacity to exchange ions, as it is high in pH, ash, nitrogen sources, and inorganic components like phosphorus (P), potassium (K), calcium (Ca), and magnesium (Mg). Biochar has been produced from fresh water macroalga *Cladophora glomerata,* as a biosorbent of heavy metals that have high toxicity. Other algal biochars include *Chlorella vulgaris, Dunaliella salina, Haematococcus pluvialis, Gracilaria,* and *Oedogoniumsps* has high affinity toward oxyanions like arsenic, molybdenum, and selenium that are difficult to be removed using conventional techniques. Algal biochar is a boon for amending soil for agricultural purposes especially applicable to acidic soils increasing the crop productivity and as absorbents for removing organic and inorganic contaminants in waste water treatment.

Keywords Algal biochar · Remediation · Macroalga

S. B. Mayegowda
Department of Psychology, CHRIST University, Bengaluru, Karnataka, India

M. Yashwant · G. Sarma · N. Reddy · N. G. Manjula (✉)
Department of Microbiology, School of Basic and Applied Sciences (SBAS), Dayananda Sagar University, Bengaluru, Karnataka, India
e-mail: manjulang-sbas@dsu.edu.in

A. Roy et al. (eds.), *Biochar Production Engineering*,
https://doi.org/10.1007/978-3-031-92594-8_5

1 Introduction

Biomass has been considered as a potential renewable energy resource, abundance in nature with carbon neutral properties (Morgan et al. 2019; Hannon et al. 2010). Generally, algae represent as one of the biochar productions for its various applications (Fig. 1). Since millions of years, most single-celled organisms like green and blue-green algae have existed on earth that predominantly live in the aquatic ecosystems, with the components necessary for propagation exist within the single cell which constitute their entire being. Unlike higher plants, that are lignocellulosic in composition such as soybean, cotton seed, agricultural and forest residues that are made up of three major polymers, i.e., hemicellulose, cellulose, and lignin that do not exist in algae. These algal biomass and cyanobacteria contain elements such as lipids, proteins, nucleic acid, and non-cellulosic carbohydrates, saccharides however, only certain algae contains significant amounts of lignin and cellulose (Heilmann et al. 2010; He et al. 2018). With many intrinsic advantages, like short growth cycles, high yield/hectare and most importantly growth in both waste water and salt water makes them more significant (Yuan et al. 2015; Cerciello et al. 2021; Brindhadevi et al. 2021; Gautam and Vinu 2020). These attractive features enable us to harness them efficiently at an earlier stage to use in innovative techniques (Bach and Chen 2017).

Algae are economical and ecologically important for providing essential needs that includes foods as biomass, animal feeds, soil additives, phycocolloids, and nutraceuticals (Stengel and Connan 2015). Certain species of algae have been commercially used in the production of food and phycocolloids. With rapid growth rate, they assimilate high amounts of nutrients such as nitrogen, phosphorous and

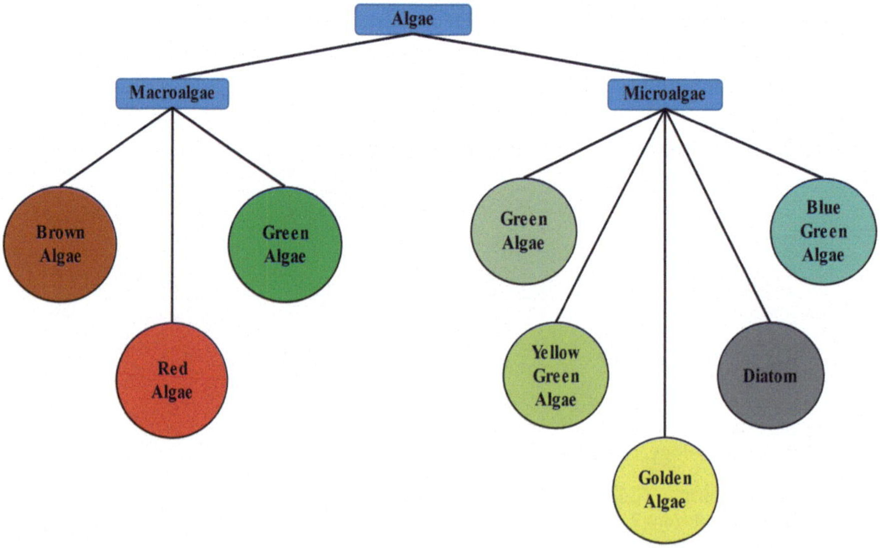

Fig. 1 Classification of algae into macroalgae and microalgae

also sequester heavy metals helping in bioremediation and waste water treatment (Joshi et al. 2024; Barrington et al. 2009). Additionally, algae with increased diversity in water ecosystem, observed in all types of water sources as marine, freshwater, and brackish commending as best for eutrophication treatment. However, it also leads to anthropogenic activities particularly with nutrients uptake in these enclosed waters (Liu et al. 2013; Cohen and Fong 2006). Examples include genera from green algae like *Cladospora, Ulva, prolifera, Chaetomorpha, and Rhizoclonium* that are fast growing with rapid colonization in nutritional rich environment (Liu et al. 2009; Pang et al. 2010).

The advantage of using algal biomass as a thermochemical feedstock takes limited days for biochar conversion and it can be cultured on arable and non-arable land with usage of water salinity and nutritional regime (Nwobaet al. 2020; Taylor et al. 2001; de Paula Silva et al. 2008; Nelson et al. 2008). In turn, acts as an attractive solution to sequester carbon by photosynthesis using of both microalgae and microalgae being a cost-effective process. There are three types of biofuels as solid, liquid, and gas, that has been gained immense popularity by algal biomass thus reducing dependence on fossil fuels (Bird et al. 2011). Algal biomass has a higher growth rate comparatively to plants making their harvest and processing faster, simple, flexible with suitability to deliver biofuels production rate in an increased amount (Xu et al. 2014; Huang et al. 2017; Neveux et al. 2014).

These algal species have been used for bioremediation process and one potential use of such biomass includes biochar production (charcoal). Biochar a probable tool to sequester carbon as a source of soil ameliorant to help improvise the water holding capacity and soil nutrients. Additionally, it also provides substrates to enhance the microbial activity in soil with high nutrients, including minerals (Ruperez 2002; Ross et al. 2008). Hence, algal biochar has been a potential carbon sequestrating and soil ameliorating source with production of high value end products as shown in Fig. 2.

2 Biochar and Its Uses

Biochar is a solid residue which is carbonaceous in nature and can be obtained when biomass is heated in conditions of oxygen deficiency (Roy and Bharadvaja 2021). It has multiple uses which include sequestration of carbon from the environment, recycling of nutrients, management of various waste systems, conditioning of soil, etc. (Brewer 2012). In recent times, algal biomass has been a source of renewable energy that is gaining attraction as an alternative sustainable substitute to other methods for the sequestration of carbon dioxide from the soil in order to reduce the everyday consequences of greenhouse effect and also global warming. Algal biomass has shown to be a suitable candidate for its rapid growth, wide availability, capability to rapidly fix atmospheric CO_2. Although, biomass can be used by itself, conversion of algal biomass to biochar provides a better opportunity for further utilization in biorefineries and other applications too. The production of biochar from biomass is done using the processes of pyrolysis, torrefaction, or hydrothermal liquefaction. With

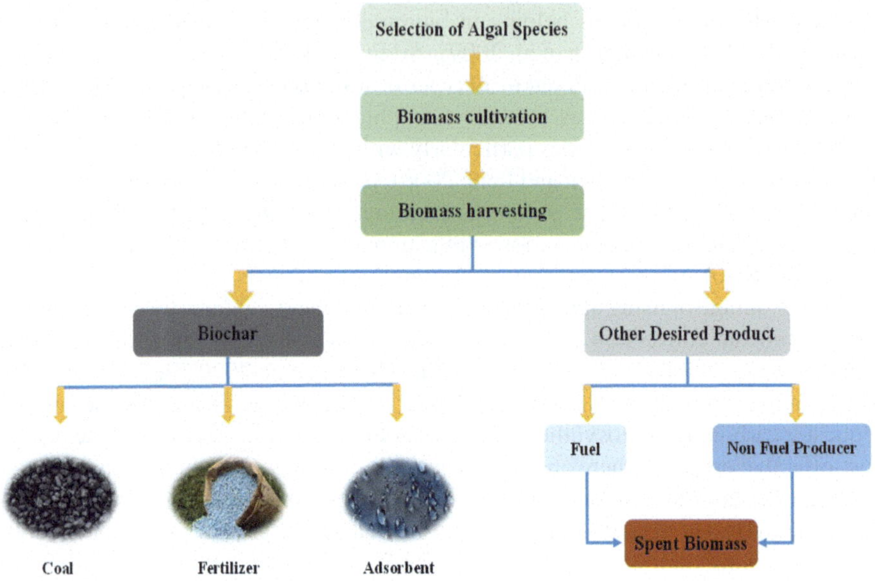

Fig. 2 Selection of algal biomass and transforming them into biochar as well as other important products

its higher potentiality toward wider applications as compared to the biomass, it can be used for its benefits as fertilizer, adsorbing environmental pollutants, source for biological source of medical applications, etc. (Yu et al. 2017). Biochar is a carbonaceous material and produced when forestry residues, wood waste, animal residues, algal biomass is thermally decomposed either incomplete absence of oxygen or by partial absence of oxygen. However, the nutritional components vary between the algal species, and research is being carried out on algal biochar feedstock during pyrolysis for varied applications and its potential in generating energy. There are various types of pyrolysis techniques employed to produce biochar from biomass as in Fig. 3 (Ross et al. 2008; Brennan and Owende 2010). However, more investigation pertaining to the issue, needs to be addressed on different algal species for their sustainability toward carbon sequestration, enhancing toward soil fertility by both microalgae and macroalgae (Ross et al. 2008; Grierson et al. 2009).

Algae are economically and ecologically important as animal feed, soil additives, food, phycocolloids, and nutraceuticals, as their growth rates are rapid that serve to assimilate nitrogen (N), phosphorous (P), and sequester heavy metals to help solve eutrophication and remediation from soil and waste waters (Mata et al. 2010; Archana et al. 2021). With recent times, agricultural and aquaculture waste water streams is a good nutrient source for algal biomass, either used as reusable source or value-added products (Prein 2002; Barrington et al. 2009). Macroalgae being feasibly used in waste water remediation, industrial moderation or sequestration of carbon source (C) through biological carbon factors in capturing and storage,

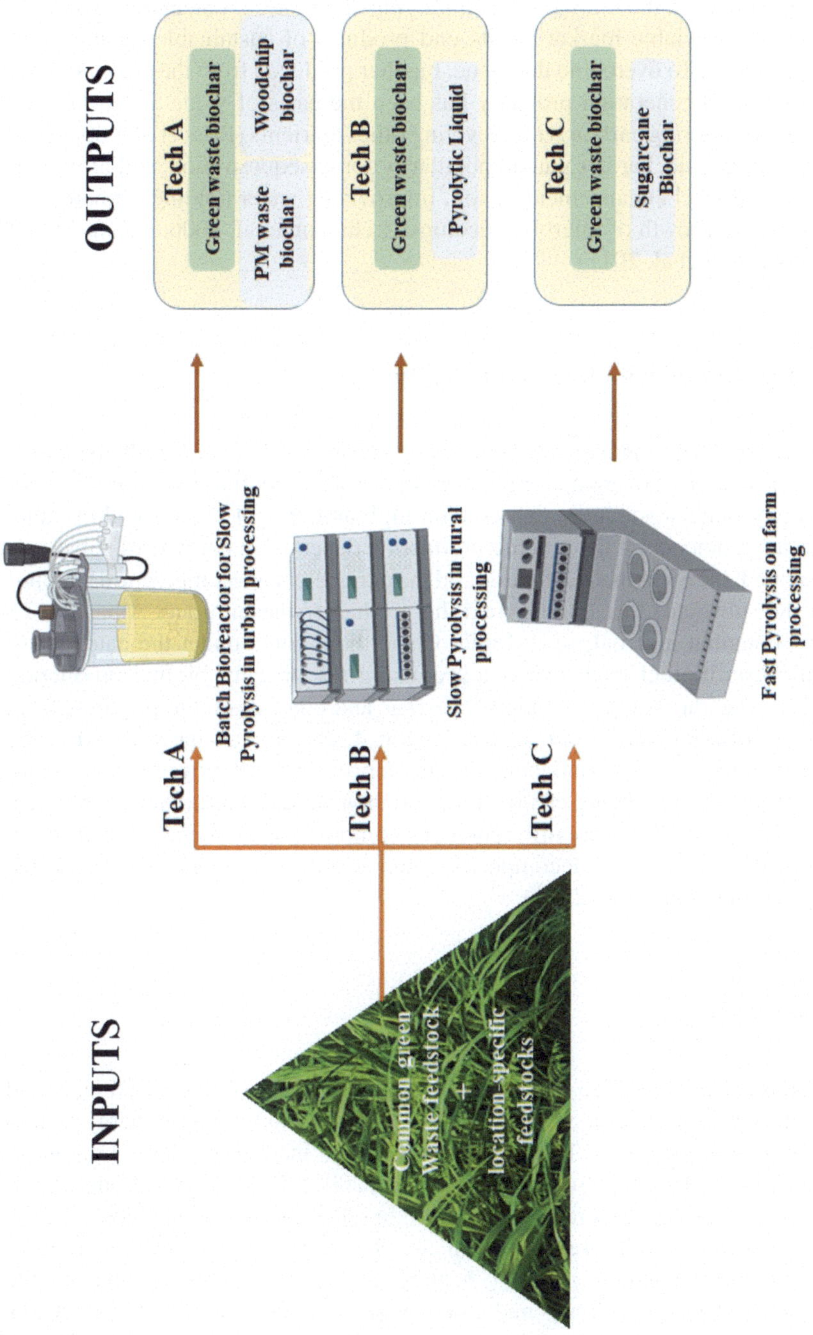

Fig. 3 Types of pyrolysis techniques employed when making biochar from biomass

similar to those of power stations run by coal or coal seam gas production (Wang et al. 2020). However, the limitation in employing algal-based remediation and CCS system requires a viable market for its end products of sustainable quantities of biomass produced. To overcome this issue, biochar produced from the algal biomass with its energy co-generation probably has been the basis of value-added product along with sequestering carbon with recycling other nutrients (Ross et al. 2008; Bird et al. 2011). Algal biochar has gained popularity for C-sequestration, soil amelioration, nutritional level enhancement in soil, improvising water holding capacity and also favoring the growth of microbial ecology (Lehmann et al. 2006; Lehmann and Joseph 2009; Bird et al. 2011).

3 Production of Biochar from Algae

The biochar production is typically done using substrates that are readily available. Basically, it includes the abundant resources such as agricultural residues, animal waste, woody plant waste, etc., that contain high amounts of carbon within them. However, the efficiency of the biochar production depends on the type of substrates used along with the combined technique and the parameters like temperature, composition of the substrates that finally yield the product (Alhashimi and Aktas 2017). Following potential macroalgae and microalgae, both come under the category of third generation biofuel feedstock is a promising resource for biofuel production (Das 2021). The chronology of making biochar and other desirable products from different type of algal biomass are depicted in Fig. 4. The biomass has been subjected to various techniques of thermochemical decomposition, with physical and chemical state of the biomass being changed irreversibly. With different thermochemical techniques, the yield results are different in the composition of biochar. The different methods of thermochemical decomposition are as follows: torrefaction, pyrolysis, and gasification (Mona et al. 2021).

3.1 Torrefaction

A pre-treatment method also referred as mild or slow pyrolysis, roasting, wood cooking, drying in high temperature for conversion of biomass into energy (Amin 2009; López Barreiro 2013). The process involves an inert atmosphere with atmospheric pressure and temperature varying between 200 and 300 °C. Additional classification is based on the temperatures used as light (200–235 °C), mild (235–275 °C), and high (275–300 °C). There is a significant change in the physiochemical properties of the fuel, including grinding, heating, ash content, moisture and volatile substances (condensable volatiles and gases) along with the density (Yu et al. 2017). The four main parts of torrefaction system includes, the reaction unit also known as torrefactor, the gas cleaning unit and inert gases cylinder with nitrogen (N_2) as

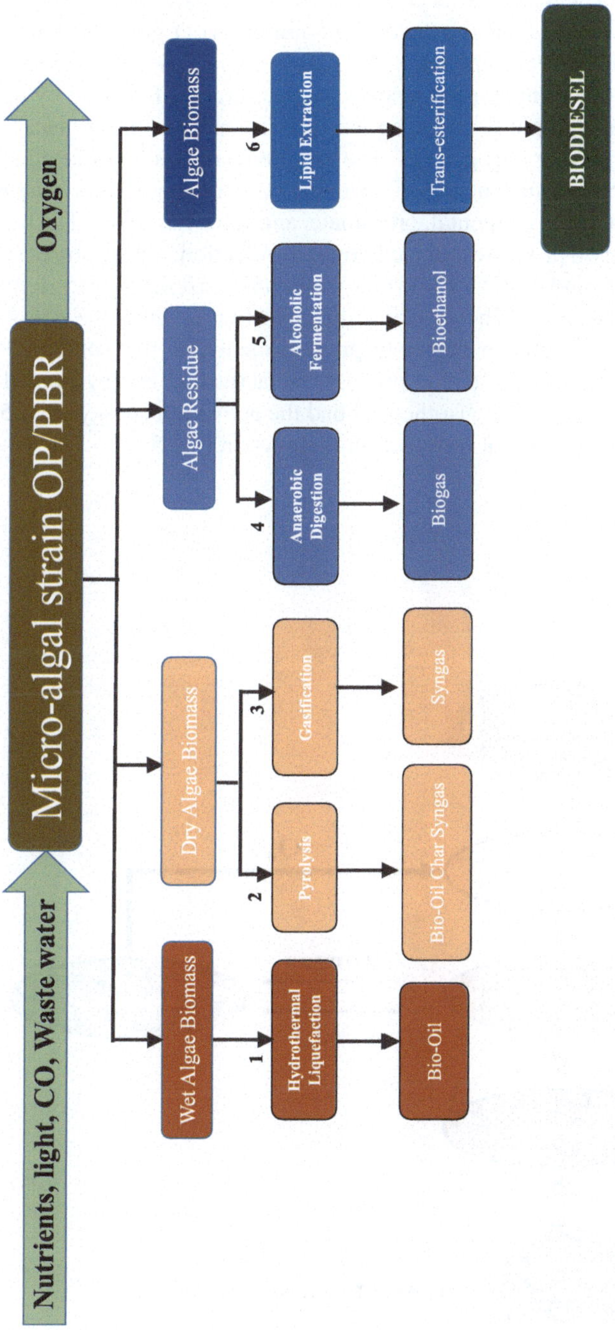

Fig. 4 Flowchart depicting the chronology of making biochar and other desirable products from different type of algal biomass

inert gas to prevent oxygen contamination, a rota-meter that regulates the flow of gas to the torrefactor. The treatment becomes thermally unstable when oxygen quantity increases with nitrogen gas being the most common carrier gas as it is abundant and cost-effective. The power is applied to the reaction unit and it is measured by watt-hour meter. A desired temperature reaches to tube furnace, followed by quartz tube filled with biomass sample to create an inert environment in the furnace connected by a quartz tube carrying nitrogen gas. The torrefied biomass is collected after being cooled in the reaction unit. The physiochemical properties of biomass are measured by analysis that includes elemental, proximate, and calorific values. The basic difference between the two processes: in isothermal torrefaction, temperatures range from 200, 235, 250, 275, and 300 °C, while non-isothermal torrefaction 0.5, 1, and 1.5 °C/min are taken in account. When torrefaction condition is same, the non-isothermal torrefaction produce a greater degree of pre-treatment than the isothermal torrefaction. In non-isothermal torrefaction process, the activation energy ranged between 44.14 and 88.41 kJ/mol, while isothermal had the activation energy at 57.52 kJ/mol (Das 2021). The procedure of torrefaction is shown in Fig. 5.

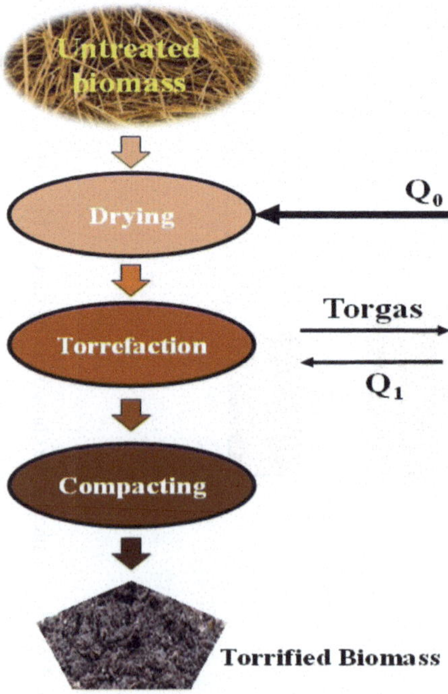

Fig. 5 Procedure of torrefaction (torgas is a by-product)

Types of Torrefactions

Conventional Torrefaction

The most common algal production used for solid biofuels with temperatures ranging from 200 to 300 °C with a time duration of 10–60 min. The process involves pre-treatment along with the existence of an inert or carrier gas (primarily nitrogen), carbon loss is slight as the temperature is raised from 200 to 300 °C and extended from 15 to 60 min, while the loss of hydrogen and oxygen alter noticeably. Following which the product is made up of carbon (74%), hydrogen (35%), and oxygen (28%). The weight loss increment is maximum reference operation given by torrefaction severity index (TSI) varying from 0 to 1 at 300 °C with a 60mins residence period. The analysis of thermal degradation demonstrates that higher torrefaction temperatures with shorter residence times are required for improvising the biomass quality to save energy as opposed to lower torrefaction temperature for a period of 60 min. Accordingly, N_2 is adequately reliable in the form of a carrier gas, as compared to the carbon dioxide in terms of energy production. The high levels of reliability, stability, and controllability are the main benefits while, difficulties are the longer residency period and operational cost (Das 2021).

Oxidative Torrefaction

The process is identical to the conventional torrefaction, except that air as carrier gas is utilized rather than nitrogen gas and occasionally, different oxidizing gases are utilized. The primary mechanisms are biomass oxidation and devolatilization resulting in a larger ash content. The oxidative torrefaction is more efficient and practical compared to non-oxidative process as it requires a significant shorter residence time and produces superior energy efficiency. Oxidative biochar produces more energy and solid yields than in non-oxidative forms as observed in *Nanochloropsis oceanica, Chlorella sps* ranging from 49.59 to 96.58% and 50.12 to 93.26%, respectively. Maximum higher heating value (HHV) of 9.48 MJ/kg was detected under the optimal torrefaction conditions of 200 °C and 60 min of residence time. The amount of processing time and energy needed is greatly decreased and additionally, inexpensive air is employed as a carrier gas. However, oxidative gases and air that is used as carrier gas gives a lower solid yield and a larger formation of ash content. The air has 21% oxygen, which impact on the process performance and increases the amount of solid ash produced resulting in somewhat lower HHV (Das 2021).

Microwave Torrefaction

The algal biomass can be pyrolyzed from the material present inside the cells during the process using microwave irradiation. The reaction unit for microwave heating is composed of a quartz bottle, a microwave absorbing media, and a microwave furnace. The bottle is made up of quartz, retained inside the reaction unit which is loaded with biomass and cannot reach the proper temperature if it does not absorb enough microwave radiation. To enhance the microwave torrefaction, effectiveness and its adsorption, a layer of media that can help to absorb microwave can be covered on the quartz bottle. The primary distinction between conventional and microwave heating is that with traditional heating, heat is transferred from the biomass surface to its interior. While, microwave heating, a phenomenon known as reverse heat transmission takes place that is more effective (Das 2021).

Wet Torrefaction

The algal biomass can be torrefied with varying temperatures in between 180 and 260 °C for about 10–24 min, with high pressure higher than the saturated vapor pressure of water with 5–20 MPa to be suggested subcritical pressure range. Compared to traditional torrefaction, this technique greatly improved the quality of the solid yield. For algal feedstocks with a higher moisture content, it works better, improves fuel quality by lowering the biomass ash level (Bach and Skreiberg 2016).

3.2 Pyrolysis

It is carried out in the absence of oxygen at higher temperatures, in which the catalyst may or may not be present. The biomass that is being thermochemically decomposed, can be done with diverse temperatures at different times in order to yield biofuels that possess low to medium calorific values. Ultimately, the end product obtained can be made of three types, viz. biochar, bio-oil, or gaseous substances (Fig. 6) (Sirajunnisa and Surendhiran, 2016; Xiong et al. 2017). While, the conversion of bio-oil or liquid requires temperature of variation between 400 and 550 °C, followed by a transitory residence time and the end products or biochar is produced at low temperature. However, a wide pyrolytic temperature is beneficial to produce biogas. Further, this process is classified into four modes of operation based on the heating rates: fast, slow, catalytic, and microwave pyrolysis. The procedure lasts 10–120 min at a temperature between 300 and 800 °C. In contrast to the catalytic pyrolysis, which requires reaction temperatures as low as 300 °C, microwave pyrolysis requires a maximum reaction temperature of 800 °C. However, pyrolysis often occurs at a high temperature and without oxygen (Yu et al. 2017).

The three main processes in the conversion of microalgae to bioenergy includes dehydration, devolatilization, and breakdown. One of the main pyrolysis processes includes the devolatization occurring at 200–550 °C, while dehydration occurs below 200 °C, and the ultimate stage disintegration of solid biomass occurs above 550 °C. Algae have a higher concentration of lipids, proteins, and carbohydrates. During the process, proteins and carbohydrates decompose at temperatures at 400 °C, while, the decomposition of lipids needs an exceeding high temperature of 550 °C (Das 2021).

Types of Pyrolysis

Fast Pyrolysis

A very high heating rate of 600 to 36,000 °C/min with a short holding/residence duration of 0.003–0.05 min is used. The process occurs in the absence of oxygen at

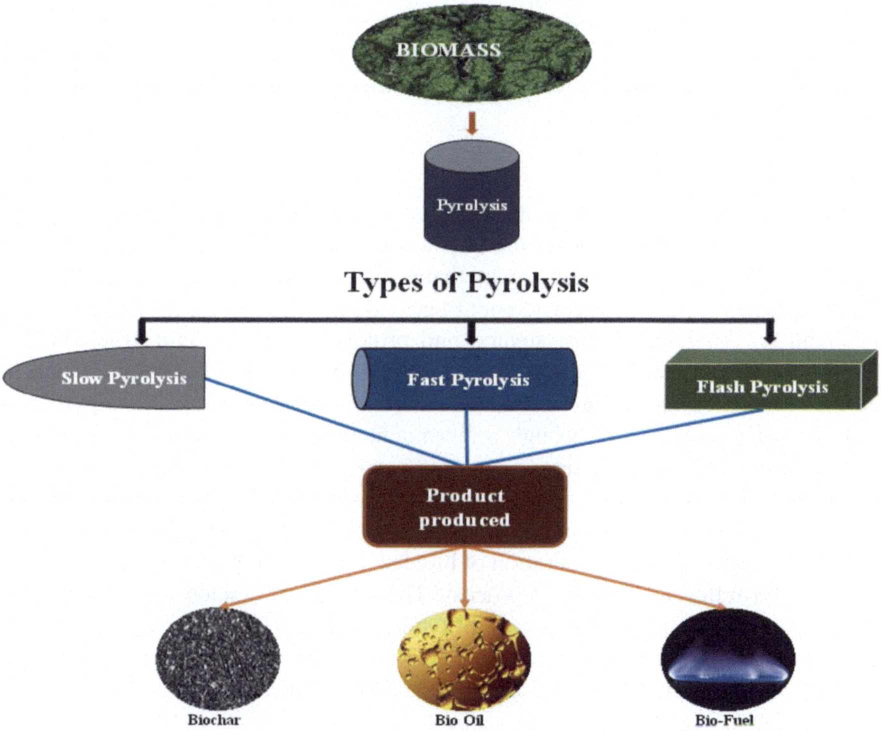

Fig. 6 Using pyrolysis technique to make biochar, bio-oil, and bio-fuel

a high temperature of 400–700 °C. This algal biomass is first dried and ground into particles that are no larger than 1 mm and added to the reactor, rapid heat breaks down the biomass into gases and molecules with shorter chains of small particle size, then cooled to create liquid biofuels. The efficiency of biomass with greater heating rate, a high temperature, and a shorter residence time along with the quick cooling increase the yield of oil from 60 to 75%, whereas solids and gaseous products range from 15 to 25% and 10 to 20%. A study estimated oil yield and algal biomass conversion for HHV ranged from 17.5 to 72% of weight and 18.4 to 41 MJ/kg, respectively (Yu et al. 2017).

An extreme HHV of the algal biomass that produces bio-oil was 29 MJ/kg that has been almost 1.4 times greater than the wood. Microalgae with low lipid content like *Chlorella vulgaris* reported 42.2% bio-oil at 400 °C. The biomass with a higher protein content causes the formation of nitrogen gas lowering the content of nitrogen in the bio-oil, a crucial task accomplished by using a quick process involving pyrolysis. It was discovered that biochar contains inorganic components that could be effective in boosting crop nutritional levels and its production. When *Chlorella vulgaris* were pyrolyzed at 500 °C, the biomass production, biogas, and bio-oil were assessed to be 31, 10, and 53% of the total weight, respectively. Correspondingly, recovered energy from bio-oil, bio-char, and gas was 57.4%, 36.8%, and 2.69%. Algal biochar from pyrolysis is utilized as fertilizer, since it contains significantly enhanced nutrients as sodium (Na), calcium (Ca), magnesium (Mg), potassium (K), and phosphorous (P) (Das 2021).

Slow Pyrolysis

Low heating rates of 3.5–10 °C/min and same lengthy holding times are used for slow pyrolysis (20–120 min) for converting biomass to more biochar and non-condensable gases like carbon dioxide, hydrogen, methane, etc. This technique is extremely slow rate, the preheat treatment is carried out to get rid of algal biomass moisture content and bond-breaking into lipids, sugars, and proteins, with resultant products like biochar, liquids, or gases. Finally, biochar breaks down into solid residues rich in carbon. *Spirulina platensis* biomass is used to produce bio-oil using a slow pyrolysis method, and producing oil through a thermochemical liquefaction (TCL) method demonstrating TCL is more significant than pyrolysis. However, the production of bio-oil in TCL is 41% with comparison to pyrolysis with 23–29%. Similar to this, oil yield for TCL and pyrolysis processes when solids were converted was 93% and 60–72%, respectively. The generation of biochar was at 500 °C along with the liquid and gases form the slow pyrolysis process. The making of biochar from microalgae was 37, 34, 37, 53, 63, and 44% weight whereas, production of gases was 20, 25, 22, 24, and 13%. These microalgae, were estimated with the volatile gases evolution like CO_2, CO, CH_4, C_2H_4, C_2H_6, and H_2 (Das 2021; Yu et al. 2017).

Catalytic Pyrolysis

The oil by quick pyrolysis method had a higher viscosity, a lower heating value, and a larger concentration of oxygenated chemicals and water. However, the problem was solved by the addition of catalyst, which improved the oil yield and quality, a technique known as catalytic pyrolysis. During the process, the catalysts can split up heavier, higher molecular weight molecules, produceing vapors undergo secondary cracking to produce liquid and gaseous by-products. In the catalytic pyrolysis mode, catalysts are used to transform the algal biomass into biofuels with temperature ranging from 300 to 500 °C with the catalyst to biomass ratio being o.2 to 5. In this mode, yield of oil and HHV for the conversion of algal biomass differ from 20.7 to 55% weight and 21.2 to 32.2 MJ/kg. While, sodium carbonate (Na_2CO_3) and Zeolite Socony mobil-5 (ZSM-5) are the main catalysts employed during the catalytic process. The oils created by the process contain fewer molecules that are not much oxygenated and acidic. During the process, the distribution of products like biochar, oil, and gases are altered, directly resulting in the synthesis of gasoline, diesel fuel, aromatics, and other chemicals. The main parts for the bioreactor in catalytic pyrolysis consists of the biomass feeding unit, catalytic regenerator, cooling unit, cleaning system, and cyclone for separation of char and gases. Algal biomass along with catalysts are added to the bioreactor by grinding as dry or wet. The catalytic pyrolysis process is divided into two basic methods as ex-situ and in situ that is based on biomass and catalytic mixing in the bioreactor. In situ involves both catalysts and biomass combined while, ex-situ the algal biomass is pyrolyzed distinctly. The products of the pyrolysis of the biomass are then moved with the catalyst bed located downstream from the pyrolyzed. Biomass is broken down pyrolytically in two steps, first stage biomass is decomposed naturally, and second volatile products are cracked or condensed (Das 2021).

Microwave Pyrolysis

Assisted heating by microwaves is applied in the pyrolysis process with necessary heat energy being produced by electromagnetic radiation that passes through the biomass. However, biomass must be heated externally using traditional pyrolysis method. The internal heating of the biomass is uniform due to the microwave assistance. Primarily, carried out by two methods, ionic conduction and dipolar polarization with the former having a significant impact on heat production. The majority of this system's components are same as those in a normal pyrolysis system but being added by a microwave oven. Quartz or borosilicate glass that are transparent materials can be used to construct this oven. This substance gives a greater temperature for the reaction mixture and low temperature for reacting in the microwave oven (Das 2021; Yu et al. 2017).

Gasification

The method of algal biomass conversion by decomposing thermochemically converts it in gaseous fuels, by utilizing the carbon source in the algae to synthesize gases also known as syngas. This produce syngas that is burned to yield fuel in liquid form for transportation. In order to undergo gasification, these microalgae must undergo different methods like dehydration, pyrolysis, oxidation, and finally gasification (Das 2021).

4 Applications of Algal Biochar

Biochar has the potential to be used in an array of situations especially in an environmental standpoint. Some of the potential uses of algal biochar (Fig. 7) are bioremediation of soil, sequestration of carbon from the environment, improvising soil health, immobilization of contaminants, biochar-derived catalysts, tar removal and Syngas production, capturing greenhouse gases, animal feed, air filters, remediation of heavy metals from the environment, adsorption of organic pollutants from different components, especially the soil, improving the overall procedure of organic solid waste composting, decontamination and filtration of wastewater as well as water, can be used as an electrode or as materials to craft an electrode especially in microbial fuel cells (Shilpa et al. 2022; Mayegowda et al. 2024a).

Biochar derived from macroalgae can be suitably used as a tool for both carbon sequestration in the long-term and soil ameliorant. The use of algal biochar for remediation of wastewater from aquaculture industries, saline waste water sources, water sources that underwent eutrophication, etc. can be used in an array of ways help to reduce the overall carbon footprint by sequestering carbon from the soil, energy

Fig. 7 Representation of the different uses of biochar

generator, soil ameliorant, and fertilizer, and thus can be a source of revenue generation (Bird et al. 2011).

Biochar has the compatibility to be coated with various metal nanoparticles which enable to form composites with specific or special physical and chemical properties (Hu et al. 2010). The impact of algal biochar on the composition and quality of soil is positive as it can mitigate the amount of greenhouse gases emitted which is done by isolating the solid carbon in the ground. This carbon is then sequestered in the soil for hundreds and sometimes for thousands of years. This in turn helps improve the condition of the soil in regards to soil health and fertility (Ok et al. 2020).

Another important application of biochar in the environment is the removal of heavy metals from the soil by adsorption technique (Carrales-Alvarado et al. 2021). According to observations made in some studies done previously, different genera and species of algae vary in their yield and composition of biochar. The yield of biochar depends on a lot of factors such as temperature, time, pH, and elemental composition. Some of the species of algae are *Spirulina sp.* (at 300 °C for 30 min at

11–12 pH) that was found to yield about 11% weight of biochar of the original weight taken (Vardon et al. 2011). *Ulva* (at 350 °C for 5 min at 11–12 pH) yielded about 12% weight of biochar (Neveux et al. 2014). *Laminaria saccharina* (at 350 °C for 15 min at 11–12 pH) yielded 19% weight of biochar (Anastasakis and Ross 2015). Similarly, different species of algae in accordance to the temperature, time, and pH produce different yields of biochar. Other examples of biochar yield are *Chlamydomonas reinhardtii* around 2–21% biochar was provided at a temperature of 230 °C for 60 min at a pH of 9–13 (Hognon et al. 2015). *Nanochloropsis sp.* with a yield of 17–62%, and *Chlorella* with a yield of 13–55% was denoted by some resaerchers (Reddy et al. 2016).

5 Future Perspective

Currently, there are certain limitations in the development of algal biochar with respect to the parameters, methodology, and techniques as it has certain limitations to determine the quality, quantity, and microscopic conformation within the soil. In recent times, there has not been any major breakthroughs that point out any amendments on biochar and substances derived such as biochar fertilizers. So far, the studies done on biochar and its effects on crops, soil, and the environment is only on simulated, experimental manner, and short-term basis. Keeping all of these hurdles in mind, it is essential to develop methods of carbonization technology to improve the efficiency and quality of biochar. The utilization of waste substrates for production of algal biochar and formation of by-products will enhance us to provide economic revenues and help in cutting down cost (Chen et al. 2020). Algal biochar, is an excellent agent of bioremediation, fertilizer, composting, enhanced microbial diversity, reduction in the emission of greenhouse gases, and increased degradation of organic matter (Manjula et al. 2023; Mayegowda et al. 2024b). The future lies in the application of biochar as energy source and adsorbent in progressive way with the fact that biochar N-doped surface chemistry with electrochemical properties will be finest choice to store energy and conserve it (Wan et al. 2020). Its application can be used for energy stored supercapacitors, microbial fuel cells (MFC), biochar-based composite production, and organic fertilizers for effective soil improvement (Czekała et al. 2019; Chakraborty et al. 2020). Furthermore, genetically modified algae enriched with lipid content help reduce the greenhouse emanation with the help of biochar production (Kumar et al. 2017). Accumulation of toxic elements onto macroalgae from wastewater streams and then encompassing them in an immobilized form within the biochar is another aspect to look out for in the future (Mehta 2008). Since this biochar contains toxic waste and hence is classified as "contaminated," it can then be used as ameliorants underneath the soil under plant life which is used for the production/extraction of biofuels, or as a bioremediation tool in another contaminated site (Bird et al. 2011).

6 Conclusion

Algal biomass can be used for biochar production along with the feedstock by using various techniques. Algae has been researched for biochar production that helps for the plant growth, soil fertility, and pollutants bioremediation while, industrial biochar used in carbon electrodes in supercapacitors for energy storage, waste water treatment, air purifier, food packaging silage agent, etc. Biochar can be amended in both agricultural and environmental sectors. With the depletion of natural resources, it becomes obligation to device techniques to restore and use alternative ways such as the biochar derived from algal biomass as a main sustainable and green energy resource that can help to fulfill the worlds energy demand, bioremediation of various heavy metals to maintain soil fertility and plant growth, co-application of compost and biochar that has potential to extend the benefits and increase the multiple cropping cycles. However, more research on the algal species for the production of algal biomass need to be carried on with their use on applications.

References

Alhashimi HA, Aktas CB (2017) Life cycle environmental and economic performance of biochar compared with activated carbon: a meta-analysis. Resour Conserv Recycl 118:13–26. https://doi.org/10.1016/j.resconrec.2016.11.016

Amin S (2009) Review on biofuel oil and gas production processes from microalgae. Energy Convers Manage 50(7):1834–1840. https://doi.org/10.1016/j.enconman.2009.03.001

Anastasakis K, Ross AB (2015) Hydrothermal liquefaction of four brown macro-algae commonly found on the UK coasts: an energetic analysis of the process and comparison with bio-chemical conversion methods. Fuel 139:546–553. https://doi.org/10.1016/j.fuel.2014.09.006

Archana S, Jayanna BK, Ananda A, Shilpa BM, Pandiarajan D, Muralidhara HB, Kumar KY (2021) Synthesis of nickel oxide grafted graphene oxide nanocomposites—a systematic research on chemisorption of heavy metal ions and its antibacterial activity. Environ Nanotechnol Monit Manage 16:100486. https://doi.org/10.1016/j.enmm.2021.100486

Bach Q-V, Chen W-H (2017) Pyrolysis characteristics and kinetics of microalgae via thermogravimetric analysis (TGA): a state-of-the-art review. Biores Technol 246:88–100. https://doi.org/10.1016/j.biortech.2017.06.087

Bach Q-V, Skreiberg Ø (2016) Upgrading biomass fuels via wet torrefaction: a review and comparison with dry torrefaction. Renew Sustain Energy Rev 54:665–677. https://doi.org/10.1016/j.rser.2015.10.014

Barrington K, Chopin T, Robinson S (2009) Integrated multi-trophic aquaculture (IMTA) in marine temperate waters. FAO Fisheries and Aquaculture Technical Paper (FAO)

Bird MI, Wurster CM, de Paula Silva PH, Bass AM, de Nys R (2011) Algal biochar—production and properties. Biores Technol 102(2):1886–1891. https://doi.org/10.1016/j.biortech.2010.07.106

Brennan L, Owende P (2010) Biofuels from microalgae—a review of technologies for production, processing, and extractions of biofuels and co-products. Renew Sustain Energy Rev 14(2):557–577. https://doi.org/10.1016/j.rser.2009.10.009

Brewer CE, Brown RC (2012) Biochar. In: Comprehensive renewable energy. Elsevier, pp 357–384

Brindhadevi K, Anto S, Rene ER, Sekar M, Mathimani T, Thuy Lan Chi N, Pugazhendhi A (2021) Effect of reaction temperature on the conversion of algal biomass to bio-oil and biochar through pyrolysis and hydrothermal liquefaction. Fuel 285:119106. https://doi.org/10.1016/j.fuel.2020.119106

Carrales-Alvarado DH, López-Olmos C, Dongil AB, Kubacka A, Guerrero-Ruiz A, Rodríguez-Ramos I (2021) Effect of N-doping and carbon nanostructures on NiCu particles for hydrogen production from formic acid. Appl Catal B 298:120604. https://doi.org/10.1016/j.apcatb.2021.120604

Cerciello F, Apicella B, Russo C, Cortese L, Senneca O (2021) Effects of pressure on lignocellulosic biomass fast pyrolysis in nitrogen and carbon dioxide. Fuel 287:119604. https://doi.org/10.1016/j.fuel.2020.119604

Chakraborty I, Sathe SM, Dubey BK, Ghangrekar MM (2020) Waste-derived biochar: applications and future perspective in microbial fuel cells. Biores Technol 312:123587. https://doi.org/10.1016/j.biortech.2020.123587

Chen Y-D, Liu F, Ren N-Q, Ho S-H (2020) Revolutions in algal biochar for different applications: state-of-the-art techniques and future scenarios. Chin Chem Lett 31(10):2591–2602. https://doi.org/10.1016/j.cclet.2020.08.019

Cohen RA, Fong P (2006) Using opportunistic green macroalgae as indicators of nitrogen supply and sources to estuaries. Ecol Appl 16(4):1405–1420. https://doi.org/10.1890/1051-0761(2006)016[1405:UOGMAI]2.0.CO;2

Czekała W, Jeżowska A, Chełkowski D (2019) The use of biochar for the production of organic fertilizers. J Ecol Eng 20(1):1–8. https://doi.org/10.12911/22998993/93869

de Paula Silva PH, McBride S, de Nys R, Paul NA (2008) Integrating filamentous 'green tide' algae into tropical pond-based aquaculture. Aquaculture 284(1–4):74–80. https://doi.org/10.1016/j.aquaculture.2008.07.035

Gautam R, Vinu R (2020) Reaction engineering and kinetics of algae conversion to biofuels and chemicals via pyrolysis and hydrothermal liquefaction. React Chem Eng 5(8):1320–1373. https://doi.org/10.1039/D0RE00084A

Grierson S, Strezov V, Ellem G, Mcgregor R, Herbertson J (2009) Thermal characterisation of microalgae under slow pyrolysis conditions. J Anal Appl Pyrol 85(1–2):118–123. https://doi.org/10.1016/j.jaap.2008.10.003

Hannon M, Gimpel J, Tran M, Rasala B, Mayfield S (2010) Biofuels from algae: challenges and potential. Biofuels 1(5):763–784. https://doi.org/10.4155/bfs.10.44

He C, Tang C, Li C, Yuan J, Tran K-Q, Bach Q-V, Qiu R, Yang Y (2018) Wet torrefaction of biomass for high quality solid fuel production: a review. Renew Sustain Energy Rev 91:259–271. https://doi.org/10.1016/j.rser.2018.03.097

Heilmann SM, Davis HT, Jader LR, Lefebvre PA, Sadowsky MJ, Schendel FJ, von Keitz MG, Valentas KJ (2010) Hydrothermal carbonization of microalgae. Biomass Bioenerg 34(6):875–882. https://doi.org/10.1016/j.biombioe.2010.01.032

Hognon C, Delrue F, Boissonnet G (2015) Energetic and economic evaluation of *Chlamydomonas reinhardtii* hydrothermal liquefaction and pyrolysis through thermochemical models. Energy 93:31–40. https://doi.org/10.1016/j.energy.2015.09.021

Hu B, Wang K, Wu L, Yu S-H, Antonietti M, Titirici M-M (2010) Engineering carbon materials from the hydrothermal carbonization process of biomass. Adv Mater 22(7):813–828. https://doi.org/10.1002/adma.200902812

Huang Y, Finell M, Larsson S, Wang X, Zhang J, Wei R, Liu L (2017) Biofuel pellets made at low moisture content—influence of water in the binding mechanism of densified biomass. Biomass Bioenerg 98:8–14. https://doi.org/10.1016/j.biombioe.2017.01.002

Joshi VS, Gowda N, Mayegowda SB, Manjula NG (2024) Algae-based nanoparticles for contaminated environs nanoremediation. In Sustainable Nanoremediation, by Fernanda Maria Policarpo Tonelli, Arpita Roy, and Khalid Rehman Hakeem, pp 327–351, 1st ed. New York: Apple Academic Press, 2024. https://doi.org/10.1201/9781003468950-12

Kumar A, Kumar A, Sharma G, Naushad M, Stadler FJ, Ghfar AA, Dhiman P, Saini RV (2017) Sustainable nano-hybrids of magnetic biochar supported g-C_3N_4/$FeVO_4$ for solar

powered degradation of noxious pollutants—synergism of adsorption, photocatalysis and photo-ozonation. J Clean Prod 165:431–451. https://doi.org/10.1016/j.jclepro.2017.07.117
Lehmann J, Gaunt J, Rondon M (2006) Bio-char sequestration in terrestrial ecosystems—a review. Mitig Adapt Strat Glob Change 11(2):403–427. https://doi.org/10.1007/s11027-005-9006-5
Lehmann J, Joseph S (eds) (2009) Biochar for environmental management: science and technology. Earthscan, London ; Sterling, VA
Liu D, Keesing JK, Xing Q, Shi P (2009) World's largest macroalgal bloom caused by expansion of seaweed aquaculture in China. Mar Pollut Bull 58(6):888–895. https://doi.org/10.1016/j.marpolbul.2009.01.013
Liu D, Keesing JK, He P, Wang Z, Shi Y, Wang Y (2013) The world's largest macroalgal bloom in the Yellow Sea, China: formation and implications. Estuar Coast Shelf Sci 129:2–10. https://doi.org/10.1016/j.ecss.2013.05.021
Lobban CS, Harrison PJ (1994) Seaweed ecology and physiology. Cambridge University Press, Cambridge; New York, NY, USA
López Barreiro D, Prins W, Ronsse F, Brilman W (2013) Hydrothermal liquefaction (HTL) of microalgae for biofuel production: state of the art review and future prospects. Biomass Bioenerg 53:113–127. https://doi.org/10.1016/j.biombioe.2012.12.029
Manjula NG, Tajunnisa, Mamani V, Meghana CA, Mayegowda SB (2023) Fungal-based synthesis to generate nanoparticles for nanobioremediation. In Tonelli FMP, Roy A, Ananda Murthy HC (eds) Green nanoremediation: sustainable management of environmental pollution. Springer International Publishing, Cham, pp 83–108. https://doi.org/10.1007/978-3-031-30558-0_4
Mata L, Schuenhoff A, Santos R (2010) A direct comparison of the performance of the seaweed biofilters, *Asparagopsis armata* and *Ulva rigida*. J Appl Phycol 22(5):639–644. https://doi.org/10.1007/s10811-010-9504-z
Mayegowda SB, Chikkud V, Barua S, Manjula NG (2024a) Heavy metal detection by nanotechnology-based sensors. In: Nanotechnology-based Sensors for Detection of Environmental Pollution. Elsevier, pp 237–263. https://doi.org/10.1016/B978-0-443-14118-8.00013-9
Mayegowda SB, Nithin Gowda BK, Chandan Gowda U, Joshi V, Manjula NG (2024b) 27—Sustainability and green nanomaterials on nanotechnology-based sensors. In Policarpo Tonelli FM, Roy A, Ozturk M, Murthy HCA (eds) Nanotechnology-based sensors for detection of environmental pollution, pp 553–572. Elsevier, Amsterdam. https://doi.org/10.1016/B978-0-443-14118-8.00027-9
Mona S, Malyan SK, Saini N, Deepak B, Pugazhendhi A, Kumar SS (2021) Towards sustainable agriculture with carbon sequestration, and greenhouse gas mitigation using algal biochar. Chemosphere 275:129856. https://doi.org/10.1016/j.chemosphere.2021.129856
Morgan TJ, Youkhana A, Turn SQ, Ogoshi R, Garcia-Pérez M (2019) Review of biomass resources and conversion technologies for alternative jet fuel production in Hawai'i and tropical regions. Energy Fuels 33(4):2699–2762. https://doi.org/10.1021/acs.energyfuels.8b03001
Nelson TA, Haberlin K, Nelson AV, Ribarich H, Hotchkiss R, Alstyne KLV, Buckingham L, Simunds DJ, Fredrickson K (2008) Ecological and physiological controls of species composition in green macroalgal blooms. Ecology 89(5):1287–1298. https://doi.org/10.1890/07-0494.1
Neori A, Msuya FE, Shauli L, Schuenhoff A, Kopel F, Shpigel M (2003) A novel three-stage seaweed (*Ulva lactuca*) biofilter design for integrated mariculture. J Appl Phycol 15(6):543–553. https://doi.org/10.1023/B:JAPH.0000004382.89142.2d
Neveux N, Yuen AKL, Jazrawi C, Magnusson M, Haynes BS, Masters AF, Montoya A, Paul NA, Maschmeyer T, de Nys R (2014) Biocrude yield and productivity from the hydrothermal liquefaction of marine and freshwater green macroalgae. Biores Technol 155:334–341. https://doi.org/10.1016/j.biortech.2013.12.083
Nwoba EG, Ogbonna CN, Ishika T, Vadiveloo A (2020) Microalgal pigments: a source of natural food colors. In: Alam MdA, Xu J-L, Wang Z (eds) Microalgae biotechnology for food, health and high value products. Springer Singapore, Singapore, pp 81–123

Ok YS, Bhatnagar A, Hou D, Bhaskar T, Mašek O (2020) Advances in algal biochar: production, characterization and applications. Biores Technol 317:123982. https://doi.org/10.1016/j.biortech.2020.123982

Pang SJ, Liu F, Shan TF, Xu N, Zhang ZH, Gao SQ, Chopin T, Sun S (2010) Tracking the algal origin of the Ulva bloom in the Yellow Sea by a combination of molecular, morphological and physiological analyses. Mar Environ Res 69(4):207–215. https://doi.org/10.1016/j.marenvres.2009.10.007

Prein M (2002) Integration of aquaculture into crop–animal systems in Asia. Agric Syst 71(1–2):127–146. https://doi.org/10.1016/S0308-521X(01)00040-3

Reddy HK, Muppaneni T, Ponnusamy S, Sudasinghe N, Pegallapati A, Selvaratnam T, Seger M, Dungan B, Nirmalakhandan N, Schaub T, Holguin FO, Lammers P, Voorhies W, Deng S (2016) Temperature effect on hydrothermal liquefaction of *Nannochloropsis gaditana* and *Chlorella* sp. Appl Energy 165:943–951. https://doi.org/10.1016/j.apenergy.2015.11.067

Ross A, Jones J, Kubacki M, Bridgeman T (2008) Classification of macroalgae as fuel and its thermochemical behaviour. Biores Technol 99(14):6494–6504. https://doi.org/10.1016/j.biortech.2007.11.036

Roy A, Bharadvaja N (2021) Efficient removal of heavy metals from artificial wastewater using biochar. Environ Nanotechnol Monitor Manag 16:100602

Ruperez P (2002) Mineral content of edible marine seaweeds. Food Chem 79(1):23–26. https://doi.org/10.1016/S0308-8146(02)00171-1

Shilpa BM, Rashmi R, Manjula NG, Sreekantha A (2022) Bioremediation of heavy metal contaminated sites using phytogenic nanoparticles. In Shah MP, Roy A (eds) Phytonanotechnology. Springer Nature, Cham, pp 227–253. https://doi.org/10.1007/978-981-19-4811-4_11

Sirajunnisa AR, Surendhiran D (2016) Algae—a quintessential and positive resource of bioethanol production: a comprehensive review. Renew Sustain Energy Rev 66:248–267. https://doi.org/10.1016/j.rser.2016.07.024

Stengel DB, Connan S (2015) Marine algae: a source of biomass for biotechnological applications. In: Stengel DB, Connan S (eds) Natural products from marine algae. Springer, New York, NY, pp 1–37

Taylor R, Fletcher RL, Raven JA (2001) Preliminary studies on the growth of selected 'green tide' algae in laboratory culture: effects of irradiance, temperature, salinity and nutrients on growth rate. Bot Marina. https://doi.org/10.1515/BOT.2001.042

Vardon DR, Sharma BK, Scott J, Yu G, Wang Z, Schideman L, Zhang Y, Strathmann TJ (2011) Chemical properties of biocrude oil from the hydrothermal liquefaction of Spirulina algae, swine manure, and digested anaerobic sludge. Biores Technol 102(17):8295–8303. https://doi.org/10.1016/j.biortech.2011.06.041

Wang D, Jiang P, Zhang H, Yuan W (2020) Biochar production and applications in agro and forestry systems: A review. Sci Total Environ 723:137775. https://doi.org/10.1016/j.scitotenv.2020.137775

Xiong X, Yu IKM, Cao L, Tsang DCW, Zhang S, Ok YS (2017) A review of biochar-based catalysts for chemical synthesis, biofuel production, and pollution control. Biores Technol 246:254–270. https://doi.org/10.1016/j.biortech.2017.06.163

Yu KL, Lau BF, Show PL, Ong HC, Ling TC, Chen W-H, Ng EP, Chang J-S (2017) Recent developments on algal biochar production and characterization. Biores Technol 246:2–11. https://doi.org/10.1016/j.biortech.2017.08.009

Yuan T, Tahmasebi A, Yu J (2015) Comparative study on pyrolysis of lignocellulosic and algal biomass using a thermogravimetric and a fixed-bed reactor. Biores Technol 175:333–341. https://doi.org/10.1016/j.biortech.2014.10.108

Chapter 6
Production of Biochar from Plant Biomass and Their Applications

Jayath de Silva, E. R. M. Kushani Senanayake, Manujith Waduge, Dewni Mindula Liyanage, and Gobika Thiripuranathar

Abstract Biochar is a carbonaceous solid substance produced due to the thermal decomposition of various types of biomasses. Highly porous stable biochar mainly contains carbon (C) and could be produced from the thermal combustion of crop residues and municipal solid trash in an oxygen-limited atmosphere. The methodologies used for the production are pyrolysis, gasification, hydrothermal carbonization, and torrefaction. However, the production methods differ depending on the type of biomass, temperature, residence time, heating rate, and several other factors. Though biochar could be produced from several sources, plant biomass is an environmentally friendly method that attracted immense attention as a low-cost amendment due to its many potent advantages for agriculture and the environment. Also, it is utilized as a soil amendment, boosts soil fertility, which indirectly influences plant development, and reduces climate change by sequestering C to slow down global warming. Further, it could be utilized in a wide spectrum of applications, including agriculture, environment, energy, and many more. Besides, it could be used as a solution for reducing the carbon dioxide content of air due to its good adsorption properties. However, biochar can also have many negative impacts on the natural environment. For instance, the addition of harmful chemicals such as heavy metals, Polycyclic aromatic hydrocarbons, Environmentally persistent free radicals, and dioxins to the environment, changing physicochemical properties of soil, eutrophication in aquatic environments, and the reduction in microorganism population in soil environments are some of them. Hence, it is necessary to use risk avoidance methods for biochar and understand its future challenges to maximize the gain obtained from biochar without harming the environment and protecting the balance of the environmental systems. Hence, this chapter summarizes the production methods and potential applications of biochar produced from plant biomass sources and their upcoming challenges.

Keywords Carbon content · Municipal waste · Plant biomass · Cellulose pyrolysis · Ecotoxicological impacts

J. de Silva · E. R. M. Kushani Senanayake · M. Waduge · D. M. Liyanage ·
G. Thiripuranathar (✉)
College of Chemical Sciences, Institute of Chemistry Ceylon, Welikada, Rajagiriya, Sri Lanka
e-mail: tgobika@ichemc.edu.lk

A. Roy et al. (eds.), *Biochar Production Engineering*,
https://doi.org/10.1007/978-3-031-92594-8_6

1 Introduction

Biochar is a carbon-rich, charcoal-like solid matter which could be formed from various types of biomasses namely forestry along with agriculture residues, urban waste, and animal manures, and it is the outcome of the thermolysis of resources in restricted oxygenate and warmth pyrolysis surroundings. It is utilized as a soil amendment to better soil physicochemical, and biological effects and upgrade soil fertility which would thereby indirectly affect plant growth as well as lessen climatic variation by sequestering carbon (C) to slow down global warming (Yaashikaa et al. 2020).

Enriched valuables of biochar include large surface area, excessive porosity, bulk density, elemental composition, high cation exchange capacity (CEC), stability, water-retention capability, hydrophobic nature and enhanced openness for plant nourishment, further, which are eco-friendly, affordable, and can be prepared simply and quickly (Gabhane et al. 2020).

Feedstocks are woody and nonwoody biomasses consisting of natural resources derived from life or non-life forms. In contrast to organisms with excessive moistness, biochar synthesis with less moistness is more economically feasible due to the notable depletion in thermal radiation along with the shortening of the pyrolysis procedure' time requirements (Liu et al. 2023a).

Carboxylic (–COOH), hydroxyl (–OH), amine, amide, and lactonic groups are crucial moiety that enhance its surface assimilation capabilities. Since additional moiety is located at the surface, the surface area is vital for biochar's catalytic function. Biochar's carbonates, bicarbonates, and silicates interact with hydrogen ions to neutralize the soil. Hydroxyl and carboxylic moiety are mostly important for removing harmful pollutants such as pesticides, hydrocarbons and heavy metals like chromium (Cr) as well as lead (Pb) (Kocsis et al. 2022).

The excessive surface area-to-volume ratio as well as affinity for the hydrophobic nature of pollutants that biochar has for surface assimilation of contaminants are key factors. It is seen as a renewable supply because it is a vital part in eradicating impurities and pollutants from soil, aqueous as well as gaseous mediums. Numerous studies are focusing on biochar as they try to determine how effective it is at removing various impurities (Zhou et al. 2022).

Biochar's high porosity may improve supplemented soils' physicochemical and biological qualities directly or indirectly by sorption of metals, enhancing potential uptake of meso- and micronutrients, greatening soil organic matter content, increasing pH, bettering Ca, K, N, and P usability and improving water-retaining ability. These are associated with the benefit of increased soil oxygen levels. By expanding the soil's surface area for adsorbing more cations, this activity improves the soil's capability to exchange positive ions. Biochar is an interesting choice for treating wastewater due to its excessive pore fraction and immense surface area (Gupta et al. 2022).

Microorganisms may be indirectly protected by the biochar's enormous porous fraction and lengthening the continuance of wetness, and air in addition to supplements. The growth, existence, and activity of bacteria and fungi could be helped by the adsorbed organic molecules, which could benefit plant growth as well. Generally, biochar is suggested for acidic soils as it raises pH levels, enhances nutrition retention through surface assimilation of cations and increases soil effectiveness. Additionally, raising the pH also raises the soil's CEC. Compared to biochar created at lower temperatures, which is more easily degradable, biochar produced at higher temperatures remains in the soil for an extensive duration (Hossain et al. 2022).

As a result of its ability to store C within the soil, biochar research has recently attracted immense attention. Crop productivity improved as a result of its capacity to enhance soil qualities. Lessening CO_2 emissions into the atmosphere is becoming more and more important due to global warming. The carbon cycle, which directly affects climate change, depends on soil. A viable strategy for lowering CO_2 emissions from the soil is carbon sequestration. Additionally, it was discovered that biochar is persistent due to its aromatic arrangement structure, which encourages slow degradation in the soil as microbes can only sporadically utilize such compounds (Zhang et al. 2023).

It was discovered that low fertile soils had a higher rate of mineralization of organic matter than high fertile soils while high carbon content soils had a higher rate of mineralization of carbon than low carbon content soils (Liu et al. 2023b).

Labile and recalcitrant C are the two categories into which the C in biochar can be divided. Liable C is easily consumed by microbes following the application of biochar, which increases C mineralization right away. Therefore, using biochar re-establish C mineralization. Recalcitrant C, on the other hand, stays in the soil for a longer period. As a result, more C is fixed as a consequence of using the application of biochar than is released as an outcome of suitable carbon mineralization (Shafiq et al. 2023).

2 Types of Plant Biomass Used in the Production Process

There are many potential feedstocks available for biochar production. One such source is agricultural biomass. Large quantities of agro-waste, including primary cellulose, hemicellulose, and lignin, are produced each year and thus could be diverted toward biochar production. Biochar produced from rice straw has been applied to soil that was contaminated by heavy metals, and afterward, its impact on the microorganism population and their enzyme activity was investigated. Accordingly, when applied, it caused an increment in the pH and the organic matter content of the soil while decreasing the extractable fraction of Zn and Cd. Besides, a higher concentration of this biochar lowered enzyme activity and microbial abundance by changing the properties of the soil to which it was applied (Huang et al. 2017).

Another type of agro-waste used to produce biochar is maize stover, the waste of maize left after harvesting. Biochar made of maize stover was added to the soil alongside unprocessed maize stover, and the subsequent effects on the soil were studied. Accordingly, when biochar was added to soil, the C content in the soil increased (Yang et al. 2017).

Slow pyrolysis of wheat straw, poplar wood, and spruce wood was carried out to obtain biochar, and the products obtained were tested and characterized. The biochar produced from straw contained not only the highest salt (4.92 mS cm^{-1}) and ash content (12.7%) but also more soluble major and trace elements within it. The straw biochar had greater amounts of water-soluble cations and anions inside them, thus may make them useful to supply plant nutrients such as K, P, and S to plants (Kloss et al. 2012).

In contrast, some other elements, such as water-extractable boron, are highly present in wood-derived biochar. The biochar derived from spruce and poplar wood had greater surface area and thermal stability when compared. Further, different pyrolysis temperatures (400, 460, and 525 °C) influence the final biochar production. Though polycyclic aromatic hydrocarbon (PAH) content was also studied, no consistent trends with regard to pyrolysis temperature was observed. Besides, the amounts present in some biochar exceeded the USEPA and European limits (Kloss et al. 2012).

Biochar has been produced using biomass obtained from municipal waste too. In the research published by Mitchell et al., various types of waste are categorized under institutional waste (paper towels, cardboard, and the like), demolition waste (particle board, plywood, and demolition wood), and garden waste (spruce chips, grass and maple twigs, and leaves) were used to produce biochar. All the types of waste considered contained cellulose, lignin, and hemicellulose in varying compositions. Further, there were also inorganic components present in them. Hence, the inorganic components in the biomass could affect the final biochar product obtained. For instance, the biochar obtained from paper and cardboard products contained higher ash concentrations. Further, biochar produced from maple leaves and grass also had a higher ash content compared to spruce chips and maple twigs. Such high-ash biochar could be advantageous for soil remediation purposes. Besides, pH values also varied depending on the source used, with biochar produced using white paper having the lowest pH value (Mitchell et al. 2013).

Plant biomass such as wood chips has also been mixed with biosolids and cattle manure to make feedstocks and to produce biochar. The feedstock was pyrolyzed, and the produced biochar was applied to soil that was used to grow ryegrass (*Lolium multiflorum* Lam.) The grass was afterward tested for phosphorous (P) content and found that the manure wood mixture biochar had a greater extractability and bioavailability of P than the biosolid wood mixture biochar (Wang et al. 2012).

3 Biochar Production Processes

The growing interest in utilizing biochar for countless applications has demanded the conversion of biomass to biochar. Biochar is commonly produced by subjecting the biomass to thermochemical conversion. Such methods include pyrolysis, hydrothermal carbonization, gasification, and torrefaction (Pang 2019; Lin et al. 2016). Each method has differences that affect the yield of biomass obtained. Conditions such as temperature are of utmost importance in the production process, and the other important factors include the rate of heating and the time the material resides in the reactor. Generally, during the conversion process of biomass to biochar, water loss occurs at around 100 °C, followed by the thermal degradation of cellulose, hemicellulose, and lignin at around 220 °C. Finally, the burning of residual carbonaceous material acts as the final weight-loss stage (Yaashikaa et al. 2020).

3.1 Production Methods

For the highest yield, the procedure picked for production should be suitable, relying upon the type of biomass and the process conditions. For instance, residence time, temperature, heating rate, and other factors should be ideal. These circumstances are essential since they might influence the physicochemical conditions during the production cycle. Since weight reduction in the biomass occurs during the process, the form of biochar produced from plant biomass differs according to the applied conditions (Table 1) (Yaashikaa et al. 2020).

3.2 Pyrolysis

It is a chemical process where a solid or a liquid undergoes thermal degradation without interacting with O_2 or other oxidants (Qin et al. 2023). Many variables affect the process, including the substrate type, the rate of temperature rise, the temperature reached, and the presence of O_2 or other chemicals (Qin et al. 2023). During this process, thermal decomposition of organic materials occurs at the temperature range of 250–900 °C (Osayi et al. 2018). Further, the absence of O_2 allows biomass heating beyond its thermal stability limit and avoids combustion of the biomass, causing it to thermally degrade into other products (Armah et al. 2022). During this process, heating of biomass produces gasses such as CO, CO_2, CH_4, and H_2. It also produces a volatile matter which can be further condensed into a liquid phase and produces biochar as a solid by-product (Chatterjee et al. 2020; Manmeen et al. 2023; Alonso-Gómez et al. 2024).

Biomass contains lignin, cellulose, hemicellulose, and other extracts such as tannins, fatty acids, and inorganic salts (John et al. 2011). These components pyrolyze

Table 1 The thermochemical conversion procedures and their associated process conditions

Technique	Temperature (°C)	Residence time	The yield of biochar (%)	The yield of bio-oil (%)	Syngas production (%)	References
Pyrolysis	300–550 (slow)	5–30 min up to 25–35 h (slow)	35 (slow)	30 (slow)	35 (slow)	Mitchell et al. (2013)
	850–1250 (fast)	< 2 s (fast)	12 (fast)	75 (fast)	13 (fast)	Ahmad et al. (2014a)
Hydrothermal carbonization	150–300	1–16 h	50–80	5–20	2–5	Lehmann et al. (2011), Prakoso et al. (2018) and Kruse et al. (2013)
Gasification	750–900	10–20 s	10	5	85	Turner et al. (2008) and Mohan et al. (2006)
Torrefaction	150–300	10–60 min	80	0	20	Wilk and Magdziarz (2017)

at different temperatures (Armah et al. 2022), with cellulose pyrolyzing at around 350 °C and lignin being well above 350 °C. Besides, the biochar production is favored by relatively lower temperatures and longer residence times (Alonso-Gómez et al. 2024). Generally, the pyrolysis of biomass occurs in three steps: initial moisture present in the solid biomass begins to evaporate, followed by the degradation of the more unstable polymers, and this primary decomposition step occurs at the 200–400 °C temperature range. Afterward, the char formed in the primary step will undergo aromatization in the secondary pyrolysis step as temperatures continue to increase above 400 °C (John et al. 2011). Many reactions, including dehydration, isomerization, depolymerization, aromatization, decarboxylation, and charring, are observed during the pyrolysis process sequentially or in tandem (Wang et al. 2020b).

3.2.1 Pyrolysis of Different Types of Biomass

Cellulose Pyrolysis

Cellulose is one of the major components of biomass, and as such, its pyrolysis is important in biochar production. Cellulose pyrolysis produces mostly levoglucosan, glycolaldehyde, and 5-hydroxyfurfural, with other products being compounds like acetone and formic acid (Wang et al. 2020a). Cellulose decomposition occurs as a depolymerization process in the form of two reactions; slow and fast pyrolysis. In slow pyrolysis, cellulose decomposition is facilitated by a longer residence time and lesser heating rate, whereas, fast pyrolysis, requires a higher heating rate with swift volatilization, which causes the formation of levoglucosan (Yaashikaa et al. 2020; Hu et al. 2018; Maduskar et al. 2018). Levoglucosan can undergo dehydration and produces hydroxymethyl furfural, which could break down to form either liquid or gaseous products, including bio-oil and syngas. Besides, it can undergo reactions such as aromatization, condensation, and polymerization to produce solid biochar (Yaashikaa et al. 2020).

Hemicellulose Pyrolysis

Hemicellulose consists of countless glycosyl groups, such as xylose, mannose, glucose, and galactose. It is typically represented as Xylan in research studies (Dengyu et al. 2022). Hemicellulose decomposition is similar to cellulose decomposition with some differences. Hemicellulose decomposition begins around 250 °C, similar to cellulose; however, other factors differentiate it. Among the several products formed during xylan pyrolysis, the most abundant was hydroxy acetaldehyde. The amount of hydroxy acetaldehyde released increases with higher temperature and time (Ansari et al. 2019). The major products of xylan decomposition through pyrolysis were glycolaldehyde, acetaldehyde, furfural, acetone, and CO (Piazza et al. 2024). Xylan pyrolysis yields xylose and pyranose molecules that could be used for char production through multistep dehydration reactions (Dengyu et al. 2022). It has also been found that pyrolysis of xylan-based hemicellulose in a fixed bed reactor showed that char yield decreased with increasing temperature (Gao et al. 2020).

Lignin Pyrolysis

Lignin can be considered generally as a group of three polymers that vary in different plants and even different parts of the plant. They are guaiacyl lignins, and guaiacyl-p-hydroxyphenyl lignin (Adler 1977). Pyrolysis of lignin is a more complicated process; and it is generally divided into two: primary and secondary pyrolysis. Primary pyrolysis involves the direct formation of products from the decomposition of original biomass material whereas secondary pyrolysis involves the further reaction of volatiles evolved in primary pyrolysis undergoing coking, cracking, and other reactions involving free radicals (Ferdous et al. 2002). Particularly, the β-O-4 lignin linkage breaks, resulting in the formation of free radicals (Yaashikaa et al. 2020). This linkage thermally decomposes at a temperature between 250 and 350 °C as per the ^{1}H-NMR analysis conducted by Chu et al. (2013).

Char resulting from lignin pyrolysis was reported to have a condensed aromatic structure, and stored up to half of the energy present in the starting biomass (Ben and Ragauskas 2012). Higher pyrolysis temperature reduced char yield and resulted in the cleavage of aliphatic OH, carboxyl, and methoxyl groups according to the FTIR and NMR results reported by Sharma et al. Further, around 40% of lignin was obtained as char after pyrolysis at 750 °C. The presence of inorganic components was also theorized to enhance char yields (Sharma et al. 2004). Research conducted by Hosoya et al. (2009) have shown that methoxyl groups are important for char formation when pyrolyzing lignin, where o-quinone methide is a key intermediate in the process.

Pyrolysis of cellulose mixed with hemicellulose and milled wood lignin at high temperatures (800 °C) showed a few interactions between lignin pyrolysis and cellulose, and hemicellulose and cellulose pyrolysis, although the latter is weaker. Lignin inhibited the thermal polymerization of levoglucosan formed from cellulose and improved the formation of products from cellulose having a lower molecular weight which reduced the char faction yield and enhanced the tar yield. In turn, cellulose helped reduce the secondary char formation from lignin and enhanced the formation of some lignin-derived products such as guaiacol, 4-methyl-guaiacol, and 4-vinyl-guaiacol (Sharma et al. 2004).

3.2.2 Fast and Slow Pyrolysis

Pyrolysis could be classified depending on residence time, peak temperature, and heating rate, such as slow, intermediate, and fast pyrolysis (Manmeen et al. 2023). Slow pyrolysis is conducted at moderate temperatures, low heating rates, and longer residence time. The reaction temperature range is about 300–550 °C while the rate of heating is around < 50 °C per minute. The residence time is about 5–30 min up to 25–35 h. Slow pyrolysis yields biochar, bio-oil, and gas, and the yield of each is depending on the feedstock properties, processing temperature, heating rate, and the pyrolysis environment (Mitchell et al. 2013; Roy and Dias 2017; Crombie and Mašek 2014; Pilon and Lavoie 2012). Slow pyrolysis promotes biochar formation giving a higher percent product yield. A lower pyrolysis temperature produces more biochar than when higher temperatures are used. This effect could be due to the organic material combustion, greater volatilization of organic compounds, and partial gasification of char at higher temperatures (Muradov et al. 2012). Slow pyrolysis results in greater biochar yield from lignocellulosic and lignin biomasses as the low heating rate diminishes the chances of secondary pyrolysis and thermal cracking process (Rangabhashiyam and Balasubramanian 2019).

Fast pyrolysis can yield more bio-oil than biochar (Maaoui et al. 2023). Biochar forms approximately 12% of the product composition due to fast pyrolysis. Increasing pyrolysis temperatures causes the production of more volatiles while lowering biochar yield (Tripathi et al. 2016) Biochar could contain about a quarter of the energy present in the biomass feed (Bridgwater 2012). In fast pyrolysis, biomass is heated up to a temperature of 850–1250 °C where the heating rate is about 10–200 °C per second with a residence time of < 2 s (Ahmad et al. 2014a).

It has been found that when increasing the particle size of the biomass used for fast pyrolysis, it could reduce the volatile product yield while increasing the biochar yield. This is attributed to the lower rate of heat transfer inside the reactor (Sharma et al. 2015). The presence of inorganic salts can also impact the char yield. Hemicellulose subject to fast pyrolysis produced biochar that comprised 10 wt% when mineral-free hemicellulose was used (Patwardhan et al. 2011).

The yield of biochar obtained via pyrolysis can be increased by reducing the temperature and heating rate while increasing the sample particle size. Particle size affects the heating rate which in turn affects the products formed through pyrolysis. A higher lignin content has also been shown to give a greater amount of biochar as a product (Demirbas 2004).

3.3 Hydrothermal Carbonization

Hydrothermal carbonization (HTC) is a thermochemical conversion methodology that could be used to produce hydrochar, a coal-like product. Hydrochar is similar to biochar, and carbonaceous products are obtained through thermochemical conversion methods; however, it is named in this manner to differentiate it from char produced from processes, including pyrolysis and gasification, and they are considered 'dry'. HTC is a valuable process as it uses wet biomass, which has an elevated moisture content of 75–90%, as its feedstock, unlike other conversion methods, forgoes an additional step of drying the biomass prior to the commencement of the process (Lehmann et al. 2011; Prakoso et al. 2018; Kruse et al. 2013).

HTC is a pressurized thermal conversion process carried out in moderate subpyrolysis temperature (150–300 °C) and self-generated pressure (10–15 bar) for a few minutes to several hours. Feedstock includes agricultural residues, industrial and animal wastes, and municipal wastes are used for this purpose (Chen et al. 2021; Kossińska et al. 2023; Rosas-Mendoza et al. 2024; Lee et al. 2018; Wang et al. 2019a). The reaction is conducted under high pressure in vessels with biomass present in liquid water (Cui et al. 2024).

The reduced feedstock particle size enables a larger surface area which increases the reactivity of the solids and uniform heat and matter transfer. The main reactions occurred in hydrothermal carbonization process are dehydration, decarboxylation, aromatization, and condensation (Prieto et al. 2024). The products formed from hydrothermal carbonization, their characteristics, relative proportions in the gas/liquid/solid phases, and the energy requirements needed are dependent on the input material and the reaction conditions used.

During the dehydration process, biomass undergoes a loss of O and C, thus, lowering their H/C and O/C ratios. Elimination of hydroxyl groups appears to be the main process that occurs. The products of dehydration of glucose include hydroxymethyl furfural (HMF) or 1,5–anhydro glucose (Kobayashi et al. 2016). Cellulose degradation by hydrothermal carbonization results in significant amounts of formic acid as well. The formic acid can further degrade to yield CO_2 and H_2 as well (Yang et al. 2017). In the decarboxylation process, carboxyl groups are eliminated, and these carboxyl and carbonyl groups degrade rapidly, resulting in CO_2 and CO, respectively (John and Omorogie 2022). Both dehydration and decarboxylation processes result in structures with high aromaticity. Lowering of H/C and O/C ratios due to carbonization and the evolution of water and CO_2 was observed (Rosas-Mendoza et al. 2024). Decarboxylation follows dehydration, but only after a significant amount of water has been generated, does the decarboxylation process initiate fully (Funke and Ziegler 2010). Important intermediates such as 5-hydroxymethylfurfural and its derivatives eventually undergo polymerization, aromatization, and condensation to produce the hydrochar (Yaashikaa et al. 2020; He et al. 2021).

It was found by Kang et al. that the amount of hydrochar produced by the hydrothermal carbonization of lignin, cellulose, D-xylose, and the wood meal was highest with lignin followed by wood mass, cellulose, D-xylose. Hydrochar yield at a temperature between 225 and 265 °C was between 45 and 60% and C content, C recovery, energy recovery, the ratio of C/O, and the ratio of C/H in all of these hydrochars are among 63–75%, 80–87%, 78–89%, 2.3–4.1, and 12–15%, respectively. A possible reason stated for the higher yield from lignin is its stable phenolic structure that can positively influence char formation as a result of a condensation reaction (Kang et al. 2012). Higher temperatures also resulted in accelerated hydrothermal carbonization of the biomass, which gave a lower yield, a lesser ion exchange capacity and volatile matter content, lesser number of O-containing functional groups but also resulted in a higher C content for hydrochars. Avoiding relatively higher temperatures can limit liquefaction and gasification to yield more solid hydrochar products (Kruse et al. 2013; Kang et al. 2012; Falco et al. 2011).

Funke and Ziegler describe three main reaction pathways that result in hydrochar as summarized by Zhao et al.; (1) direct solid to solid conversion of biomass materials (termed "char") undergoing devolatilization, condensation, dehydration, and decarboxylation; (2) polymerization of dissolved intermediates that were derived from the degradation of the original biomass, toward polyfurans; (3) and finally a carbonization step via further inter molecules. These pathways are relevant to pure carbohydrates such as glucose, amylopectin, lignin, cellulose xylose, maltose, and hemicellulose (Wüst et al. 2020).

Hydrothermal carbonization offers several advantages over other production methods. As mentioned previously, the wet feedstocks could be used for production, and thus energy-intensive drying processes to pretreat the biomass is not required. Biomass particle size can also influence the hydrothermal carbonization process as proposed by Wüst et al. who showed that particles of 0–250 μm resulted in hydrochar having the smallest carbon percentage while particles of sizes 250–500 μm had a constant carbon percentage in hydrochar formed. The usage of residual process water also increased the efficiency of hydrothermal carbonization (Wüst et al. 2020; Chen et al. 2024).

3.4 Gasification

It is a usual method for producing syngas from different sustainable fuel assets. In correlation with other ordinary strategies, including pyrolysis, combustion, and fermentation, gasification offers larger syngas volume and lower levelized discharges. The significant byproduct of gasification is H_2 (Yang et al. 2023). But during this process, a substantial amount of biochar can also be produced. Moreover, biochar generated is regarded as a waste, though it has numerous useful applications, including waste water dye removal/decoloration, chemical adsorption, carbon sequestration, and as soil stabilization agents (Al-Rahbi and Williams 2017; González et al. 2008).

When carbonaceous material is exposed to high temperatures and combustion agents like O_2, air, and steam, a thermochemical process called gasification process take place, where it converts the carbonaceous material into gaseous products like syngas, which comprises CO, CO_2, CH_4, H_2 and residues of hydrocarbons. It should be mentioned that the production of syngas is heavily impacted by the reaction temperature (Hu et al. 2017). It was discovered that when the temperature increases, the production of CO and H_2 increases while other substances such as CH_4, CO_2, and hydrocarbons decrease (Prabakar et al. 2018). Syngas is the main output of this process, while char is considered as a by-product with a lower yield.

The gasification mechanism can be subdivided into a variety of stages, including certain:

Drying

During this process, the biomass's moisture content entirely evaporates without any energy being recovered. The moisture content varies according to the type of biomass materials. Also, drying is utilized as a discrete process during gasification when the biomass comprises high moisture content (Hu et al. 2017). Additionally, heat transfer within a particle raises the biomass's localized temperature, which causes water to evaporate and then gradually releases pyrolytic volatiles. The forerunners of essential volatiles are cellulose, hemicellulose, lignin, and extractives, including super durable gases (e.g., CO_2, CO, and CH_4).

Oxidation/Combustion

The oxidation and combustion of gasification agents are the main energy sources for this process. These gasification specialists respond with the combustible species present in the gasifier to produce CO_2, CO, and H_2O. Here, the reaction is takes place at high temperatures (>700 °C) under a limited supply of vapor and O_2 (Turner et al. 2008). Different gasifying agents could be used in the process depending on the specifications of the intended gas composition (Shen et al. 2015). By supplying steam, which raises syngas's heating value, a gas with a higher concentration of H_2 can be produced (Luo et al. 2009). Commonly, carbonaceous materials obtained from natural petroleum products could be converted into H_2, CO, and CO_2 utilizing gasification.

The overall degradation completes almost in the range of 400–500 °C, where biochar is a significant item, but biomass degrades to some extent at different temperatures. At a higher temperature, where secondary decomposition and a number of secondary reactions take place, biochar is further transformed into fuel. Syngas composed from H_2 and CH_4 is generated (Mohan et al. 2006). Sewage sludge is another group that is taken into consideration for the production of H_2 and biochar by gasification with steam at 800 °C, which showed 35–45 vol% H_2 with a considerable amount of singe (Sattar et al. 2014).

Petroleum/oil coke and coal are utilized as the essential feedstock for some huge gasification plants around the world. Metropolitan strong waste, agricultural and industrial waste, sewage slop, and so on are likewise being utilized as feedstock while performing gasification operations. Similar discussions have focused on producing biochar through gasification, using sunflowers, miscanthus, sewage sludge, and wood pellets in a quartz cylinder furnace while providing steam. Likewise, steam gasification for pine sawdust results in further cracking at 850 °C, leading to an increase in H_2 volume fraction with rising temperature (Yan et al. 2010).

3.5 Torrefaction

Torrefaction differs from slow pyrolysis as the latter uses temperatures in between 350 and 700 °C with low heating rates, and this temperature is maintained for 15–60 min. Torrefaction mainly targets the hemicellulose component of the biomass, which degrades above 200 °C (Wilk and Magdziarz 2017). Several reasons can be given for the upper limit of 300 °C for torrefaction. Above 300 °C, higher devolatilization and carbonization occurs with the thermal degradation of cellulose and lignin being the major reaction taking place (Chen et al. 2018a).

The degradation of hemicellulose is the key weight-loss process that occurs in torrefaction. According to the published literature, about 30% of the mass of biomass is lost during the process, while 90% of the initial lower heating value is retained. As torrefaction targets the hemicellulose component of biomass primarily, herbaceous biomass such as grasses (which have greater amounts of hemicellulose) undergoes greater changes rather than woody biomass (Bridgeman et al. 2008). Torrefaction gives a solid product that is hydrophobic and more brittle. Further, it removes volatile matter and moisture from biomass and reduces H/C and O/C ration while increasing the mass density, calorific value, and grindability among other characters (Piersa et al. 2022).

The change in C, H, and O percentages of miscanthus biomass has been reported. Further, the values of temperature and residence time selected for the torrefaction process influences the properties of the final biochar obtained. Raw miscanthus has a C mass percentage of 46.30%. Using residence times of 30 min to 1.5 h and different temperatures ranging from 230 to 290 °C, it was shown that the C wt% increased from 50.3 up to 62.2%. Besides, the mass percentage of O was also reduced by roughly 13% with the highest residence time and heating temperature, 1.5 h and 290 °C, respectively (Wilk and Magdziarz 2017). Torrefaction of crude oil kernel at temperatures between 150 to 300 °C resulted in an increase of C% of the resultant product up to around 75%, a linear decrease in O_2 to around 10%, while H content remained about constant throughout with slight variation (Fokaides 2017).

Three main types of torrefactions are present; dry torrefaction, wet torrefaction and steam torrefaction. Dry torrefaction is the thermal treatment of biomass in an inert environment at atmospheric pressure and at a temperature in between 200 and 300 °C. Dry torrefaction gives three main products; solid biochar, a liquid faction containing water, organic compounds and lipids, and permanent gasses such as H_2, CO_2, CO, and hydrocarbons. A prerequisite for dry torrefaction is a completely dry feedstock. The biochar product obtained also has a higher ash content. It must be noted that the larger amount of volatile matter and lesser moisture content of the biochar obtained has a risk of self-ignition (Chen et al. 2018a).

Wet torrefaction is also referred to as hydrothermal carbonization. Factors that affect wet torrefaction include temperature, pressure, atmosphere, residence time, and the liquid medium (Bach et al. 2015). Here the biomass feedstock used will contain high moisture content, and the process occurs in a subcritical pressurized water vessel maintained at a pressure between 1 and 250 MPa. Further, the temperature range and residence time of this process are 180–265 °C and 5 min to several hours. The key product of wet torrefaction is hydrochar. The ash content of the resulting solid is lesser, while the C content is higher. Wet torrefaction requires a postdrying step after the product is obtained (Chen et al. 2018a).

Steam torrefaction is another alternative method. The research conducted by Mustaza et al. used superheated steam in the torrefaction process of mangrove wood as it is a less lengthy process than the traditional one used to produce char, which takes many days. Superheated steam can remove volatile matter from solid biomass, which is then removed by the carrier gas. The effect of treatment time on the resulting char with a longer operating time reduces the sulfur present inside the biomass. Further, with the upsurge in operating time, the C content of the biomass decreased as more hemicellulose and cellulose were thermally degraded. Interestingly, it was theorized that as the operating time increases, a reduction in mass yield is expected. However, practically, when the operating time increases, it causes a temperature overshoot due to cellulose and hemicellulose decomposition (which is an exothermic process). Thus, the degradation of volatile matter was reduced, affecting the final mass of the char and condensed volatile matter product, increasing the mass yield by a few percentage points (Mustaza et al. 2021).

4 Characterization of Biochar

The techniques used to characterize biochar are based on physicochemical, surface, and structural analysis and are performed mainly to determine the ability to remove pollutants. pH, CEC, and electrical conductivity are physical methods and surface area, particle size distribution, density and porousness are chemical methods used to characterize biochar.

The surface area of biochar could be studied from Brunauer–Emmett–Teller (BET) technique. While temperature has a substantial influence on biochar synthesis, the surface area is the basic factor in analyzing the surface assimilation ability of biochar. Between processed and unprocessed primal matter, the surface area could vary. Therefore, throughout the process of making biochar, activation is used to improve the porosity and surface area of the biochar. The activation procedure might require physicochemical activation actions. Generally, biochar with enormous surface area as well as excessive porousness will have high sorption properties. Scanning electron microscopy (SEM) can examine morphology, such as biochar's pore size, while SEM with energy dispersive X-ray spectroscopy (EDX) could be used to analyze surface elements.

The surface functional groups that increase their sorption properties are characterized using Fourier transform infrared spectroscopy (FTIR) and Raman spectrograph alongside nuclear magnetic resonance (NMR) spectrograph (Rudra et al. 2024). Temperature and biomass affect the surface moiety of biochar, while there is a chance of a decrease in biochar moiety when some factors, including pH, surface area, and porousness increase.

FTIR spectrograph is an oscillation analysis for examining the moiety on the surface of biochar. Huge changes happen to the biochar in the mixture along with additional ordering when increasing temperature. A nondamaged FTIR instrument could effectively observe these progressions. Raman scattering depends on the atoms' progress in vibration while emitting electromagnetic radiation. Raman spectrograph is appropriate in characterizing biochar due to its excessive sensitivity, littlest sampling composition as well as less involvement; however, its price limits its usage. Besides, Raman spectrograph is also utilized to examine aromaticity.

The constitutional composition of biochar such as the general quantity of carbon moiety, the surmised level of the build-up of aromatic rings along with the basic arrangement of char particles can be analyzed using NMR spectroscopy. A strong magnetic field and radio recurrence (RF) pulses are used in NMR to analyze the arrangement of fragments through the resonance frequencies of unique atomic cores.

Thermo-gravimetric analysis (TGA) is used in thermic study to note the compound's physicochemical qualities as they change with temperature (El-Sayed et al. 2023). TGA is used to examine the igniting characteristics of biochar and the biomass mixtures. Additionally, the usual mass mean of each constituent could be studied to sight if there was synergistic interaction between the blends' parts. Hence, the effects could be useful for both the analysis of tests and for a greater concern of the warmth methods as well as properties of the examples from a broader perspective.

X-ray diffraction (XRD) is a widely applicable method to decide the crystallinity and arrangement of biochar. The polarizer, radiation source, and stepping motor are incorporated into the digital XRD setup. Further, to produce high-quality biochar quickly and without causing any damage, XRD patterns are helpful. Transmission electron microscopy (TEM) could also be used to examine biochar's structural arrangements (Yaashikaa et al. 2020).

5 Applications and Mechanisms Related to Biochar

5.1 *Soil Remediation*

Negative charge and huge surface area are just two of the biochar's main qualities for cleaning up contaminated soils. The use of biochar improves soil crop productivity by improving soil nutrient availability, microbial activity, soil organic matter availability, and water-retention capacity. Ion exchange, physical adsorption, electrostatic interaction, precipitation, and complexation are among the biochar's remediation mechanisms for soil pollution (Ahmad et al. 2014b).

5.1.1 Ion Exchange

It refers to the method by which acidic O_2-containing functional groups on the surface of biochar, like carbonyl groups, carboxyl groups, and hydroxyl groups, can ionize H^+ or surface base ions like Na^+, K^+, Ca^{2+}, Mg^{2+}, etc., to exchange with heavy metal ions or cationic organic pollutants (Hassan and Carr 2018).

5.1.2 Physical Adsorption

Biochar physically adsorbs pollutants like heavy metals or chemical compounds on its surface or disperses them into its micropores as a result of its surface properties, such as its porosity and enormous specific surface area. The average pore diameter of the biochar is larger than the diameter of the heavy metal ions. In general, the deeper the biochar's pores are accessed by heavy metals, the greater it's capacity for adsorption (Ko et al. 2004; Ngah et al. 2008).

The characteristics and exposed area of biochar, the characteristics and concentration of pollutants, and the temperature during the adsorption process all have a huge impact on the intensity of physical adsorption. The Langmuir and Freundlich model often fit physical adsorption, which can be either single- or multi-layered. Pseudo-first-order and pseudo-second-order kinetic models are typically used to fit physical adsorption kinetics (Qiu et al. 2009; Chun et al. 2004).

5.1.3 Electrostatic Interaction

The term "electrostatic interaction" describes the electrostatic adsorption of heavy metal ions onto the surface charge of biochar. The negative charge on the biochar surface and the heavy metal with a positive charge produces an electrostatic adsorption when the pH value of the solution is larger than the charge point of the biochar (pH pzc) (Pan et al. 2013; Uchimiya et al. 2010; Sohi et al. 2010; Tang et al. 2013; Xu et al. 2011).

5.1.4 Precipitation

Heavy metal ions combine with mineral constituents in biochar, such as CO_3^{2-}, SiO_3^{4-}, PO_4^{3-}, Cl^-, SO_4^{2-}, SO_3^{2-}, and OH^-, to form water-insoluble compounds (metal oxides, metal carbonates, and metal phosphates) that aid in the adsorption and immobilization of heavy metals. According to Xu, et al. (2013), the precipitation of CO_3^{2-} and PO_4^{3-} was the major reason of the Cu, Zn, and Cd adsorption on fertilizer biochar, whereas the amount of electron surface complexation via OH^- groups or delocalization was little.

5.1.5 Complexation

Complexation is the process by which O_2-containing functional groups on the surface of biochar interact with heavy metals to create complexes that may be permanently established. In the study on the aluminum phytotoxicity of cow manure biochar to wheat, Qian et al. (2013) discovered that the main mechanism by which aluminum (Al) was adsorbed by biochar rather than electrostatically attracted to sites with negative charges was the complexation of the carboxyl group with $[Al(OH)]^{2+}$ and its monomer surface. According to Jia et al. (2013), surface complexation was the primary mechanism by which oxytetracycline was adsorbed onto biochar, whereas metal bridges and possible cationic exchange were secondary.

5.2 Adsorption Mechanism of Biochar for Pollution Remediation

Adsorption often involves more than one process, but rather a combination of them. The following is a description of the adsorption mechanism of biochar for pollution remediation: municipal sewage sludge was employed as a starting point for the pollutant Cd through a process called surface precipitation in an alkaline environment and the swapping of exchangeable cations with Cd (Chen et al. 2015). Using fertilizers as a source of pollution Cu, Zn, and Cd are precipitated as CO_3^{2-} and by partially complexing with the OH^- group or a delocalized electron, PO_4^{3-} can be found on the biochar surface.

Bamboo biomass was employed to bind the sulfonamide pollutants sulfathiazole, sulfamethoxazole, and sulfamethazine. The neutralization of OH^- groups by –H freed from functionalized biochar surface and electron–acceptor–acceptor (EAA) interaction were the further steps in the adsorption of negative species, which were mostly occurred by proton exchange with water and the formation of negative charge assisted hydrogen bonding (Ahmed et al. 2017). Since the constituents of OH^-, CO_3^{2-}, and Si in biochar can precipitate with Zn^{2+}, wheat straw and pine needles are employed for pollutants like Zn (Qian et al. 2016). Tetracycline is a contaminant that is adsorbable to rice straw and swine manure using the principal mechanisms of hydrogen bonding, electrostatic attraction, and electron donor–acceptor (EDA) interactions (Chen et al. 2018b).

5.3 *Biochar in Soil Remediation for Removing Heavy Metals*

The exclusion of heavy metals by biochar is largely occurred in two aspects: (1) the adsorption of heavy metals in the pores of biochar to minimize the remaining amount in the soil; (2) ion interchange or redox reaction between the effective constituents in biochar and heavy metal ions to fixed the formation of heavy metal precipitates or to reduce toxicity by converting them into low-valent states. Boostani, et al. (2019) found that the effect of sheep and earthworm manure biochars on Pb immobilization in contaminated calcareous soil. The presence of biochars significantly increased the amount of Pb in the residual state, which decreased the Pb activity in the soil.

Despite the fact that biochar decreases the concentration of producible heavy metals in soil that has been contaminated by complex heavy metals, due to competitive adsorption, biochar has varying adsorption effects on various heavy metals. Yang et al. (2016) found that biochar made from straw and bamboo decrease the production of Cu and Pb more effectively than Zn. Interestingly, Zhou et al. also came to the same conclusion.

Accordingly, in the single metal adsorption test, sludge biochar had the maximum adsorption capacity to Zn, however, in the polymetallic adsorption test, Zn, Mn, and Cu had decreased adsorption capacities while Cr had higher adsorption capacity (Zhou et al. 2017).

The Cu absorption in roots is declined by 15%, 35%, and 26%, respectively, by biochar made from bamboo, rice straw, and Chinese walnut shells (Wang et al. 2019b). Additionally, the solubility of Cu and Pb in industrially polluted soil is decreased by rice straw biochar (Khan et al. 2018). Further, rice husks increase Cd adsorption by 21–41% on saturated soil and by 38–54% on dry land soil. Besides, sewage sludge decreases the concentration and bioavailable levels of Cd, Pb, and Zn in the leachates of Brazil soil. Moreover, paddy soil samples taken near to Zn and Pb mines revealed that the levels of acid-soluble Cd were 8–10% lower in soils treated with poultry litter biochar than in the contaminated control soil (Lu et al. 2018).

5.4 *Biochar in Agriculture and Environment*

The production of biochar from agro-environmental waste biomass is generating immense attention as a low-cost amendment due to its many beneficial effects for agriculture and the environment, in addition its capacity to retain soil moisture. Its usage in the agricultural environment causes observable changes in the physicochemical, biological, and carbon balance. Some mechanistic evidence exhibits biochar's ability for increasing crop productivity, soil water availability, biochar co-composting, and nutrient use efficiency (Ayaz et al. 2021). Its application in agricultural soils is an intriguing new technique with exciting possibilities for improving soil fertility, long-term C storage, and sustainable waste disposal. Information on biochar's tremendously beneficial properties is widely available in the literature (Montanarella and Lugato 2013a; Kwapińska et al. 2023).

In order to increase soil fertility and ensure long-term C storage, which has positive impacts on the global C balance as a measure to prevent climate change, using biochar as a soil amendment is ideal (Sombroek 1966). The scientific explanation for such a different use of charcoal to the traditional usage as fuel was motivated by the discovery of the "Terra Preta de Indios," or carbon-rich soils, in the Central Amazon (Lehmann 2007).

According to Jeffery et al., a quantitative investigation of the outcomes of adding biochar to soils displays a modest but overall advantageous effect to biochar amendment on crop productivity. Also, the liming effect, the enhanced ability of the soil to retain water, and the availability of nutrients were most likely the main processes behind these reactions as these effects were more apparent in acidic and coarse-textured soils (Jeffery et al. 2011).

The potential of soil to function as a carbon sink and so aid to prevent climate change is one of the significant responsibilities identified. To fulfill the objectives of this plan, biochar is obviously a proven product for C storage in soils for a long period of time (Montanarella and Lugato 2013b). Using biochar as a soil supplement may have some other benefits in addition to this. As a result of its porous composition and large surface area, biochar has been proven to have beneficial effects on soil field capacity, fertilizer use efficiency, pH ("liming effect"), nutrient availability, and CEC (Zhao et al. 2024; Atkinson et al. 2010; Chan et al. 2008).

Nevertheless, to completely comprehend the interactions between various types of soil and biochar, more study is necessary (in terms of feedstock and processing conditions) particularly in light of some crucial issues like the alteration of heavy metal availability, the sorption of pesticides, and the introduction of metal contaminants (Kookana et al. 2011).

Biochar can prevent herbicide dispersion and degradation because of its high sorption capability. The use of insecticides that are less likely to contaminate water and are administered externally may have some advantages (Jones et al. 2011). However, biochar may reduce the effectiveness of pesticides applied to the soil, requiring the usage of further herbicides to reach the appropriate standard of weed control (Nag et al. 2011). Other studies have found that adding charcoal to soil might have conflicting effects on pesticide penetration, depending on the substance's ability for adsorption (Larsbo et al. 2013).

A recent study suggests that high-ash biochar with maximum P contents may be a great source of P for agriculture (Wang et al. 2012). It will aid to create agricultural systems with a more continuous loop by minimizing the significant reliance on external input, especially that obtained from scarce resources. The CEC of biochar was improved by composting it with organic manures, although this process also changed the surface characteristics of the biochar (Prost et al. 2013).

The surface area of the biochar was reduced during the composting process as the sorbed components from compost-derived materials blocked the micropores. The biochar selectively absorbs organic matter and nutrients. In addition, composting biochar with poultry manure accelerated up the rate of manure degradation relative to the control treatment while reducing total N losses by up to 52% (Steiner et al. 2010).

5.5 Industrial Applications

Biochar is a feasible replacement for the currently used fossil fuels due to its less emission nature of greenhouse gasses, increased effectiveness of manufacturing of electricity as well as thermal energy, economic benefit of reducing waste disposal costs, and scarcity of fossil fuels on the global energy requirements (Yaashikaa et al. 2020; Gabhane et al. 2020).

The calorific value, carbon content, moisture, ash, and volatile components of biochar are important to the power industry, which depend on factors such as the kind of natural resources utilized to make biochar and the pyrolysis process conditions. Biochar's high calorific value and less release characteristics led to its excessive capability as a renewable fuel in the power sector in the future (Liu et al. 2023c).

It is used to enhance atmosphere standard, soak up water along with protection from electromagnetic radiation due to its great insulating properties, high specific surface, excellent water-retaining ability, and very low thermal conductivity. These effects of biochar are advantageous in the housing business.

Besides, biochar is utilized as a precursor in activated carbon manufacture along with specially benefitting for the chemisorption from gasses as well as fluid that are carbon materials with a strongly advanced specific surface and high porosity. Therefore, it has extremely varied and expanding applications (Yaashikaa et al. 2020).

Biochar can be widely beneficial in coloring manufacture due to its excellent selective adsorptive qualities, which enable its use as adsorbents for the cleaning of effluent or to draw off gasses as well as in water filtration. Moreover, it can be employed in the pharmaceutical industry to eliminate toxicants effectively (Gabhane et al. 2020). Biochar's qualities build it an effective and potentially useful catalyst. As a catalyst, biochar has various uses such as compost, to adsorb pollutants, treating soil and cleaning up contaminated soil in agricultural fields, the manufacture of biodiesel, energy, the elimination of tar, the management of trash, the generation of syngas, electrodes for microbial fuel cells, the creation of chemicals, and the elimination of environmental toxins (Yaashikaa et al. 2020; Zhang et al. 2023).

5.5.1 Applicability of Biochar as a Building Material

Due to the energy-intensive procedures required and the significant amount of CO_2 released during the process, the building sector is generating environmental issues on a global scale (Gupta et al. 2018). One of the newest methods for combating climate change is the direct air absorption of CO_2, and one of the possible alternatives is the development of materials that can do so by adsorption. Adsorption is a relatively inexpensive and straightforward way to collect C than industrial procedures (Gupta and Kua 2017).

The inclusion of biochar in building materials appeared to enhance the characteristics of those materials and increase the toughness of composites. However, the amount of biochar used ended up being crucial. The following were the key conclusions about changes in material properties (Gupta et al. 2018; Gupta and Kua 2017; Legan et al. 2022).

The biochar dosage of 0.5–24% by weight of cement improved the strength of cementitious materials and also the higher content of biochar up to 40% diminished the material's flexural strength. The substitution of sand by biochar in concrete caused a decrease in compressive strength. When the biochar dosage is between 2 and 40%, it reduces the bulk density of fresh mortar. When Biochar dosage is between 0.5 and 2%, water absorption into the material and water penetration through the material is decreased (Legan et al. 2022).

When adding biochar, some properties have increased while some properties have decreased. Therefore, it is not advisable to apply biochar in every construction process as the strength of the cement could be decreased. However, the properties could be improved when the best proportion of biochar is used for different cement mixtures in various construction processes. Utilizing biochar in construction materials can directly or indirectly impact lowering C footprints and preventing climate change. This biochar will adsorb CO_2 from the air. However, it is necessary to add CO_2-saturated biochar; otherwise, the adsorption process will not occur well (Legan et al. 2022).

According to preliminary research, this biochar cement combination offers good insulation and could keep the optimum indoor humidity levels. The composite may avert condensation on the exterior walls, which could otherwise result in growth of mold, and the too-dry indoor air, which is one of the main causes of respiratory issues (Gupta and Kua 2017). Low heat conductivity, good chemical stability, and low flammability are among biochar's best qualities that make it a suitable building material. The occurrence of a large variety of nanopores, mesopores, and micropores on the surface of biochar particles contributes to its low thermal conductivity. However, the feedstock utilized, and the pyrolysis temperature affects how the holes are distributed. The pores in biochar prevent thermal bridging in concrete, which improves building insulation. This is crucial for lowering the amount of energy used to heat or cool buildings (Gupta and Kua 2017).

6 Biochar's Negative Effects and Upcoming Challenges

The previous analyses and studies on biochar have been on how biochar is modified, how reactions occur, and how biochar actively participates in environmental rehabilitation. However, the drawbacks and dangers of biochar have just lately come to light. Given the toxic elements, structure, and particle size of biochar, it is vital to evaluate the negative effects of applying it to the environment. Heavy metals, Polycyclic aromatic hydrocarbons (PAHs), dioxins, environmentally persistent free radicals (EPFR), perfluorochemicals (PFC), and volatile organic compounds (VOCs) can be considered as the major environmentally damaging pollutants that can be found in biochar (Xiang et al. 2021).

The type of biomass influences the concentration and bioavailability of heavy metals in biochar. When producing biochar from biomass that contains a high level of heavy metals, such as through the process of leaching, the amount of heavy metals in the environment may rise. Given that wood-derived biochar has a large surface area (180–270 m^2/g, which is sometimes greater than 50% of the total), it may have a higher concentration of heavy metals in the exchangeable/acid-soluble fraction. Hence, the pyrolysis temperature must be managed carefully to control the concentration of heavy metals in the biochar (Xiang et al. 2021). This happens mostly as a result of the decomposition of the organic matter (OM) contained in the biomass at higher temperatures, which release heavy metals attached to the organic matter (OM). It has been observed that the bioavailability of heavy metals in biochar may decline when the pyrolysis temperature increases. The pH, present forms of heavy metals, mineral structures, and application environment are other elements that affect the environmental risk of heavy metals in biochar in addition to the heavy metal content and pyrolysis temperature (Xiang et al. 2021; Liu et al. 2020).

Due to the low pH settings, typically promoting metal dissolution, the biochar's heavy metals demonstrated their greatest capacity to leach at a pH of 3. Leaching decreased when the pH of the solution increased from 3 to 7. As the pH of the solution continued to rise from 7 to 13, the leaching of heavy metals particularly Cr, increased. This might have happened because, during carbonate decomposition, the leached Cr interacted with CaO to generate $CaCrO_4$ (Xiang et al. 2021).

Polycyclic Aromatic Hydrocarbons (PAHs)

Highly biotoxic PAHs have the ability to affect plant and microbial life in various environments. The PAH content of biochar made from biomass varied depending on the production circumstances according to reports in the literature. For instance, the type of biomass feedstocks utilized to generate biochar as well as the production temperature can both affect the variation of PAH level (Xiang et al. 2021).

Environmentally Persistent Free Radicals (EPFR)

Biochar exhibits a very potent EPFR signal. These EPFRs play a significant role in many environmental processes that take place during the creation and extensive use of biochar (Xiang et al. 2021). Since the organic components of biomass are thermally degradable during pyrolysis, EPFRs are produced as a result of the carbonization conditions and the utilization of the raw material in the process. Lignin, cellulose, and hemicellulose serve as the key precursors for the formation of EPFR in biochar. The cellulose chain has two potential cleavage locations, which means that the chain's homogeneous cleavage reaction could result in the formation of free radicals. Lignin has a tighter structure than cellulose and hemicellulose. Thus, cellulose suffers a powerful breakdown reaction, which includes the slow reaction of EPFRs. Besides, environmental biochar residues could also be used to make EPFRs (Xiang et al. 2021). Fe^{2+} and other transitional metals are the main culprits in this process. Typically, transition metals chemically adsorb biomass during pyrolysis, continuing to carry electrons from the polymer to the metal center to produce EPFRs. When lignin and cellulose in biomass are heated during pyrolysis, they may disintegrate to produce aromatic molecule precursors that, when exposed to air, develop into EPFRs (Xiang et al. 2021).

Dioxins

The creation of harmful substances like dioxins can occur during the biochar preparation. The main variables impacting the amount of dioxins in biochar are the preparation conditions, such as temperature (Xiang et al. 2021). Due to the many kinds of raw materials used to produce biochar, there may be other environmentally damaging compounds in addition to the conventional pollutants described above. For instance, biochar generated from plants did not include PFCs, which are persistent pollutants that exhibit high resistance to chemical and thermal degradation. Besides, biochar may also contain biotoxic VOCs, which are possible environmental contaminants (Xiang et al. 2021).

6.1 Adverse Effects of Biochar on the Soil Environment

The pH, structure, porosity, mobility, the bioavailability of dangerous chemicals, and other properties of the soil could be altered by biochar. The pH of the biochar steadily changes from neutral or acidic to alkaline due to the amount of acidic functional groups on the surface of the biochar declines, where the loss of O_2 takes place as the pyrolysis temperature rises. The biochar-induced rise in soil pH may limit the availability of some nutrients, such NH_4^+, to the original soil (Xiang et al. 2021; Devi and Saroha 2014).

Availability of biochar in the soil can have an impact on the microorganism population in the soil environment. The increase in amount of biochar can cause an increment in pH of the soil. This may decrease the soil nutrient content. The increment of soil pH can increase the hydrolysis rate of N-acyl-homoserine lactone (AHL), which is an intercellular signaling molecule used by gram-negative bacteria. A decreasing in the bioavailability of AHL may arise from the soil pH elevation brought on by biochar will affect the growth and existence of this gram-negative bacterial population. This will have a negative effect on soil mycobacterial activity, which will also affect the decomposition of organic components and plant growth (Xiang et al. 2021).

Biochar can contain potentially dangerous organic or inorganic substances like heavy metals and PAHs. Therefore, when it is added to the soil, it may directly affect plants. In the process of making biochar, raw materials' cellulose or hemicellulose is broken down to release gaseous hydrocarbon groups, which are then put through a series of processes to create aromatic rings. It was discovered that the soil supplemented with biochar has a higher concentration of PAHs than the soil without biochar (Brtnicky et al. 2021).

6.2 Adverse Effects of Biochar on the Aquatic Environment

Aquatic environments may also be negatively impacted by the incorporation of biochar to the natural environment. Biochar may contain a huge amount of N and P content with them which is dependent on their biomass source. Hence, these could release into the water resources and increase their nutrient content (NH_4^+, PO_4^{3-}) (Xiang et al. 2021). This will cause eutrophication in the water sources. Then the plant and algae growth in the water will increase. Eutrophication can increase the biochemical oxygen demand of the water and cause aquatic organism pollution to decrease because of a lack of O_2. The growth of algae will reduce the sunlight which comes to the bottom of the water resources, and it will cause the inhibition of plant growth.

Biochar can migrate pollutants from their adsorption potential and also can carry its surface adsorbed pollutants through surface runoff of excess water, ditches, and irrigation work (Xiang et al. 2021). Therefore, this biochar can go through wells that were uncovered at ground level or can infiltrate through the cracks in the surface of the ground and contaminate the groundwater sources also. This will pollute the entire water system and can harm all the organisms which survive by using the particular water.

6.3 Risk Avoidance Measures of Biochar

According to research findings, a pressing need to lessen or get rid of toxicity risks in biochar materials is essential. This would improve the potent value of biochar and increase the range of uses of it. Biochar's toxicity is primarily a result of the feedstock and manufacturing processes; hence, it is required to use feedstock with low levels of hazardous compounds. Biomass feedstock which has a low amount of PAH precursor molecules can be used to make biochar. Further, slow pyrolysis is advisable in terms of pyrolysis rate (Xiang et al. 2021; Devi and Saroha 2014).

The major causes of slow-growing biochar's lower ecological dangers than fast-growing biochar are its lower content of hazardous compounds, which is created at low temperatures and slow rates, the limited soil nutrient immobilization ability, and the biochar's stronger mineralization capacity. When biochar is produced at low temperatures, it contains fewer toxic compounds (such as PAHs) and is less ecotoxic than when produced at high temperatures. Additionally, when pyrolysis time and temperature increase, the concentration of PAHs often falls (Xiang et al. 2021).

Before incorporating biochar to the soil, the physicochemical properties of the medium (such as soil moisture and aeration) should be determined. For instance, it has been discovered that after applying biochar, soil dust emissions increase as a result of the low water content in the soil, suggesting that tilling after wetting plots modified with biochar effectively reduces exposure to both soil and the biochar particles. For maximum dust reduction in coarse-grained soils, high water levels close to saturation are necessary, although this is not the case for fine-grained soils. Researchers advise using large particles of biochar when adding it to the soil since they have less sorption (owing to a lower surface area-to-volume ratio) and are less likely to be ingested or transferred by plants or animals (Xiang et al. 2021).

7 Upcoming Challenges of Biochar

Generally, most of the biochar research has been conducted on the laboratory scale and pilot scale; therefore, more field research needs to be conducted to identify the conditions and effects. When considering sustainability, controlling the pyrolysis process might also be quite important. The electricity required to run the pyrolysis process could produce greenhouse gas emissions, especially if it is generated from fossil and non-renewable resources. Hence, it is necessary to develop green energy generation technologies to supply the energy required for turning biomass into biochar. It is possible to explore several cutting-edge technologies as a potential means of promoting the sustainability of biochar production, such as solar energy concentration in the form of concentrated solar towers (Kamali et al. 2022; Mohammadi et al. 2020).

Although utilizing biochar in soil applications may have some benefits, there are still some uncertainties regarding how it will interact with the environment. To prevent significant environmental issues, such as ecotoxicological impacts, it is imperative to look at the long-term impacts of biochar in soil and monitor its fate. Thereby, this can guarantee the longevity of biochar's soil applications. In this regard, it is vital to create and standardize analytical procedures that may be used to look at how biochar is affected by the environment (Kamali et al. 2022; Mohammadi et al. 2020).

Find the best ways to make biochar that is less hazardous while also protecting the farms and workers that handle the materials is an essential need. Also, it is mandatory to thoroughly research the potentially toxic effects of biochar on human health. Comprehensive research is still needed on the fate of biochar when it is applied in the environment, as well as its long-term effects on the properties of the soil. Additionally, the possibility for biochar discharge into water bodies requires thorough analysis of how mobile it is in the soil (such as underground water sources).

Biochar might be a ground-breaking material in many industries, including global agriculture, if these difficulties could be resolved in the future (Kamali et al. 2022; Mohammadi et al. 2020).

8 Conclusion

The use of biochar is gaining attention in a variety as a means of addressing the most severe ecological problems. Due to its lower greenhouse gas emissions, higher efficiency in producing heat and electricity, financial gain in lowering waste disposal costs, and scarcity of fossil fuels concerning the world's energy needs, biochar is a viable alternative to currently used fossil fuels. Since it is so effective at removing impurities and toxins from soil, aqueous, and gaseous mediums, it is regarded as a renewable resource. The growth, survival, and activity of microorganisms may be helped, as biochar enhances nutrient retention through cation adsorption and improves soil productivity. Further, biochar is used as a component in cement materials, which can act as a carbon dioxide adsorber. The properties of the material could be changed according to the percentage of biochar in the cement. Besides, they can reduce plant growth and add pollutants to the surface and groundwater systems. However, to use biochar as an efficient and safe material for its applications, it is necessary to make the production process environmentally friendly and use biomass feedstocks with less harmful precursor compounds to produce biochar with minimum harmful components.

References

Adler E (1977) 'Lignin chemistry—past, present and future. Wood Sci Technol 11:169–218. https://doi.org/10.1007/BF00365615

Ahmad M, Rajapaksha AU, Lim JE, Zhang M, Bolan N, Mohan D, Vithanage M, Lee SS, Ok YS (2014) Biochar as a sorbent for contaminant management in soil and water: a review. Chemosphere 99:19–33. https://doi.org/10.1016/j.chemosphere.2013.10.071

Ahmad M, Rajapaksha AU, Lim JE, Zhang M, Bolan N, Mohan D, Vithanage M, Lee SS, Ok YS (2014b) Biochar as a sorbent for contaminant management in soil and water: a review. Chemosphere 99:19–33. https://doi.org/10.1016/j.chemosphere.2013.10.071

Ahmed MB, Zhou JL, Ngo HH, Guo W, Johir MAH, Sornalingam K (2017) Single and competitive sorption properties and mechanism of functionalized biochar for removing sulfonamide antibiotics from water. Chem Eng J 311:348–358. https://doi.org/10.1016/j.cej.2016.11.106

Alonso-Gómez LA, Celis-Carmona DD, Rodríguez-Sánchez YF, Castro-Ladino JR, Solarte-Toro JC (2024) Biochar production from cassava waste biomass: a techno-economic development approach in the Colombian context. Bioresour Technol Rep 26:101872. https://doi.org/10.1016/j.biteb.2024.101872

Al-Rahbi AS, Williams PT (2017) Hydrogen-rich syngas production and tar removal from biomass gasification using sacrificial tyre pyrolysis char. Appl Energy 190:501–509. https://doi.org/10.1016/j.apenergy.2016.12.099

Ansari KB, Arora JS, Chew JW, Dauenhauer PJ, MushrifS H (2019) 'Fast pyrolysis of cellulose, hemicellulose, and lignin: effect of operating temperature on bio-oil yield and composition and insights into the intrinsic pyrolysis chemistry. Indus Eng Chem Res 58(35):15838–15852. https://doi.org/10.1021/acs.iecr.9b00920

Armah EK, Chetty M, Adedeji JA, Estrice DE, Mutsvene B, Singh N, Tshemese Z (2022) Biochar: production, application and the future. In: Biochar—productive technologies, properties and application. IntechOpen. https://doi.org/10.5772/intechopen.105070

Atkinson C, Fitzgerald J, Hipps N (2010) Potential mechanisms for achieving agricultural benefits from biochar application to temperate soils: a review. Plant Soil 337:1–18. https://doi.org/10.1007/s11104-010-0464-5

Ayaz M, Feiziene D, Tilvikiene V, Akhtar K, Stulpinaitė U, Iqbal R (2021) Biochar role in the sustainability of agriculture and environment. Sustainability 13:1330. https://doi.org/10.3390/su13031330

Bach QV, Tran KQ, Skreiberg O (2015) Accelerating wet torrefaction rate and ash removal by carbon dioxide addition. Fuel Process Technol 140:297–303. https://doi.org/10.1016/j.fuproc.2015.09.013

Ben H, Ragauskas AJ (2012) Torrefaction of Loblolly pine. Green Chem 14(1):72–76. https://doi.org/10.1039/C1GC15570A

Boostani HR, Najafi-Ghiri M, Hardie AG, Khalili D (2019) Comparison of Pb stabilization in a contaminated calcareous soil by application of vermicompost and sheep manure and their biochars produced at two temperatures. Appl Geochem 102:121–128. https://doi.org/10.1016/j.apgeochem.2019.01.013

Bridgeman TG, Jones JM, Shield I, Williams PT (2008) Torrefaction of reed canary grass, wheat straw and willow to enhance solid fuel qualities and combustion properties. Fuel 87(6):844–856. https://doi.org/10.1016/j.fuel.2007.05.041

Bridgwater AV (2012) Review of fast pyrolysis of biomass and product upgrading. Biomass Bioenerg 38:68–94. https://doi.org/10.1016/j.biombioe.2011.01.048

Brtnicky M, Datta R, Holatko J, Bielska L, Gusiatin Z, Kucerik J, Hammerschmiedt T, Danish S, Radziemska M, Mravcova L, Fahad S, Kintl A, Sudoma M, Ahmed N, Pecina V (2021) A critical review of the possible adverse effects of biochar. Soil Environ Sci Total Environ 796:148756. https://doi.org/10.1016/j.scitotenv.2021.148756

Chan KY, Van Zwieten L, Meszaros I, Downie A, Joseph S (2008) Using poultry litter biochars as soil amendments. Aust J Soil Res 46:437–444. https://doi.org/10.1071/SR08036

Chatterjee R, Sajjadi B, Chen W-Y, Mattern DL, Hammer N, Raman V, Dorris A (2020) Effect of pyrolysis temperature on physicochemical properties and acoustic-based amination of biochar for efficient CO_2 adsorption. Front Energy Res 8:85. https://doi.org/10.3389/fenrg.2020.00085

Chen T, Zhou Z, Han R, Meng R, Wang H, Lu W (2015) Adsorption of cadmium by biochar derived from municipal sewage sludge: impact factors and adsorption mechanism. Chemosphere 134:286–293. https://doi.org/10.1016/j.chemosphere.2015.04.052

Chen Z, Wang M, Ren Y, Jiang E, Jiang Y, Li W (2018a) Biomass torrefaction: a promising pretreatment technology for biomass utilization. IOP Conf Ser Earth Environ Sci 113:012201. https://doi.org/10.1088/1755-1315/113/1/012201

Chen T, Luo L, Deng S, Shi G, Zhang S, Zhang Y, Deng O, Wang L, Zhang J, Wei L (2018) Sorption of tetracycline on H_3PO_4 modified biochar derived from rice straw and swine manure. Bioresour Technol 267:431–437. https://doi.org/10.1016/j.biortech.2018.07.074

Chen C, Liang W, Fan F, Wang C (2021) The effect of temperature on the properties of hydrochars obtained by hydrothermal carbonization of waste *Camellia oleifera* shells. ACS Omega 6(25):16546–16552. https://doi.org/10.1021/acsomega.1c01787

Chen Z, Guo Y, Luo L, Liu Z, Miao W, Xia Y (2024) A critical review of hydrochar based photocatalysts by hydrothermal carbonization: synthesis, mechanisms, and applications. Biochar. https://doi.org/10.1007/s42773-024-00364-9

Chu S, Subrahmanyam AV, Huber GW (2013) The pyrolysis chemistry of a β-O-4 type oligomeric lignin model compound. Green Chem 15(1):125–136. https://doi.org/10.1039/C2GC36332A

Chun Y, Sheng G, Chiou CT, Xing B (2004) Compositions and sorptive properties of crop residue-derived chars. Environ Sci Technol 38:4649–4655. https://doi.org/10.1021/es035034w

Crombie K, Mašek O (2014) Investigating the potential for a self-sustaining slow pyrolysis system under varying operating conditions. Biores Technol 162:148–156. https://doi.org/10.1016/j.biortech.2014.03.134

Cui H-M, Liu Y, Bian J, Li M-F (2024) Catalytic hydrothermal carbonization of corn and rice straws with citric acid: implications for charcoal production and combustion performance. Ind Crops Prod 219:119162. https://doi.org/10.1016/j.indcrop.2024.119162

Demirbas A (2004) Effects of temperature and particle size on bio-char yield from pyrolysis of agricultural residues. J Anal Appl Pyrol 72(2):243–248. https://doi.org/10.1016/j.jaap.2004.07.003

Dengyu C, Kehui C, Xiaozhuang Z, Ziyu G, Jianbin Z, Yimeng Z, Hong Z (2022) Insight into biomass pyrolysis mechanism based on cellulose, hemicellulose, and lignin: evolution of volatiles and kinetics, elucidation of reaction pathways, and characterization of gas, biochar and bio-oil. Combus Flame 242:112142. https://doi.org/10.1016/j.combustflame.2022.112142

Devi P, Saroha AK (2014) Risk analysis of pyrolyzed biochar made from paper mill effluent treatment plant sludge for bioavailability and eco-toxicity of heavy metals. Bioresour Technol 162:308–315. https://doi.org/10.1016/j.biortech.2014.03.093

El-Sayed SA, Khass TM, Mostafa ME (2023) Thermal degradation behaviour and chemical kinetic characteristics of biomass pyrolysis using TG/DTG/DTA techniques. Biomass Convers Biorefin. https://doi.org/10.1007/s13399-023-03926-2

Falco C, Baccile N, Titirici M-M (2011) Morphological and structural differences between glucose, cellulose and lignocellulosic biomass derived hydrothermal carbons. Green Chem 13(11):3273. https://doi.org/10.1039/C1GC15742F

Ferdous D, Dalai AK, Bej SK, Thring RW (2002) Pyrolysis of lignins: experimental and kinetics studies. Energy Fuels 16(6):1405–1412. https://doi.org/10.1021/ef0200323

Fokaides PA (2017) Energy recovery alternatives for the sustainable management of olive oil industry. In: Olive mill waste, pp 79–96. https://doi.org/10.1016/b978-0-12-805314-0.00004-2

Funke A, Ziegler F (2010) Hydrothermal carbonization of biomass: a summary and discussion of chemical mechanisms for process engineering. Biofuels Bioprod Bioref 4:160–177. https://doi.org/10.1002/bbb.198

Gabhane JW, Bhange VP, Patil PD, Bankar ST, Kumar S (2020) Recent trends in biochar production methods and its application as a soil health conditioner: a review. SN Appl Sci 2(7):1–21. https://doi.org/10.1007/s42452-020-3121-5

Gao Z, Li N, Wang Y, Niu W, Yi W (2020) Pyrolysis behavior of xylan-based hemicellulose in a fixed bed reactor. J Anal Appl Pyrol 146:104772. https://doi.org/10.1016/j.jaap.2020.104772

González J, Román S, Bragado D, Calderon M (2008) Investigation on the reactions influencing biomass air and air/steam gasification for hydrogen production. Fuel Process Technol 89:764–772. https://doi.org/10.1016/j.fuproc.2008.01.011

Gupta S, Kua H (2017) Factors determining the potential of biochar as a carbon capturing and sequestering construction material: critical review. J Mater Civ Eng. https://doi.org/10.1061/(ASCE)MT.1943-5533.0001924

Gupta S, Kua HW, Low CY (2018) Use of biochar as carbon sequestering additive in cement mortar. Cement Concr Compos 87:110–129. https://doi.org/10.1016/j.cemconcomp.2017.12.009

Gupta S et al (2022) Biochar for enhancing soil health and crop productivity: a comprehensive review. Environ Sci Pollut Res 29(2):1935–1954. https://doi.org/10.1007/s11356-021-16990-3

Hassan MM, Carr CM (2018) A critical review on recent advancements of the removal of reactive dyes from dyehouse effluent by ion-exchange adsorbents. Chemosphere 209:201–219. https://doi.org/10.1016/j.chemosphere.2018.06.043

He Q, Yu Y, Wang J, Suo X, Liu Y (2021) Kinetic study of the hydrothermal carbonization reaction of glucose and its product structures. Indus Eng Chem Res 60(12):4552–4561. https://doi.org/10.1021/acs.iecr.0c06280

Hosoya T, Kawamoto H, Saka S (2009) Role of methoxyl group in char formation from lignin-related compounds. J Anal Appl Pyrol 84(1):79–83. https://doi.org/10.1016/j.jaap.2008.10.024
Hossain MA, Saha M, Islam MR (2022) Biochar for enhancing soil health and crop productivity: a comprehensive review. Environ Sci Pollut Res 29(2):1935–1954. https://doi.org/10.1007/s11356-021-16990-3
Hu X, Xu J, Wu M, Xing J, Bi W, Wang K, Ma J, Liu X (2017) Effects of biomass pre-pyrolysis and pyrolysis temperature on magnetic biochar properties. J Anal Appl Pyrol 127:196–202. https://doi.org/10.1016/j.jaap.2017.08.006
Hu B, Lu Q, Jiang X, Dong X, Cui M, Dong C, Yang Y (2018) Pyrolysis mechanism of glucose and mannose: the formation of 5-hydroxymethyl furfural and furfural. J Energy Chem 27(2):486–501. https://doi.org/10.1016/j.jechem.2017.11.013
Huang D, Liu L, Zeng G, Xu P, Huang C, Deng L, Wang R, Wan J (2017) The effects of rice straw biochar on indigenous microbial community and enzymes activity in heavy metal-contaminated sediment. Chemosphere 174:545–553. https://doi.org/10.1016/j.chemosphere.2017.01.130
Jeffery S, Verheijen FGA, van der Velde M, Bastos AC (2011) Quantitative review of the effects of biochar application to soils on crop productivity using meta-analysis. Agric Ecosyst Environ 144:175–187. https://doi.org/10.1016/j.agee.2011.08.015
Jia M, Wang F, Bian Y, Jin X, Song Y, Kengara FO, Xu R, Jiang X (2013) Effects of pH and metal ions on oxytetracycline sorption to maize-straw-derived biochar. Bioresour Technol 136:87–93. https://doi.org/10.1016/j.biortech.2013.02.098
John KI, Omorogie MO (2022) Biomass-based hydrothermal carbons for catalysis and environmental cleanup: a review. Green Chem Lett Rev 15(1):162–186. https://doi.org/10.1080/17518253.2022.2028017
John E, White W, James C, Benjamin LL (2011) Biomass pyrolysis kinetics: a comparative critical review with relevant agricultural residue case studies. J Anal Appl Pyrol 91(1):1–33. https://doi.org/10.1016/j.jaap.2011.01.004
Jones DL, Edwards-Jones G, Murphy DV (2011) Biochar mediated alterations in herbicide breakdown and leaching in soil. Soil Biol Biochem 43(4):804–813. https://doi.org/10.1016/j.soilbio.2010.12.015
Kamali M, Sweygers N, Al-Salem S, Appels L, Aminabhavi T, Dewil R (2022) Biochar for soil applications—sustainability aspects, challenges and future prospects. Chem Eng J 428:131189. https://doi.org/10.1016/j.cej.2021.131189
Kang S, Li X, Fan J, Chang J (2012) Characterization of hydrochars produced by hydrothermal carbonization of lignin, cellulose, d-xylose, and wood meal. Ind Eng Chem Res 51(26):9023–9031. https://doi.org/10.1021/ie300565d
Khan MA, Khan S, Ding X, Khan A, Alam M (2018) The effects of biochar and rice husk on adsorption and desorption of cadmium on to soils with different water conditions (upland and saturated). Chemosphere 193:1120–1126. https://doi.org/10.1016/j.chemosphere.2017.11.110
Kloss S, Zehetner F, Dellantonio A, Hamid R, Ottner F, Liedtke V, Schwanninger M, Gerzabek HM, Soja G (2012) Characterization of slow pyrolysis biochars: effects of feedstocks and pyrolysis temperature on biochar properties. J Environ Qual 41(4):990. https://doi.org/10.2134/jeq2011.0070
Ko DC, Cheung CW, Choy KK, Porter JF, McKay G (2004) Sorption equilibria of metal ions on bone char. Chemosphere 54:273–281. https://doi.org/10.1016/j.chemosphere.2003.08.004
Kobayashi T, Khuwijitjaru P, Adachi S (2016) Decomposition kinetics of glucose and fructose in subcritical water containing sodium chloride. J Appl Glycosci 63(4):99–104. https://doi.org/10.5458/jag.jag.JAG-2016_013
Kocsis T, Ringer M, Biró B (2022) Characteristics and applications of biochar in soil-plant systems: a short review of benefits and potential drawbacks. Appl Sci 12(8):4051. https://doi.org/10.3390/app12084051
Kookana R, Van ZL, Krull E, Singh B (2011) Biochar application to soil: agronomic and environmental benefits and unintended consequences. Adv Agron 112:103–143. https://doi.org/10.1016/B978-0-12-385538-1.00003-2

Kossińska N, Krzyżyńska R, Ghazal H, Jouhara H (2023) Hydrothermal carbonisation of sewage sludge and resulting biofuels as a sustainable energy source. Energy 275:127337. https://doi.org/10.1016/j.energy.2023.127337

Kruse A, Funke A, Titirici M-M (2013) Hydrothermal conversion of biomass to fuels and energetic materials. Curr Opin Chem Biol 17(3):515–521. https://doi.org/10.1016/j.cbpa.2013.05.004

Kwapińska M, Pisanó I, Leahy JJ (2023) Preliminary assessment of pyrolysis biochar derived from milk/dairy processing sludge as a potential component of fertilizers. ACS Sustain Chem Eng 11(6):2345–2353. https://doi.org/10.1021/acssuschemeng.2c06093

Larsbo M, Löfstrand E, de Veer DV, Ulén B (2013) Pesticide leaching from two Swedish topsoils of contrasting texture amended with biochar. J Contam Hydrol 147:73–81. https://doi.org/10.1016/j.jconhyd.2013.01.003

Lee J, Lee K, Sohn D, Kim YM, Park KY (2018) Hydrothermal carbonization of lipid extracted algae for hydrochar production and feasibility of using hydrochar as a solid fuel. Energy 153:913–920. https://doi.org/10.1016/j.energy.2018.04.112

Legan M, Gotvajn AŽ, Zupan K (2022) Potential of biochar use in building materials. J Environ Manage 309:114704. https://doi.org/10.1016/j.jenvman.2022.114704

Lehmann J (2007) Bio-energy in the black. Front Ecol Environ 5:381–387. https://doi.org/10.1890/1540-9295(2007)5[381:BITB]2.0.CO;2

Lehmann J, Rillig MC, Thies J, Masiello CA, Hockaday WC, Crowley D (2011) Biochar effects on soil biota—a review. Soil Biol Biochem 43(9):1812–1836. https://doi.org/10.1016/j.soilbio.2011.04.022

Lin Y, Ma X, Peng X, Yu Z, Fang S, Lin Y, Fan Y (2016) Combustion, pyrolysis and char CO_2-gasification characteristics of hydrothermal carbonization solid fuel from municipal solid wastes. Fuel 181:905–915. https://doi.org/10.1016/j.fuel.2016.05.031

Liu T, Yang L, Hu Z et al (2020) Biochar exerts negative effects on soil fauna across multiple trophic levels in a cultivated acidic soil. Biol Fertil Soils 56:597–606. https://doi.org/10.1007/s00374-020-01436-1

Liu Q, Wu L, Zhang A (2023a) Biochar as a carbon sink and soil amendments: potential and challenges. Environ Sci Pollut Res 30(15):4141–4155. https://doi.org/10.1007/s11356-023-23390-2

Liu X, Zhang Y, Wang Z (2023b) Recent advances in biochar applications for carbon sequestration and soil improvement. Sci Total Environ 844:157059. https://doi.org/10.1016/j.scitotenv.2022.157059

Liu Y, Weng Z, Han B, Guo Z, Tian H, Tang Y, Cai Y, Yang Z (2023c) Recent studies on the comprehensive application of biochar in multiple environmental fields. J Clean Prod 421:138495–138495. https://doi.org/10.1016/j.jclepro.2023.138495

Lu HP, Li ZA, Gascó G, Méndez A, Shen Y, Paz-Ferreiro J (2018) Use of magnetic biochars for the immobilization of heavy metals in a multi-contaminated soil. Sci Total Environ 622:892–899. https://doi.org/10.1016/j.scitotenv.2017.12.056

Luo S, Xiao B, Hu Z, Liu S, Guo X, He M (2009) Hydrogen-rich gas from catalytic steam gasification of biomass in a fixed bed reactor: influence of temperature and steam on gasification performance. Int J Hydrogen Energy 34(5):2191–2194. https://doi.org/10.1016/j.ijhydene.2008.12.075

Maaoui A, Ben A, Abdallah AB, Chagtmi R, Lopez G, Cortazar M, Olazar M (2023) Assessment of pine wood biomass wastes valorization by pyrolysis with focus on fast pyrolysis biochar production. J Energy Inst 108:101242–101242. https://doi.org/10.1016/j.joei.2023.101242

Maduskar S, Maliekkal V, Neurock M, Dauenhauer PJ (2018) On the yield of levoglucosan from cellulose pyrolysis. ACS Sustain Chem Eng 6(5):7017–7025. https://doi.org/10.1021/acssuschemeng.8b00853

Manmeen A, Kongjan P, Palamanit A, Jariyaboon R (2023) Biochar and pyrolysis liquid production from durian peel by using slow pyrolysis process: regression analysis, characterization, and economic assessment. Indus Crops Prod 203:117162–117162. https://doi.org/10.1016/j.indcrop.2023.117162

Mitchell PJ, Dalley TSL, Helleur RJ (2013) Preliminary laboratory production and characterization of biochars from lignocellulosic municipal waste. J Anal Appl Pyrol 99:71–78. https://doi.org/10.1016/j.jaap.2012.10.025

Mohammadi A, Khoshnevisan B, Venkatesh G, Eskandari SA (2020) Critical review on advancement and challenges of biochar application in paddy fields: environmental and life cycle cost analysis. Processes 8:1275. https://doi.org/10.3390/pr8101275

Mohan D, Pittman CU, Steele PH (2006) Pyrolysis of wood/ biomass for bio-oil: a critical review. Energy Fuels 20(3):848–889. https://doi.org/10.1021/ef0502397

Montanarella L, Lugato E (2013a) The application of biochar in the EU: challenges and opportunities. Agronomy 3(2):462–473. https://doi.org/10.3390/agronomy3020462

Montanarella L, Lugato E (2013b) The application of biochar in the EU: challenges and opportunities. Agronomy 3:462–473. https://doi.org/10.3390/agronomy3020462

Muradov N, Fidalgo B, Gujar AC, Garceau N, T-Raissi A (2012) Production and characterization of Lemna minor bio-char and its catalytic application for biogas reforming. Biomass Bioenergy 42:123–131. https://doi.org/10.1016/j.biombioe.2012.03.003

Mustaza M, Mizan M, Yoshida H, Izhar S (2021) Torréfaction of mangrove wood by introducing superheated steam for biochar production. IOP Conf Ser Earth Environ Sci 765:012027. https://doi.org/10.1088/1755-1315/765/1/012027

Nag SK, Kookana R, Smith L, Krull E, Macdonald LM, Gill G (2011) Poor efficacy of herbicides in biochar-amended soils as affected by their chemistry and mode of action. Chemosphere 84(11):1572–1577. https://doi.org/10.1016/j.chemosphere.2011.05.052

Ngah WW, Hanafiah M, Hanafiah MAKM (2008) Removal of heavy metal ions from wastewater by chemically modified plant wastes as adsorbents: a review. Bioresour Technol 99:3935–3948. https://doi.org/10.1016/j.biortech.2007.06.011

Osayi JI, Iyuke S, Daramola MO et al (2018) Pyrolytic conversion of used tyres to liquid fuel: characterization and effect of operating conditions. J Mater Cycles Waste Manag 20:1273–1285. https://doi.org/10.1007/s10163-017-0690-5

Pan J, Jiang J, Xu R (2013) Adsorption of Cr(III) from acidic solutions by crop straw derived biochars. J Environ Sci 25:1957–1965. https://doi.org/10.1016/S1001-0742(12)60305-2

Pang S (2019) Advances in thermochemical conversion of woody biomass to energy, fuels and chemicals. Biotechnol Adv 37(4):589–597. https://doi.org/10.1016/j.biotechadv.2018.11.004

Patwardhan PR, Brown RC, Shanks BH (2011) Product distribution from the fast pyrolysis of hemicellulose. Chemsuschem 4(5):636–643. https://doi.org/10.1002/cssc.201000425

Piazza V, Batisda da Silva Junior R, Frassoldati A, Lietti L, Gambaro C, Rajendran K, Chen D, Faravelli T, Beretta A (2024) Unravelling the complexity of hemicellulose pyrolysis: quantitative and detailed product speciation for xylan and glucomannan in TGA and fixed bed reactor. Chem Eng J 497:154579. https://doi.org/10.1016/j.cej.2024.154579

Piersa P, Unyay H, Szufa S, Lewandowska W, Modrzewski R, Ślężak R, Ledakowicz S (2022) An extensive review and comparison of modern biomass torrefaction reactors vs. biomass pyrolysis—part 1. Energies 15(6):2227. https://doi.org/10.3390/en15062227

Pilon G, Lavoie JM (2012) Pyrolysis of switchgrass (*Panicum virgatum* L.) at low temperatures within N_2 and CO_2 environments: product yield study. ACS Sustain Chem Eng 1(1):198–204. https://doi.org/10.1021/sc300098e

Prabakar D, Manimudi VT, Suvetha S, Sampath S, Mahapatra DM, Rajendran K, Pugazhendhi A (2018) Advanced biohydrogen production using pretreated industrial waste: outlook and prospects. Renew Sust Energ Rev 96:306–324. https://doi.org/10.1016/j.rser.2018.08.006

Prakoso T, Nurastuti R, Hendriansyah R, Rizkiana J, Suantika G, Guan G (2018) Hydrothermal carbonization of seaweed for advanced biochar production. MATEC Web Conf 156:05012. https://doi.org/10.1051/matecconf/201815605012

Prieto M, Yue H, Brun N, Ellis GJ, Naffakh M, Shuttleworth PS (2024) Hydrothermal carbonization of biomass for electrochemical energy storage: parameters, mechanisms, electrochemical performance, and the incorporation of transition metal dichalcogenide nanoparticles. Polymers 16(18):2633–2633. https://doi.org/10.3390/polym16182633

Prost K, Borchard N, Siemens J, Kautz T, Séquaris JM, Möller A, Amelung W (2013) Biochar is affected by composting with farmyard manure. J Environ Qual 42(1):164–172. https://doi.org/10.2134/jeq2012.0064
Qian L, Chen B, Hu D (2013) Effective alleviation of aluminum phytotoxicity by manure-derived biochar. Environ Sci Technol 47:2737–2745. https://doi.org/10.1021/es3047872
Qian T, Wang Y, Fan T et al (2016) A new insight into the immobilization mechanism of Zn on biochar: the role of anions dissolved from ash. Sci Rep 6:33630. https://doi.org/10.1038/srep33630
Qin X, Cheng S, Xing B, Qu X, Shi C, Meng W, Zhang C, Xia H (2023) Preparation of pyrolysis products by catalytic pyrolysis of poplar: application of biochar in antibiotic wastewater treatment. Chemosphere 338:139519. https://doi.org/10.1016/j.chemosphere.2023.139519
Qiu Y, Xiao X, Cheng H, Zhou Z, Sheng GD (2009) Influence of environmental factors on pesticide adsorption by black carbon: pH and model dissolved organic matter. Environ Sci Technol 43:4973–4978. https://doi.org/10.1021/es900573d
Rangabhashiyam S, Balasubramanian P (2019) The potential of lignocellulosic biomass precursors for biochar production: Performance, mechanism and wastewater application—a review. Indus Crops Prod 128:405–423. https://doi.org/10.1016/j.indcrop.2018.11.041
Rosas-Mendoza ES, Alvarado-Vallejo A, Vallejo-Cantú NA, Velasco-Santos C, Alvarado-Lassman A (2024) Valorization of the complex organic waste in municipal solid wastes through the combination of hydrothermal carbonization and anaerobic digestion. Renew Energy 231:120916. https://doi.org/10.1016/j.renene.2024.120916
Roy P, Dias G (2017) Prospects for pyrolysis technologies in the bioenergy sector: a review. Renew Sustain Energy Rev 77:59–69. https://doi.org/10.1016/j.rser.2017.03.136
Rudra A, Petersen HI, Sanei H (2024) Molecular characterization of biochar and the relation to carbon permanence. Int J Coal Geol 291:104565–104565. https://doi.org/10.1016/j.coal.2024.104565
Sattar A, Leeke GA, Hornung A, Wood J (2014) Steam gasification of rapeseed, wood, sewage sludge and miscanthus biochars for the production of a hydrogen-rich syngas. Biomass Bioenerg 69:276–286. https://doi.org/10.1016/j.biombioe.2014.07.025
Shafiq M, Latif M, Mahmood R (2023) Role of biochar in improving soil quality and mitigating climate change. J Environ Manage 314:114104. https://doi.org/10.1016/j.jenvman.2022.114104
Sharma RK, Wooten JB, Baliga VL, Lin X, Geoffrey CW, Hajaligol MR (2004) Characterization of chars from pyrolysis of lignin. Fuel 83(11–12):1469–1482. https://doi.org/10.1016/j.fuel.2003.11.015
Sharma A, Wang S, Pareek V, Yang H, Zhang D (2015) Multi-fluid reactive modeling of fluidized bed pyrolysis process. Chem Eng Sci 123:311–321. https://doi.org/10.1016/j.ces.2014.11.019
Shen Y, Zhao P, Shao Q, Takahashi F, Yoshikawa K (2015) In situ catalytic conversion of tar using rice husk char/ash supported nickel–iron catalysts for biomass pyrolytic gasification combined with the mixing-simulation in fluidized-bed gasifier. Appl Energy 160:808–819. https://doi.org/10.1016/j.apenergy.2014.10.074
Sohi S, Krull E, Lopez-Capel E, Bol R (2010) A review of biochar and its use and function in soil. Adv Agron 105:47–82. https://doi.org/10.1016/S0065-2113(10)05002-9
Sombroek WG (1966) Amazon soils: a reconnaissance of the soils of the Brazilian Amazon region. Centre for Agricultural Publications and Documentation, Wageningen
Steiner C, Das KC, Melear N, Lakly D (2010) Reducing nitrogen loss during poultry litter composting using biochar. J Environ Qual 39(4):1236–1242. https://doi.org/10.2134/jeq2009.0337
Tang J, Zhu W, Kookana R, Katayama A (2013) Characteristics of biochar and its application in remediation of contaminated soil. J Biosci Bioeng 116:653–659. https://doi.org/10.1016/j.jbiosc.2013.05.035
Tripathi M, Sahu JN, Ganesan P (2016) Effect of process parameters on production of biochar from biomass waste through pyrolysis: a review. Renew Sustain Energy Rev 55:467–481. https://doi.org/10.1016/j.rser.2015.10.122

Turner J, Sverdrup G, Mann MK, Maness P-C, Kroposki B, Ghirardi M, Evans RJ, Blake D (2008) Renewable hydrogen production. Int J Energy Res 32:379–407. https://doi.org/10.1002/er.1372
Uchimiya M, Lima IM, Klasson KT, Chang S, Wartelle LH, Rodgers JE (2010) Immobilization of heavy metal ions (CuII, CdII, NiII, and PbII) by broiler Litter-derived biochars in water and soil. J Agric Chem 58:5538–5544. https://doi.org/10.1021/jf9044217
Wang T, Camps-Arbestain M, Hedley M et al (2012) Predicting phosphorus bioavailability from high-ash biochars. Plant Soil 357:173–187. https://doi.org/10.1007/s11104-012-1131-9
Wang Y, Qiu L, Zhu M et al (2019a) Comparative evaluation of hydrothermal carbonization and low temperature pyrolysis of *Eucommia ulmoides Oliver* for the production of solid biofuel. Sci Rep 9:5535. https://doi.org/10.1038/s41598-019-38849-4
Wang Y, Zhong B, Shafi M, Ma J, Guo J, Wu J, Ye Z, Liu D, Jin H (2019b) Effects of biochar on growth, and heavy metals accumulation of moso bamboo (*Phyllostachy pubescens*), soil physical properties, and heavy metals solubility in soil. Chemosphere 219:510–516. https://doi.org/10.1016/j.chemosphere.2018.11.159
Wang Q, Song H, Pan S et al (2020a) Initial pyrolysis mechanism and product formation of cellulose: an experimental and density functional theory (DFT) study. Sci Rep 10:3626. https://doi.org/10.1038/s41598-020-60095-2
Wang D, Jiang P, Zhang H, Yuan W (2020) 'Biochar production and applications in agro and forestry systems: a review. Sci Total Environ 723:137775. https://doi.org/10.1016/j.scitotenv.2020.137775
Wilk M, Magdziarz A (2017) Hydrothermal carbonization, torrefaction and slow pyrolysis of *Miscanthus giganteus*. Energy 140:1292–1304. https://doi.org/10.1016/j.energy.2017.03.031
Wüst D, Correa CR, Jung D et al (2020) 'Understanding the influence of biomass particle size and reaction medium on the formation pathways of hydrochar. Biomass Conv Bioref 10:1357–1380. https://doi.org/10.1007/s13399-019-00488-0
Xiang L, Liu S, Ye S, Yang H, Song B, Qin F, Shen M, Tan C, Zeng G, Tan X (2021) Potential hazards of biochar: the negative environmental impacts of biochar applications. J Hazard Mater 420:126611. https://doi.org/10.1016/j.jhazmat.2021.126611
Xu RK, Xiao SC, Yuan JH, Zhao AZ (2011) Adsorption of methyl violet from aqueous solutions by the biochars derived from crop residues. Bioresour Technol 102:10293–10298. https://doi.org/10.1016/j.biortech.2011.08.089
Xu X, Cao X, Zhao L, Wang H, Yu H, Gao B (2013) Removal of Cu, Zn, and Cd from aqueous solutions by the dairy manure-derived biochar. Environ Sci Pollut Res 20:358–368. https://doi.org/10.1007/s11356-012-0873-5
Yaashikaa PR, Kumar PS, Varjani S, Saravanan A (2020) A critical review on the biochar production techniques, characterization, stability and applications for circular bioeconomy. Biotechnol Rep 28:e00570. https://doi.org/10.1016/j.btre.2020.e00570
Yan F, Luo S, Hu Z, Xiao B, Cheng G (2010) Hydrogen-rich gas production by steam gasification of char from biomass fast pyrolysis in a fixed-bed reactor: influence of temperature and steam on hydrogen yield and syngas composition. Biores Technol 101(14):5633–5637. https://doi.org/10.1016/j.biortech.2010.02.025
Yang X, Liu J, McGrouther K, Huang H, Lu K, Guo X, He L, Lin X, Che L, Ye Z, Wang H (2016) Effect of biochar on the extractability of heavy metals (Cd, Cu, Pb, and Zn) and enzyme activity in soil. Environ Sci Pollut Res 23:974. https://doi.org/10.1007/s11356-015-4233-0
Yang X, Meng J, Lan Y, Chen W, Yang T, Yuan J, Liu S, Han J (2017) Effects of maize stover and its biochar on soil CO_2 emissions and labile organic carbon fractions in Northeast China. Agr Ecosyst Environ 240:24–31. https://doi.org/10.1016/j.agee.2017.02.001
Yang G, Hu Q, Hu J, Yang H, Yan S, Chen Y, Wang X, Chen H (2023) Hydrogen-rich syngas production from biomass gasification using biochar-based nanocatalysts. Biores Technol 379:129005. https://doi.org/10.1016/j.biortech.2023.129005
Zhang Y, Liu H, Wang Z (2023) Advances in biochar production and application for environmental sustainability. Environ Sci Technol 57(7):3721–3736. https://doi.org/10.1021/acs.est.3c01223

Zhao Y, Hu Z, Lu Y, Shan S, Zhuang H, Gong C, Cui X, Zhang F, Li P (2024) Facilitating mitigation of agricultural non-point source pollution and improving soil nutrient conditions: the role of low temperature co-pyrolysis biochar in nitrogen and phosphorus distribution. Biores Technol 394:130179. https://doi.org/10.1016/j.biortech.2023.130179

Zhou D, Liu D, Gao F, Li M, Luo X (2017) Effects of biochar-derived sewage sludge on heavy metal adsorption and immobilization in soils. Int J Environ Res Public Health 14:681. https://doi.org/10.3390/ijerph14070681

Zhou X, Li F, Tan J (2022) Biochar as an efficient adsorbent for organic pollutants: mechanisms and applications. Sci Total Environ 839:156318. https://doi.org/10.1016/j.scitotenv.2022.156318

Chapter 7
Biochar Is a Solution for Pharmaceutical Waste Disposal

Satyalakshmi Siragam

Abstract Nature-friendly waste disposal methods are attracting the attention of researchers and industrialists to avoid the huge damage that going to occur to the earth in the coming future. The researchers are striving to modernize the available conventional disposal methods or substances through mechanization and modeling to meet the current demand for global waste treatment approaches. Biochar (BC) is a nature compatible carbonized porous compound obtained from sustainable biomass via controlled thermal treatment. It is the most trending compound to treat solid and liquid wastes. Various studies already elaborated the BC active role in cleaning of chemical pollutants of contaminated soil and wastewater treatments. Hence, this chapter covers the disposal of pharmaceuticals by BC. As these pharmaceuticals are also the most prevailing water contaminants that can be discharged from industrial and domestic wastes. BC preparation from various biological sources, optimization of production conditions like temperature, time of exposure and pyrolysis, characterization, principal mechanism for removal of pharmaceuticals through BC, and key publications that pin the elimination of drug contaminants from water resources and biological fluids were summarized in this chapter. Finally, advanced BC compounds such as magnetic, nanoBC preparation, applications, integration of machine learning (ML), and artificial intelligence (AI) tools in BC-based waste water treatment were discussed.

Keywords Biochar · Pharmaceuticals · Micropollutants · Nanobiochar · Magnetic biochar · Machine learning

S. Siragam (✉)
Department of Pharmaceutics, Vignan Institute of Pharmaceutical Technology, Besides VSEZ, Kapujaggraju Peta, Duvvada, Visakhapatnam, Andhra Pradesh, India
e-mail: satyalaxmi148@gmail.com

A. Roy et al. (eds.), *Biochar Production Engineering*,
https://doi.org/10.1007/978-3-031-92594-8_7

1 Introduction

Globally, the pharmaceutical market is growing with a 9.1% of annual growth rate and by the end of this year; it will reach US$1,155.00 billion in 2024. (Pharmaceuticals Global Market Report 2022). Global healthcare sectors spent for medicines around $1.25 trillion in 2019, whereas $887 billion was spent in 2010. The revenue in this sector is expected to show an annual growth rate of 4.71%, leading to a market volume of US$1,454.00 billion by 2029 (Statista Market Forecast). The pharmaceutical industries are an important sector to protect the lives of human beings and animals through various healthcare products. While in its development, advanced science and technological methods have been employed. However, taking into consideration the achievements of numerous pharmaceutical industries, we should also consider the unresolved impact of pharmaceutical waste in the form of parent drugs and their metabolites which are being generated through expired drugs, unused pharmaceuticals, and drugs excreted from living beings after their pharmacological activity. A lack of balance between pharmaceutical production and its consumption may produce a huge amount of waste generation. The pharmaceutical waste-generating sources are pharmaceutical making industries, hospitals, sewage treating plants, waste controlling and reutilizing settings-dumping lands, agriculture waste peculiarly crop residues, livestock, aquaculture, and cesspool (Rajbongshi et al. 2016; Binaya and Agamuthu 2023; Han et al. 2023).

Leading pharmaceutical industries are the major contributors to global environmental issues that present crucial challenges for disposing methods of unused pharmaceutical waste.

All healthcare services generate > 3% of pharmaceutical waste globally (World health organization 2011). Today pharmaceutical disposal is an alarming problem in every nation (Kadam et al. 2016). Dangerous waste disposal in developing countries is however a significant challenge (Kaposiet al. 2024) and there were no established policies or programs available for management of medical and pharmaceutical waste (Bataduwaarachchi and Weeraratne 2018). Because of the poor management of healthcare waste, is responsible to cause serious problems for some healthcare professionals, workers in the sanitary departments, patients, and the general public. Pharmaceutical wastes are often noxious, reactive, caustic, flammable, and exhibit a deleterious impact on our ecosystem (Megan 2019). The hazardous pharmaceutical wastes are categorized into listed and characteristic waste. All commercial pharmaceuticals are covered under listed wastes on the other hand noxious, reactive, caustic, and flammable wastes are covered under characteristic wastes. Time-lapsed medicines, unused physician samples, loose pills, and spoiled or contaminated patient medication because of improper packing into the containers come under non-hazardous.

Pharmaceutical wastes found in environs are responsible for water, soil, and air pollution. Most of the pharmaceuticals reaching wastewater treatment plants are resistant to various physical, chemical, and biological degradation methods. Pharmaceutical wastes left after the treatment process are retained in wastewater treatment effluents and enter the adjoining water bodies which may affect the water quality and aquatic life. The solid sludge by-products (biosolids) of solid waste treating plants are used in land application as a fertilizer and wastewater effluent is employed for irrigation purposes. Consequently, provides a way of entry for pharmaceuticals into the terrestrial environment. A study has therefore been raised on the potential pharmaceutical uptake into soil organisms and its effects on ground-dwelling organisms and living creatures that depend on soil for nutrients. A few researchers have already demonstrated the effect of pharmaceuticals on soil organisms and investigated the impact on invertebrates such as earthworms (Carter et al. 2016).

More than a thousand samples from 258 rivers that flow through 104 nations were collected by a group of scientists for the determination of drug pollutants. They reported that except river from Iceland and another river near Venezuela, remaining all river samples have drug concentrations beyond the safe level. Pharmaceuticals and their metabolites release badly influence the ecosystem of water bodies that demands effective treatment methods for pharmaceutical industrial wastes, expired drugs as well as pharmaceutical domestic wastes (John et al. 2022).Type of waste degradation through BC are presented in Fig. 1.

Various approaches that have been in use since time immemorial such as filtration, adsorption, precipitation, coagulation, flocculation, advanced oxidation processes,

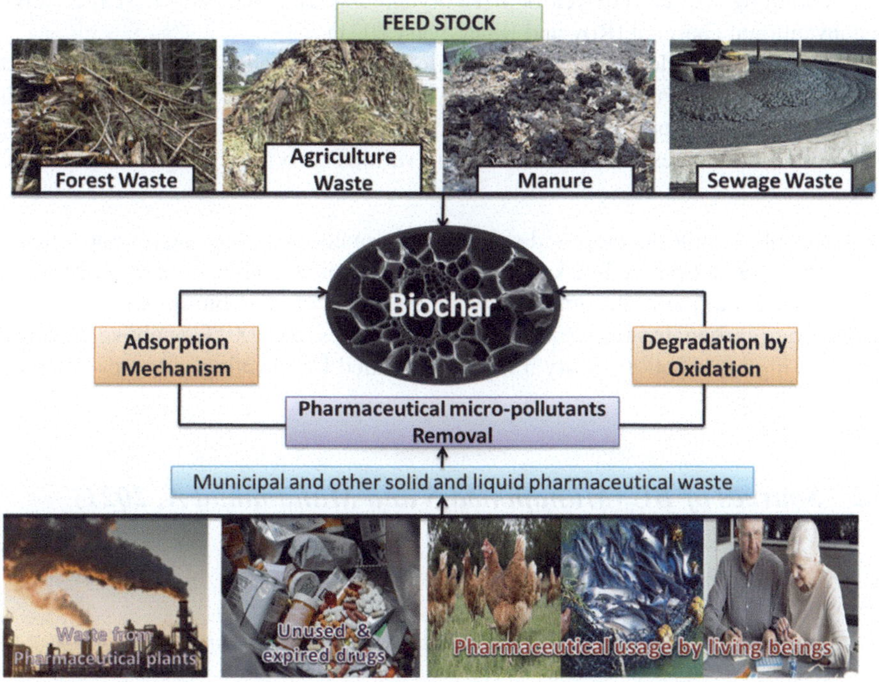

Fig. 1 Pharmaceutical waste degradation by biochar

reverse osmosis, ozonation, and fuel cells of microbial sources are integrated with the current techniques to treat large amounts of waste (Mouele et al. 2015; Ambaye et al. 2021).

High operational and capital costs are main obstacles of conventional disposal methods that limit their application in the removal of potentially low concentrations of toxic contaminants of pharmaceutical wastes in upper middle and middle income countries. The conventionally treated pharmaceutical waste remains with a low percentage of the parent compound and its metabolites; hence, the treated by-products are discharged into natural water resources. As a result, the leftover toxic pharmaceutical content is responsible for human and animal health hazards (Sayeda et al. 2021).

1.1 Pharmaceutical Waste Disposal by BC

BC is defined by the International Biochar Initiative as the solid substance of biomass that is treated thermochemically under oxygen-controlled conditions. BC term evolved from the Greek word βίος, *bios*, "life," and "char" (produced by carbonization of biomass) (Oxford University Press). BC is a porous, high surface area containing cost-effective carbonized organic material, and is the best alternative to conventional charcoal (Roy and Bharadvaja 2021).

BC contributes added benefits to the environment by remediating the contaminated soil. The basic working principle involved in BC is still poorly understood although BC has been in use for many centuries.

Some of the names used to describe the various carbon-rich organic matter are given below.

1. Char: physically decomposed solid organic material at very high temperatures such as forest fire and black carbon from incomplete combustion of fossil fuels.
2. Charcoal: thermally decomposed wood and other organic biomass.
3. Activate carbon: treating char at high temperatures like 500 °C or above. Because of its high adsorption ability it is recommended for cleaning purposes (Richika et al. 2023).

1.2 Sources of BC (Muhammad S and Muhammad A, 2021)

1. Energy crops (corn, cereals, wood, etc.)
2. Solid waste residues
3. Agricultural waste (nutshells and straw)
4. Compost (green waste)
5. Manure/animal waste
6. Kitchen waste
7. Sewage waste.

1.3 Methods of BC Production (Ibitoye et al. 2024)

BC production from wood carbonization is in practice since ancient times. Treating waste resources with enhanced BC technology can fill the gap energy needs of mankind shortly and also improves soil carbon sequestration potential. Fast pyrolysis, slow pyrolysis, carbonization, and gasification methods used for making of BC.

Pyrolysis

It is the thermochemical transformation of organic waste under a depleted oxygen environment. By considering the processing parameters pyrolysis can be categorized into the following.

Slow Pyrolysis

This method offers comparable BC production to other existing technologies. 50% of the feedstock carbon is retained. It is the most recommended method in the chemical industry, to enhance the yield of methanol, to derive carbonaceous organic substances from wood, to get vinyl chloride from ethylene dichloride, to convert biomass into BC and synthesis gas mixture, to turn waste into safely disposable substances.

Fast Pyrolysis

It occurs within a couple of seconds. Therefore, phenomena of phase conversion and processes of mass-heat transfer, chemical reaction kinetics, play a vital role. Charcoal obtained during the burning of crop residues, parts of *P. juliflora*, *P. hysterophorus*, *I. carnea*, *L. camera*, tree branches, twigs, etc.

Gasification

Gasification is the thermal transformation of biomass at > 700 °C under a controlled oxygen content. Nature-friendly waste can also be used for gasification. The high temperature induces the removal of caustic elements like chloride and potassium, thereby results pure and clean gas instead of pollutant fuels.

Carbonization

Through this process, organic materials are transformed into carbonaceous materials. It is similar to the traditional methods employed for charcoal production. The temperature ranges from ≥ 270 °C to ≥ 490 °C. The biomass degraded immediately results in production of combustible and non-combustible gases.

Torrefaction is another thermochemical method to transform biomass into BC (coal fuel). Solid coal fuel is the main product of torrefaction with other by-products like liquids and gases. Torrefaction is an enhanced energy generation process from biomass. Organic biomass is degraded under inert or nitrogen gases at one atmospheric pressure, in the range of 200–300 °C temperature for several hours. Torrefaction intensifies the properties of biomass and directly can use as fuel (Khairyet al. 2024).

Small, portable units and large immobile units are recommended for BC production. Portable units can hold 50 to 1 kg, whereas large units can hold up to 8000 kg per hour (Best Energies website 2009).

1.4 Factors Affecting BC Properties (Yaashikaa et al. 2020)

The reaction conditions during BC production are the key factors that influence its properties. These properties will decide the applicability of BC for various purposes. Some of the factors that influence BC yield and its properties were discussed here.

Feedstocks

Biomass is the feedstock in BC production derived from solid waste materials of living and non-living organisms with biological, organic, or inorganic nature. Biomass is categorized as woody biomass of plant origin and non-woody of animal, industrial, and agricultural origin. The two categories of biomass have properties like low humidness, low debris, less porous, high density, and calorific power of woody biomass. On the other hand, non-woody biomass has all the properties quite opposite to woody biomass. Among the different properties, humidness has a significant impact on the formation of biomass. The residual moisture can exist as various aqueous forms with in the biomass. Dried biomass is most preferable compared to high moisture-containing biomass as less heat energy and heating time are required.

Temperature

The range of the temperatures used in the pyrolysis process of biomass can influence the nature of the produced BC. The BC properties such as structure, composition, size and structure of pores, surface area, and functional groups are get influenced by the pyrolysis temperature.

Residence Time

Increased time of contact at a low operational temperature ($\leq$ 300 °C) could decrease BC yield, iodine adsorption number, and pH tolerance. In contrast, low contact time at high pyrolysis temperature (600 °C) had no significant impact on BC yield or pH tolerance, results in diminished iodine adsorption number.

Pre-treatment of Biomass

Different pre-treatment procedures of biomass before pyrolysis, influence the characteristics of BC described in Table 1.

Table 1 Pre-treatment methods of BC

Method of pre-treatment	Properties of BC
Dipping the biomass in solution and particle size reduction	Results in high BC yield
Nitrogen and metal doping	Influence yield of BC
Soaking or steaming	Influence chemical composition
Baking method	Decreased percentage of oxygen and moisture and improved carbon content

1.5 Characterization of BC

The ideal BC can remove pollutants from various solid and liquid residues. BC characterization involves (a) structural and elemental analysis (b) prediction of BC effect on the environment (c) Determination of various parameters like pH, adsorption capacity, total surface, and porosity. Advanced tools like scanning electron microscopy, Brunauer–Emmett–Teller, Fourier transform infrared spectroscopy, nuclear magnetic resonance spectroscopy, X-ray diffraction, thermo-gravimetric analysis, proximate and ultimate analysis, Raman spectroscopy, etc. are employed to characterize BC.

2 BC Mechanism in Removal of Pharmaceutical Pollutants (Ambaye et al. 2021)

Adsorption and persulfate-based AOPs (Advanced oxidation processes) are the basic mechanism for the disposal of pharmaceutical waste through carbon-rich chars. Among them, BC fabricated from animal and plant organic wastes is characterized as next-generation carbonic materials with their own advantages-low of cost, nature friendliness, porous structure, and reactive chemical moieties (e.g., –OH, –COOH, and $–NH_2$). Many researchers have proposed different mechanisms of BC for pharmaceutical pollutants removal. The various the pharmaceutical pollutants removal mechanisms are depicted in Fig. 2.

2.1 Adsorption Mechanism in Pharmaceutical Removal

Fluoxetine (FXT)

FXT is a most prescribed selective serotonin reuptake inhibitor (SSRI) used in the treatment of depressive disorders, obsessive–compulsive disorder (OCD), bulimia nervosa, panic disorder, and premenstrual dysphoric disorder (The American Society

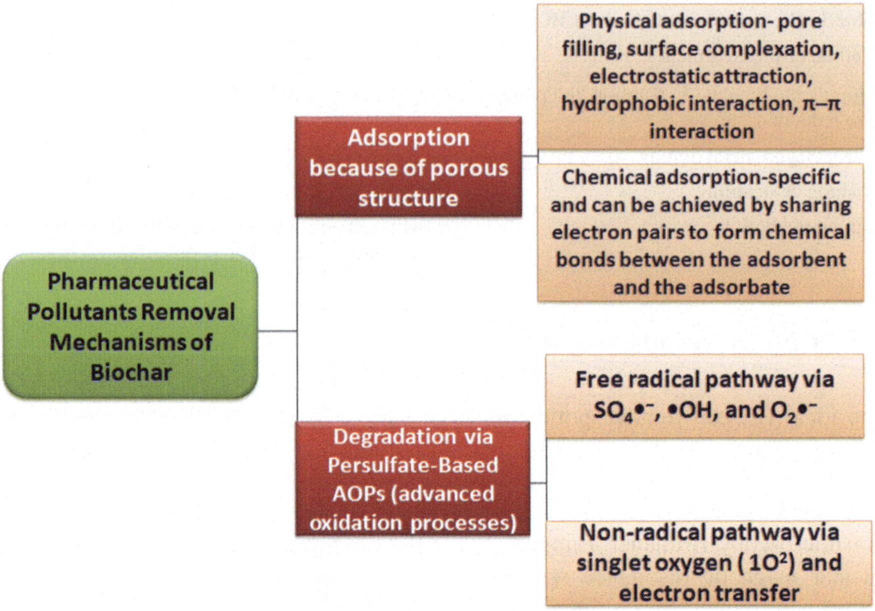

Fig. 2 Removal of pharmaceuticals through adsorption mechanism

of Health-System Pharmacists 2015). The usage of this drug is doubled in most economically developing countries because of their anxiety and stressful life. These drugs are found as micropollutants in urban wastewater treatment procedures. BC is the best alternative and inexpensive method to dispose micropollutants such as FXT from abundant wastewater to be treated. Silvia and his colleagues synthesized BC from rice production waste through thermochemical conversion. The recommended pyrolysis temperature was from 300 to 750 °C. The highest FXT removal (92.6%) was possible through the primary adsorption mechanism of electrostatic attraction with 500 °C treated biomass and obtained BC which was autoclave post-treated (Silvia et al. 2021).

Paracetamol (PML)/Acetaminophen (ACT)

The Covid-19 crisis increased the demand for over-the-counter (OTC) medicines. Among them, several antipyretics were in use and PML is the most frequently used drug. Human excreta which contribute major pharmaceutical pollutants that also have metabolic products of PML are recurrently found in water bodies. Hydrogen bonding, π–π, and electrostatic interactions are the main BC adsorption mechanisms. Adsorption of aromatic compounds is achieved with the π–electron interaction. The amide and phenolic hydroxyl groups of PML contribute their electrons in turn activate the aromatic ring and allows interaction with moving adsorbed electrons. The transfer of ionized molecules to the charged adsorbent surface is responsible for electrostatic interactions. The electrostatic interaction will be lower when the adsorbent exists in zwitterionic form or at zero net surface charge. Functional groups of proton accepting property and strong electro-negativity of nitrogen group of PML can promote firm hydrogen-bonding interactions with BC. Tran et al., removed PML from wastewater samples by his spherical BC which was developed from pure glucose through a pyrolysis process. The spherical BC exhibited improved surface area and enhanced total pore volume. The adsorption capacity of spherical BC was doubled compared to conventional BC via pore-filling, hydrogen bonding formations, n-π and π-π interactions, and van der Waals force (Tran et al. 2020).

Tetracyclines (TC)

BC derived from pyrolyzed rice straw biomass at different temperatures like 300 °C, 500 °C, and 700 °C were employed as adsorbents to remove TC from an aqueous solution. The authors fitted the obtained data in Langmuir and Freundlich models ($R^2 > 0.919$). 500 °C and 700 °C pyrolyzed BC exhibited more adsorption capacity compared to low-temperature treated biomass. The adsorption capacity increased with pyrolysis temperatures and achieved 92.8%–96.7% of TC removal with BC (700 °C at 35 °C). The π–π electron donor and relatively high surface area are the main reasons for the improved capacity of the 700 °C samples. The tight binding in between the TC aromatic ring and π electrons in BC, lead to the π–π interactions. The hydrogen bonding shows less impact on the adsorption mechanism of the BC for TC. Polar groups of every molecule of TC such as –OH and –NH_2, H-bond donors H-bond acceptors contribute to H-bond formation with the C–H groups of BC. A thermodynamic process investigated that the TC adsorption process was rapid and endothermic (Wang et al. 2017).

Sulfamethoxazole (SMX9)

The H-bond, hydrophobicity, and π–π electron interaction are the main mechanisms for the adsorption of SMX by BC. An interaction between aromatic benzene rings of SMX and BC through π–π electron bond enhances adsorption. Another adsorption mechanism is the H-bond formation between dissociated H and an electron-friendly atom such as N, O, and F (Reguyal and Sarmah 2018).

Cephalexin (CPX)

The important adsorption mechanisms of CPX on BC are H-bond interactions, electrostatic bonds, and hydrophobicity. N–H functional groups of CPX make H-bond with oxygen-containing chemical group/s of BC. Functional groups with oxygen on the surface of BC are responsible for hydrophobic interaction with CPX to enhance adsorption (Gheethi et al. 2021).

Levofloxacin (LVF)

The electrostatic reaction, H-bonding, complexation of chemical groups, and π–π electron donor–acceptor reactions are the main attributes to promote the adsorption of LVF on BC. Monolayer-homogeneous and chemical adsorption are dominating mechanisms of LVF onto BC. Neutral and weakly acidic environments are most suitable for maximum adsorption of LVF by BC. Temperature is another factor that can improve the adsorption capacities of BC. The spontaneous and endothermic adsorption process was confirmed by thermodynamic peaks. The site energy distribution analysis showed that the effectiveness of adsorption was improved by increasing BC's surface site energy through –OH, –COOH functional groups (Xu et al. 2021).

Ciprofloxacin (CFC)

CFC is the second-generation, quinolone antibiotic recommended to treat microbial infections. After the pharmacological activity, a small percentage of un-metabolized antibiotics is excreted. These will be considered micropollutants in sewage water treatment plants and responsible for environmental pollution. BC produced from rice husk at various pyrolysis temperatures was investigated to assess the adsorption capabilities of CFC from municipal wastes. BC obtained by treating biomass at 300 °C exhibited 96% CFC removal in 90 min of contact time. The drastic changes in BC surface after adsorption was evaluated by SEM images. Electrostatic forces, H-bonding, and reaction between electrons are the main reasons of adsorption (Payal and Sija 2019).

Ibuprofen (IBP)

The non-steroidal anti-inflammatory drug, ibuprofen (IBP) adsorption by straw-based BC was investigated. The two main mechanisms responsible for adsorption are the electrostatic and dispersive interaction of aromatic compounds from the liquid phase onto the BC. High IBP uptake through electrostatic interaction between ring B electrons of IBP and positive charge on weak or non-acidic oxygen groups of BC facilitates the adsorption (Nafisa and Sobhy 2016). Steam activated wood apple shell biochar reported an increased rate of IBP removal in water from 82% to 85.5% with the adsorption capacity of 12.658 mg/g (Xu et al. 2022).

Triclosan (TCS)

BioSolid BC derived from—anaerobically digested waste-activated sludge was employed for the effective removal of TCS from water and wastewater solutions. Competitive adsorption between organic micropollutants and inorganic nutrients of biosolids decreases TCS removal. Functional groups with polar oxygen-like phenolic and carboxyl groups interacted with phenyl groups on TCS molecules through hydrogen bonding, and aromatic groups present on both the pollutant and BC surface promote p-p stacking in turn responsible for adsorption (Tong et al. 2017).

Eun et al. (2022) examined the removal of triclosan using biochar derived from *Aesculus turbinata* seed shells. Raising the biochar dose from 1.7 to 10.0 g/L reduced the triclosan adsorption per unit mass of adsorbent, yet enhanced the removal percentage; the 8.3 g/L adsorbent dose eliminated over 90% of triclosan.

Caffeine (CFN)

Engineered tea-waste BC is the better choice to remove CFN from aqueous media. Tea-waste BC was obtained by pyrolyzing at 700 °C under steam activation. The determined values fit exactly to the non-linear Elovich kinetic model, to study the chemical force influence in adsorption on heterogeneous tea-waste BC surfaces. The authors identified that chemical adsorption as the main mechanism through electrostatic interactions and nucleophilic attraction. Aromatic hydroxyl and carbonyl groups of BC are responsible for the adsorption of CFN. The carbonyl groups on the tea-waste BC surface have the capability of undergoing nucleophilic attack. Physical adsorption of CFN occurs on the tea-waste BC surface, because of π-π and hydrogen bonding interactions between the BC and the heterocyclic ring of the CFN molecule (Keerthanan et al. 2020).

Deoxyribonucleic Acid (DNA) and Ribonucleic Acid (RNA)

Paula et al., investigated cleaning of DNA and RNA from municipal wastewater treatment plants. DNA and RNA are contributed from some viruses belonging to the classes of enveloped and non-enveloped responsible for contagious infections. After antibiotic treatment, the dead viral DNA and RNA pieces enter the domestic waste through excreta and urine and are responsible to acquire antibiotic resistance. The gene fragments are considered micropollutants capable to show a remarkable impact on the environment. The commercial BC (brand name, BC 4073) obtained from an aerobically treated organic waste containing more than 50% wood and remaining amounts of corn silage was used to remove the gene pollutants from treated wastewater samples. Commercial BC removed more than 90% of gene pollutants (Paula et al. 2021).

Sulfapyridine-SPY, Docusate-DCT, Erythromycin-ETM

Novel BCs derived from cotton gin waste (CG) and guayule bagasse (GB) were tested for SPY, DCT, and ETM removal from an aqueous system. Physico-chemical factors such as contact time, solution pH, BC-specific surface area, and functional groups are the main factors that influence the adsorption capability of BC. The authors studied that the hydrophobic partitioning, hydrogen bonding, and π–π electron interactions are the main mechanisms of adsorption. Decreased hydrophobicity and decreased electrostatic interaction between SPY and CG BC (both surfaces are negatively charged) at higher pH, SPY adsorption was improved from 40 to 70% with neutral and alkaline pH, respectively. Interactions of polar and non-polar functional groups of pharmaceutical pollutants with the chemical moieties of BC are responsible for the removal of micropollutants. Similar mechanisms were also observed for the remaining pollutants in this study (Marlene et al. 2021).

Amoxicillin (AMOX) and Diclofenac (DFC)

Francisco et al., have investigated the adsorption capacity of pharmaceuticals such as AMOX and DFC by hair char. A fixed bed reactor was used for the thermal treatment of animal fur waste obtained during the tanning process. BC resulted from heating at 450 °C and was post-treated with chemicals like KOH and K_2CO_3. More than 90% of AMOX and more than 80% of DFC removal were possible with chemically treated hair char (Francisco et al. 2020).

Pharmaceutical Adsorption from Urine

Pharmaceutical-free urine can be recommended to use as a nutrient for soil (Avni and Treavor 2017). Researchers have investigated the adsorption of pharmaceuticals of analgesics and anticonvulsant category from synthetic urine using BC obtained from agricultural waste materials (coconut tree materials, bamboo, Florida pine, and northern hardwood). Among the different BCs tested, activated coconut carbon, bamboo, and Florida yellow pine BCs had shown the highest pharmaceuticals clearing, i.e., greater than 90% (40 g/L) from urine. Weak interactive forces that can promote pharmaceutical removal by BC.

Surface-Modified BC in Pharmaceutical Adsorption from Synthetic Urine

Numerous researchers have studied the pharmaceuticals' adsorption capacity of surface-modified BC. Modification of the BC surface would change the chemical properties thereby enhancing the adsorption capacity. The role of surface-modified BCs in pharmaceutical removal are presented in Table 2 (Sayeda et al. 2021).

Table 2 Surface-modified BCs for adsorption of pharmaceuticals

Name of the pharmaceutical	Method of modification	Mechanism	Result
TC	Phosphoric acid	Increased pore volume and –COOH and –OH groups	Enhanced capacity of manure-based BC to absorb TC
TC	Methanol-modified rice husk BCs	Increased oxygen containing functional groups	Increase in adsorption efficiency
Sulfonamides	Hydrogen peroxide based BC	Generated hydroxyl radicals degrade pharmaceuticals because of the high oxidation potential	Increased BC doses, increase the pharmaceuticals removal
SMX	Modified by magnetic $CuZnFe_2O_4$ composite	Hydrophobicity, and p–p EDA (Electron donor Acceptor) interactions	Adsorption capacity increased at lower solution pH
TC	Silica modified BC composites	Pore-filling effect and improved p-p EDA interaction	Increased adsorption efficiency
TC	NaOH-activated BCs	High hydrophobicity, high surface area, and pore volume	Effective adsorption efficiency

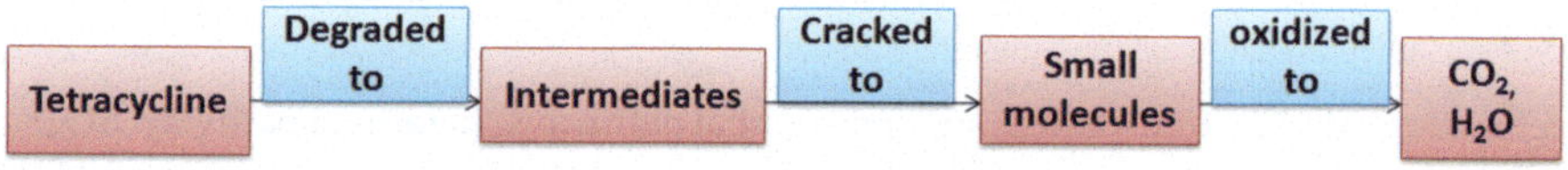

Fig. 3 Steps involved in the degradation of TC through AOPs

2.2 Degradation via Persulfate-Based AOPs

Eco-friendly and efficient AOPs have attracted the attention of researchers for organic-contaminated soil remediation and water treatment (Zhou et al. 2019). Recent reports describing the mechanism of AOPs in pharmaceutical removal are described here.

TC

The use of persulfate-based AOPs is an effective approach for disposing of TC. The flow chart provides the sequential events of TC degradation (Fig. 3).

TC under persulfate-based AOPs follows two different pathways. In pathway I (radical pathway), the dissociation of C–N bond in TC was responsible for removal of methyl group which in turn produced two intermediates. Dehydration of intermediate compounds results in the formation of ring cleavage products. The second pathway (radical pathway) utilizes reactive oxygen species such as SO4 $\bullet^-$, •OH, and $^1O_2$ and transferred electrons to attack C–C, C–N, and C = C bonds leading to hydroxylation, demethylation, cleavage, and decarboxylation. Non-radical pathway of TC degradation utilizes $^1O_{2,}$ responsible to attack amino group on the aromatic ring, leading to its demethylation (Gao et al. 2022).

SMX

As per Qi et al., study, free-radical and non-radical pathways of persulfate-based AOPs (Advanced oxidation process) were possible for the catalytic degradation of SMX. SMX was more vulnerable to $^1O_2$ attack because of electron-rich groups in the molecule. SMX is degraded to a several intermediate compounds through a series of chemical reactions. Figure 4 explains the fate of SMX in persulfate-based AOP (Qi et al. 2020).

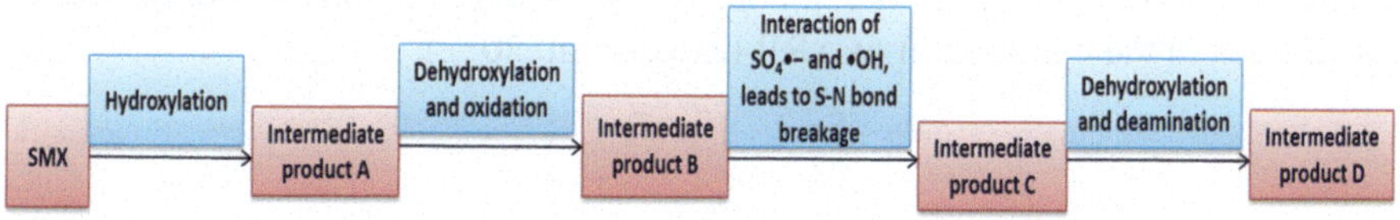

Fig. 4 Fate of SMX in persulfate-based AOP

ACT

ACT is degraded by free-radical and non-radical pathways. In radical pathway, $SO_4^{\bullet -}$ and $\bullet OH$ play a dominant role. $SO_4^{\bullet -}$ easily attacks the acetylamino group of ACT. According to previous reports, the acetylamino group of ACT was spontaneously attacked by $SO_4^{\bullet -}$, and $\bullet OH$ attacks the substituent's on the aromatic ring and supports its opening by hydroquinone or 4-benzoquinone. Radicals $O_2^{\bullet -}$ and $^1O_2$ contributed to ring-opening and hydrogen abstraction reactions (Wang and Wang 2020).

Cephalexin (CPX)

Radicals $\bullet OH$ and $SO_4^{\bullet -}$ are the important free radicals for the degradation of CPX. Dissociation of β-lactam ring and the association of a hydroxyl leads to demethylation, cleavage of the C–N bond, decarboxylation, deamination, and dehydration is the chemical reactions responsible for the degradation of the parent compound to give intermediate molecules and further degradation ends with the formation of CO_2, H_2O, and small molecular components as final products (Ali Noman et al. 2021).

Levofloxacin (LVF)

The non-radical pathway has a major role in the degradation of LVF through $^1O_2$ and electron transfer systems. To some extent $SO_4^{\bullet -}$ the mediated radical pathway is also involved in LVF degradation. Piperazinyl, quinolone moieties, and the benzene ring of LVF are the most sensitive sites of LVF for degradation through various chemical reaction mechanisms such as demethylation, depiperazinylation, and carboxylation/decarboxylation (Abukhadraet al. 2021).

Drugs for Treatment of Covid

Covid-19 crisis, in the past two years led to the unnecessary use of antibiotics. The increased usage and production of Covid treating drugs increased water pollution. Ziwei et al., studied the degradation of Covid treating drugs such as arbidol (ARB), chloroquine phosphate, hydroxychloroquine sulfate, and acyclovir by BC-supported red mud catalyst (RM-BC). RM-BC was synthesized to contain abundant free hydroxyl groups. RM-BC was removed 100% of ARB in a short span of contact. SO4 $\bullet^-$ and HO• powerful oxidants were responsible for the production of DMPOX (5,5-dimethylpyrrodidone-2(2)-oxyl-(1)) in electron spin resonance experiments. Iron oxides and oxygen-containing functional groups in RM-BC played crucial roles in the elimination of ARB (Ziwei et al. 2022).

3 Advancements in BC Application for Pharmaceutical Removal

Engineered BC fabricated by layering Mg/Al double hydroxides impregnated rice husk biomass treated at 700 °C under a controlled atmosphere in a muffle furnace. The adsorption of acetaminophen from aqueous media was enhanced with engineered BC (Abhishek et al. 2022).

Nano-Biochar (NBC)

NBC is a carbon-rich nano-material because of the provision of high specific surface area and high affinity the micropollutant adsorption capacity is increased (Naghdi et al. 2019).The removal of pharmaceuticals pollutants in from various environments are shown in Table 3.

Table 3 NBC in pharmaceuticals removal

NBC	Study report	References
NBC-EDA (Ethylenediamine)	NBC-EDA was effective in extracting more than 90% of prednisolone and Chromium VI from tap, waste, and seawater, respectively	Mahmoud et al. (2022)
Pinewood NBC	95% of carbamazepine (CBZ) (74 μg CBZ/g nanoBC) after 3 h of contact time. Increasing pH from acidic to alkaline can improve the adsorption efficiency by 2.3 folds	Naghdi et al. (2019)
Graphitic nanoBC	Oxytetracycline (OTC), glyphosate (GL), chromium (CrVI), and cadmium (CdII) removal was enhanced	Ramanayaka et al. (2019)
Dendro wood (*Gliricidia sepium*) derived nano BC	The weak Van der Waals forces, hydrogen bonding and dipolar attractions, pore-filling and film diffusion prominently govern the physisorption of oxytetracycline from synthetic human urine	Ramanayaka et al. (2020)
Rice husk BC-supported manganese dioxide composite (BC/MnO_2)	Enhanced adsorption of BC/MnO_2 may be because of improved surface area, pore volume, and MnO_2-loaded BC serving as reactive sites for tetracycline hydrochloride (TC) and doxycycline	Li et al. (2020)
Carbon nanodots and nano zero-valent iron composites-synthesized by date palm waste BC	Electrostatic, π–π electron interactions, degradation and diffusion were responsible for MB (methylthioninium chloride removal from wastewater	Munir et al. (2020)
Multi-walled carbon nanotubes	90.6% of CBZ was adsorbed after 24 h	Oleszczuk et al. (2009)

Magnetic Biochar (MBC)

MBC is an advanced BC showing an improved adsorption performance for organic pollutants such as antimicrobial agents, synthetic colorants, herbicides, and organochlorine compounds (Kumar et al. 2020). Many researchers have investigated the adsorption efficiency of MBC for antibiotics. MBC removed 33 to 300 mg/g of tetracycline and 5.19 to 212.8 mg/g of methanoxazole from contaminated water samples. MBC exhibits similar adsorption mechanisms as ciprofloxacin and also shows pore-filling.

MBC was prepared by ball milling, in which mechanically BC was mixed with iron oxides (Li et al. 2023), through direct pyrolysis achieved cross-linked BC with iron oxides (Mojiri et al. 2019). Liang et al. (2022) studied the magnetic Fe_2O_3/BC composite derived from poplar wood to remove Norfloxacin from water samples. In another study 62.7 mg/g and 94.2 mg/g of CBZ and TC were reported with ballmilled BC and iron oxide composites MBC.

Role of Artificial Intelligence (AI) and Machine Learning (ML) in Removal of Pharmaceutical Waste Using BC

The application of ML and AI to biochar is a subject gaining significant attention. Recent progress in computational science, particularly in the field of ML, has allowed researchers to use a powerful tool to reveal the intricate nonlinear connection between biochar synthesis and adsorption processes (Kumari et al. 2024). A new research trend uses machine learning to improve biochar synthesis and adsorption, leading to better pollution removal (Manatura et al. 2023). ML summarized the connections between biochar use and computational methods, including DT (Decision Trees), ANN (artificial neural network) algorithms, and RF (Random Forest) (Supraja et al. 2023).

In machine learning methods for biochar adsorption in wastewater, DT is utilized to model the intricate relationships between the properties of biochar and its adsorption efficiency in order to enhance biochar-based wastewater treatment processes. ANN algorithms were developed by comparing the processing of programming languages with the human brain system, as researchers seek to utilize artificial intelligence to replicate the functional characteristics of the human brain and neural system in learning and memory. In other words, humans develop ANN models with programming languages to replicate how the brain performs a specific task. A summary was provided of the latest advancements in machine learning for biochar production in the context of pollution removal. This encompasses the forecasting of biochar production and physical–chemical properties, determination of optimal synthesis conditions, and evaluation of economic expenses (Zhang et al. 2023).

Research initiatives in the field of ML mainly focus on two key areas as noted by showcasing ML methodologies and applying ML techniques to specific biochar-related fields. Significant uses of biochar include simulating thermochemical processes, enhancing synthesis techniques, treating wastewater, dye removal, heavy metal elimination, and forecasting biochar yield and characteristics. Nevertheless, a comprehensive and systematic review has not been carried out from the

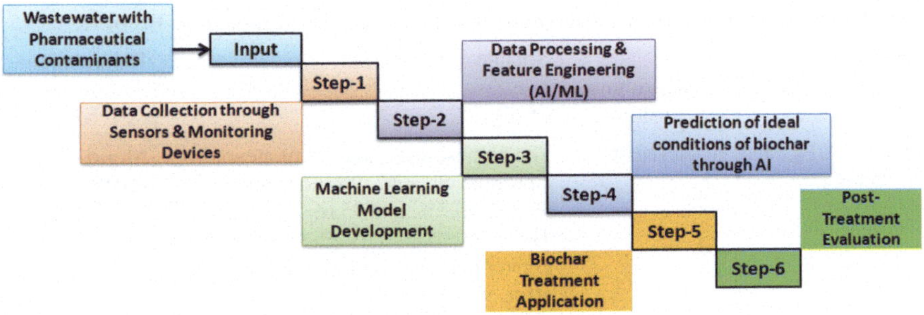

Fig. 5 Various stages of wastewater treatment using biochar

perspective of the entire regulatory process of biochar. Promptly summarizing the ML of biochar from the entire process regulation is beneficial to aid the integration of biochar and ML (Kumari et al. 2024). The following flowchart (Fig. 5) outlines the various stages of wastewater treatment using biochar, integrated with machine learning (ML) and artificial intelligence (AI) tools.

Human urine is a significant source of pharmaceutical pollutants that flow into wastewater treatment facilities. Consequently, eliminating pharmaceutical active compounds from source-separated urine is crucial for efficient nutrient recovery and reducing environmental pollution. A recent study utilized ANN to analyze the effectiveness of the granular activated carbon (GAC) adsorption process in eliminating pharmaceutical contaminants like naproxen, carbamazepine, ibuprofen, acetaminophen, and metronidazole from human urine (Almuntashiri et al. 2022). The ANN model successfully predicted a removal effectiveness of over 99% for all pharmaceutical compounds except acetaminophen (84.5%). Table 4 provides a summary of earlier research regarding the use of ML for modeling and optimizing pharmaceutical adsorption on biochar or activated carbon (AC).

4 Conclusion and Future Prospective

BC is a most useful carbon adsorbent and has a highly promising substitute for traditional activated carbon. It is an efficacious and in-expensive adsorbent that can be synthesized from enormous biomass substances, of various sources that includes forest and agricultural waste, municipal sewage sludge, hair and manure of animals, and other solid wastes. The expansion of pharmaceutical industries, the productivity of the pharmaceuticals, increased usage of drugs, and increased concentrations of parent and its drug metabolites through excreta demand effective solid and liquid waste treatment procedures. The pharmaceutical wastes deposited in the environment as micropollutants and the removal of these low concentrations of drug pollutants create complications during the application of conventional waste treatment methods. Modernization of BC with existing conventional treatment approaches

Table 4 Use of ML for modeling and optimizing pharmaceutical adsorption on BC or AC

Pharmaceutical	ML algorithms/ Adsorbent	Notable observations	References
Atenolol, CFC, and diazepam	ANN/BC	The highest removal rates of atenolol, ciprofloxacin, and ciprofloxacin were 90.3%, 95.5%, and 95.6%, respectively The ANN showed a high R^2 value of greater than 0.99 and a mean squared error of less than 1.0	Mojiri et al. (2019)
Data for 35 different pharmaceutical and personal care products including antibiotics and steroids	RF algorithm/ BC	The experimental findings indicate that machine learning is a promising approach for modeling the adsorption of pharmaceuticals onto BC The microscopic details of functionalities and pores enhanced the accuracy of predictions	Zhu et al. (2022)
AMOX antibiotics	RF algorithm/ BC	• An MCC value of 0.674 and a ROC of 0.834 indicate that the model can accurately predict the adsorption capacity of the BC. BC made from cotton seeds demonstrated a pharmaceutical removal efficiency exceeding 90%	Ram et al. (2020)
PML	ANN/AC	The comparison of various ANN architectures indicates that the Levenberg Marquardt method and the tansig-purelin function are the most promising The activated carbon made from orange peel exhibited encouraging results for paracetamol adsorption	Afolabi et al. (2020)
Human urine contains pharmaceutical contaminants such as naproxen, CBZ, IBP, ACT, and metronidazole	ANN/AC	The modeling ANN was capable of predicting a removal effectiveness exceeding 99% for all assessed compounds, except for ACT, which had a removal effectiveness of 84.5% The flow rate and the mass of the adsorbent affect the removal of pharmaceuticals through adsorption	Almuntashiri et al. (2022)

(continued)

Table 4 (continued)

Pharmaceutical	ML algorithms/ Adsorbent	Notable observations	References
Nimesulide and PML	ANN/AC	The existence of nimesulide inhibited the adsorption of PML. The ideal adsorption capacity was determined at 150 μm, 0.5 g L^{-1}, and 300 min ANN was effectively created to predict both single and binary adsorption at the same time	Pauletto et al. (2021)
SMX and TC	RF, Gradient boosting trees and ANN/BC and activated carbon	RF outperformed gradient boosting trees and ANN for both SMX and TC adsorption models The BC BET surface area, pH and initial concentrations were important factors for affecting the pharmaceutical removal efficiency	Zhu et al. (2021)

satisfies the current demand for waste extraction from polluted water and soil samples. The present chapter summarized various adsorption mechanisms of pharmaceuticals by BC. Also focused on advanced techniques of engineered BC, NBC, MBC, and ML role in the removal of hazardous pharmaceutical pollutants from water and synthetic urine samples. The morphological and chemical nature of BC supports the degradation and extensive removal of pharmaceutical pollutants through the two basic adsorption mechanisms. BC could be recommended for clearing detrimental pollutants due to its low cost, abundant surface area, and porosity.

Acknowledgements I convey my profound gratitude to Dr. L. Rathaiah the Chairman of Vignan Group of Institutions for providing his support to complete this book chapter. My heartfelt thanks to Dr. Y. Srinivasa Rao, Principal of Vignan Institute of Pharmaceutical Technology for his assistance and advice. Sincere thanks to my colleagues for their valuable suggestions.

My acknowledgment would not be complete without thanking the biggest source of my strength, my family.

References

Abhishek KC, Manvendra P, Dinesh M (2022) Synthesis of engineered biochar for aqueous pharmaceutical removal. EGU21-9358, updated on 18.07.2022

Abukhadra MR, Fathallah W, Kashief E et al (2021) Insight into the antimicrobial and photocatalytic properties of NiO impregnated MCM-48 for effective removal of pathogenic bacteria and toxic levofloxacin residuals. Microporous Mesoporous Mater 312:110769

Afolabi IC, Segun IP, Olugbenga SB et al (2020) Modelling pseudo-second-order kinetics of orange peel-paracetamol adsorption process using artificial neural network. Chemom Intell Lab Syst 203:104053. https://doi.org/10.1016/j.chemolab.2020.104053

Ali Noman E, Al-Gheethi A, Saphira RM (2021) Sustainable approaches for removal of cephalexin antibiotic from non-clinical environments: a critical review. J Hazard Mater 417:126040
Almuntashiri A, Ahmad H, Umakant B et al (2022) Removal of pharmaceutical compounds from synthetic hydrolysed urine using granular activated carbon: column study and predictive modelling. J Water Process Eng 45:102480. https://doi.org/10.1016/j.jwpe.2021.102480
Ambaye TG, Vaccari M, Van Hullebusch ED et al (2021) Mechanisms and adsorption capacities of biochar for the removal of organic and inorganic pollutants from industrial wastewater. Int J Environ Sci Technol 18:3273–3294
Avni S, Treavor HB (2017) Pharmaceutical removal in synthetic human urine using biochar. Environ Sci Water Res Technol 3:553–565
Bataduwaarachchi VR, Weeraratne CL (2018) Global medication waste management practices: challenges and opportunities in developing countries. Int J Basic Clin Pharmacol 5(6):2290–2294
Best Energies (2009) Best pyrolysis Inc. https://www.bestenergies.com/companies/bestpyrolysis.html
Binaya S, Agamuthu P (2023) Pharmaceutical waste management system—are the current techniques sustainable, eco-friendly and circular? A review. Waste Manag 168:83–37. https://doi.org/10.1016/j.wasman.2023.05.052
Carter LJ, Ryan JJ, Boxall ABA (2016) Effects of soil properties on the uptake of pharmaceuticals into earthworms. Environ Pollut 213:922–931
Desalin Water Treat 266:256–267. https://doi.org/10.5004/dwt.2022.28616
Eun JC, Joon-Kwan M, Chang GL, Seong JP (2022) Removal of triclosan from aqueous solution using biochar derived from seed shell of *Aesculus turbinate*
Fluoxetine Hydrochloride (2015) The American society of health-system pharmacists. Archived from the original on 8 December 2015. Retrieved 2 Dec 2015
Francisco R, Carolina MR, Idoia E et al (2020) Removal of drugs in polluted waters with char obtained by pyrolysis of hair waste from the tannery process. ACS Omega 5(38):24389–24402
Gao Y, Wang Q, Ji G et al (2022) Degradation of antibiotic pollutants by persulfate activated with/various carbon materials. Chem Eng J 429:132387
Gheethi A, Noman E, Saphira RM et al (2021) Cephalexin removal by a novel Cu-Zn bionanocomposite biosynthesized in secondary metabolic products of *Aspergillus arenarioides* EAN603 with pumpkin peels medium: optimization, kinetic and artificial neural network models. J Hazard Mater 419:126500
Han J, He S, Lichtfouse E (2023) Waves of pharmaceutical waste. Environ Chem Lett 21:1251–1255. https://doi.org/10.1007/s10311-022-01491-0
Ibitoye SE, Loha C, Mahamood RM et al (2024) An overview of biochar production techniques and application in iron and steel industries. Bioresour Bioprocess 11:65. https://doi.org/10.1186/s40643-024-00779-z
John LW, Alistair BAB, Dana WK et al (2022) Pharmaceutical pollution of the world's rivers. Proc Natl Acad Sci USA 119(8):e2113947119
Kadam A, Patil S, Patil S et al (2016) Pharmaceutical waste management an overview. Indian J Phar Pract 9(1):3
Kaposi A, Nagy A, Gomori G et al (2024) Analysis of healthcare waste and factors affecting the amount of hazardous healthcare waste in a university hospital. J Mater Cycles Waste Manag 26:1169–1180. https://doi.org/10.1007/s10163-024-01890-1
Keerthanan S, Amit B, Kushani M (2020) Engineered tea-waste biochar for the removal of caffeine, a model compound in pharmaceuticals and personal care products (PPCPs), from aqueous media. Environ Technol Innov 19:100847
Khairy M, Amer M, Ibrahim M et al (2024) The influence of torrefaction on the biochar characteristics produced from sesame stalks and bean husk. Biomass Conv Bioref 14:17127–17148. https://doi.org/10.1007/s13399-023-03822-9
Kumar A, Sharma G, Naushad M et al (2020) Bio-inspired and biomaterials-based hybrid photocatalysts for environmental detoxification: a review. Chem Eng J 382:122937

Kumari S, Jyoti CH, Manish K et al (2024) Machine learning (ML): an emerging tool to access the production and application of biochar in the treatment of water and wastewater. Groundw Sustain Dev 26:101243. https://doi.org/10.1016/j.gsd.2024.101243

Li J, Cai X, Liu Y et al (2020) Design and synthesis of a biochar-supported nano manganese dioxide composite for antibiotics removal from aqueous solution. Front Environ Sci 8:62

Li C, Zhang C, Zhong S et al (2023) The removal of pollutants from wastewater using magnetic biochar: a scientometric and visualization analysis. Molecules 28(15):5840. https://doi.org/10.3390/molecules28155840

Liang H, Zhu C, Ji S et al (2022) Magnetic Fe_2O_3/Biochar composite prepared in a molten salt medium for antibiotic removal in water

Mahmoud ME, El-Ghanam AM, Saad SR (2022) Sequential removal of chromium (VI) and prednisolone by nanobiochar-enriched-diamine derivative. Biomass Conv Bioref

Manatura K, Chalermsinsuwan B et al (2023) Machine learning and statistical analysis for biomass torrefaction: a review. Bioresour Technol 369:128504. https://doi.org/10.1016/j.biortech.2022.128504

Marlene CN, Herschel AE, Heather EP et al (2021) Adsorption of pharmaceuticals from aqueous solutions using biochar derived from cotton gin waste and guayule bagasse. BC 3:89–104

Megan C (2019) Consequences of improperly disposed pharmaceuticals. Danielshealth (Internet). Accessed on 23 June 2020. Available online at: https://www.danielshealth.com/knowledge-center/pharmaceutical-improper-disposal-consequences

Mojiri A, Andasht KR, Gholami A (2019) Cross-linked magnetic chitosan/activated BC for removal of emerging micropollutants from water: optimization by the artificial neural network. Water 11(3):551

Mouele ESM, Tijani OJ, Fatoba OO et al (2015) Degradation of organic pollutants and microorganisms from wastewater using different dielectric barrier discharge configurations-acritical review. Environ Sci Pollut Res 22:18345–18362

Muhammad S, Muhammad A (2021) Chapter 9—biochar as a sorbent for organic and inorganic pollutants. In: Nunez-Delgado A (ed) Sorbents materials for controlling environmental pollution. Elsevier, pp 189–208

Munir A, Mutair AA, Adel RAU et al (2020) Turning date palm waste into carbon nanodots and nano zerovalent iron composites for excellent removal of methylthioninium chloride from water. Sci Rep 10:16125

Nafisa AS, Sobhy MY (2016) Non-steroidal anti-inflammatory drug, ibuprofen adsorption using rice straw based BC. Int J Pharmacol 12:729–736

Naghdi M, Taheran M, Pulicharla R et al (2019) Pine-wood derived nanobiochar for removal of carbamazepine from aqueous media: adsorption behavior and influential parameters. Arab J Chem 12(8):5292–5301

Oleszczuk P, Pan B, Xing B (2009) Adsorption and desorption of oxytetracycline and carbamazepine by multi-walled carbon nanotubes. Environ Sci Tech 43(24):9167–9173

Oxford English Dictionary (Online ed) Oxford University Press

Paula B, Petra R, Karel K (2021) An efficient sorption material for the removal of pharmaceutically active compounds, DNA and RNA fragments from wastewater. J Environ Chem Eng 9:105746

Pauletto PS, Lutke SF, Dotto GL et al (2021) Forecasting the multicomponent adsorption of nimesulide and paracetamol through artificial neural network. J Chem Eng 412:127527. https://doi.org/10.1016/j.cej.2020.127527

Payal M, Sija A (2019) Characterisation of BC obtained from organic material and Its application for removal of ciprofloxacin. Orient J Chem 35(3):1086–1093

Pharmaceuticals Global Market Report (2022) https://www.globenewswire.com/news-release/2022/03/04/2396935/0/en/Pharmaceuticals-Global-Market-Report-2022.html. Accessed 16 Aug 2022

Qi Y, Ge B, Zhang Y et al (2020) Three-dimensional porous graphene-like biochar derived from *Enteromorpha* as a persulfate activator for sulfamethoxazole degradation: role of graphitic N and radicals transformation. J Hazard Mater 399:123039

Rajbongshi S, Yamini DS, Ahsan US (2016) Pharmaceutical waste management: a review. European J Biomed Pharm Sci 12:192–206
Ram K, Sivarajasekar N, Muthusaravanan S (2020) An ensemble-based model for the adsorptive removal of amoxicillin by microwave-biochar of waste cotton seeds. AIP Conf Proc 2270:20011
Ramanayaka S, Tsang D, Hou D et al (2019) Green synthesis of graphitic nanobiochar for the removal of emerging contaminants in aqueous media. Sci Total Environ 706:135725
Ramanayaka S, Manish K, Thusitha E et al (2020) Macro, colloidal and nanobiochar for oxytetracycline removal in synthetic hydrolyzed human urine. Environ Pollut 267:115683
Reguyal F, Sarmah AK (2018) Adsorption of sulfamethoxazole by magnetic biochar: effects of pH, ionic strength, natural organic matter and 17alpha-ethinylestradiol. Sci Total Environ 628–629
Richika G, Shveta S, Ashish K et al (2023) Activated carbon: fundamentals, classification, and properties. https://doi.org/10.1039/BK9781839169861-00001
Roy A, Bharadvaja N (2021) Efficient removal of heavy metals from artificial wastewater using biochar. Environ Nanotechnol Monit Manag 16:100602
Sayeda UM, Pavani D, Yuqing S et al (2021) Sustainable use of biochar for resource recovery and pharmaceutical removal from human urine: a critical review. Crit Rev Environ Sci Technol 51(24):3016–3048
Silvia EC, Valeria AGM, Angeles S et al (2021) Eco-approach for pharmaceutical removal: thermochemical waste valorisation, biochar adsorption and electro-assisted regeneration. Electrochim Acta 389:138694
Statistica Market Forecast. Pharmaceuticals-worldwide: https://www.statista.com/outlook/hmo/pharmaceuticals/worldwide
Supraja K, Himanshu K, Gayatri V et al (2023) Biochar production and its environmental applications: recent developments and machine learning insights. Bioresour Technol 387:129634. https://doi.org/10.1016/j.biortech.2023.129634
Tong Y, Lee KK, Brooke KM (2017) Biosolids-derived biochar for triclosan removal from wastewater. Environ Eng Sci 35(6):1–12
Tran HN, Tomul F, Nguyen HTH (2020) Innovative spherical biochar for pharmaceutical removal from water: Insight into adsorption mechanism. J Hazard Mater 394:12225
Wang J, Wang S (2020) Reactive species in advanced oxidation processes: formation, identification and reaction mechanism. Chem Eng J 401:126158
Wang H, Chu Y, Fang C et al (2017) Sorption of tetracycline on biochar derived from rice straw under different temperatures. PLoS ONE 12(8):e0182776
World Health Organization (2011) Waste from health-care activities. Fact Sheet no. 253. World Health Organization, Geneva
Xu Z, Xiang Y, Zhou H et al (2021) Manganese ferrite modified biochar from vinasse for enhanced adsorption of levofloxacin: effects and mechanisms. Environ Pollut 272:115968
Xu Y, Xinbo Z, Huu HN et al (2022) Sorptive removal of ibuprofen from water by natural porous biochar derived from recyclable plane tree leaf waste. J Water Process Eng 46:102627. https://doi.org/10.1016/j.jwpe.2022.102627
Yaashikaa PR, Senthil Kumar P, Sunita V et al (2020) A critical review on the biochar production techniques, characterization, stability and applications for circular bioeconomy. Biotechnol Rep 28:1–15
Zhang W, Wu B, Wenguang H et al (2023). Synthesis optimization and adsorption modeling of biochar for pollutant removal via machine learning. Biochar. https://doi.org/10.1007/s42773-023-00225-x
Zhou Z, Liu X, Sun K et al (2019) Persulfate-based advanced oxidation processes (AOPs) for organic-contaminated soil remediation: a review. Chem Eng J
Zhu X, Zhonghao W, Daniel CWT et al (2021) Machine learning for the selection of carbon-based materials for tetracycline and sulfamethoxazole adsorption. J Chem Eng 406:126782. https://doi.org/10.1016/j.cej.2020.126782

Zhu X, Mingjing H, Yuqing S et al (2022) Insights into the adsorption of pharmaceuticals and personal care products (PPCPs) on biochar and activated carbon with the aid of machine learning. J Hazard Mater. 423(Part B):127060. https://doi.org/10.1016/j.jhazmat.2021.127060

Ziwei G, Yue Z, Shuchai G et al (2022) Effective degradation of COVID-19 related drugs by biochar-supported red mud catalyst activated persulfate process: mechanism and pathway. JCleanProd 340:130753

Chapter 8
Removal of Inorganic Compounds from Soil Using Biochar

Mahendra Pratap Singh Tomar, Prakash B. Rathod, and Parshuram Maske

Abstract Currently, substantial quantities of water contaminated with toxic metals are being released into the environment as a result of industrial development. Various manufacturing sectors including mining, metal polishing, electroplating, textiles, ceramics, and battery production discharge heavy metals into the environment, posing significant hazards to all living organisms. Heavy metals interact with soil components through various mechanisms, including adsorption, immobilization, and surface precipitation. Consequently, it is essential to pursue the remediation of contaminated areas. Among the available technologies, adsorption is widely regarded as an effective method for removing heavy metal ions due to its high efficiency, rapid processing, simplicity, cost-effectiveness, and the availability of a wide range of adsorbents. Using adsorbents reduces the bioavailability of contaminants and limits their dispersion throughout the environment. This chapter will focus on the need for inorganic pollutant removal and the role of biochar in this process. It discusses biochar preparation and its applications, specifically examining the efficacy of biochar in removing inorganic pollutants from soil. Furthermore, this chapter presents a comparative analysis of biochar with other substrates in terms of its effectiveness for soil remediation.

Keywords Inorganic compounds · Soil remediation · Toxic metals · Health threats · Biochar efficiency

M. P. S. Tomar (✉) · P. B. Rathod
Department of Chemistry, Shri Sadguru Saibaba Science and Commerce College, Ashti, Maharashtra 442707, India

P. B. Rathod
e-mail: drmahendrapsingh1975@gmail.com

P. B. Rathod
School of Chemical Sciences, Kavayitri Bahinabai Chaudhari North Maharashtra University, Jalgaon, Maharashtra 425001, India

P. Maske
Department of Chemistry, S.G.V. & S.S.P'S, Art's Commerce and Science College, Onde, Maharashtra 401605, India

A. Roy et al. (eds.), *Biochar Production Engineering*,
https://doi.org/10.1007/978-3-031-92594-8_8

1 Introduction

Industrial expansion has resulted in substantial water contamination by toxic metals, releasing untreated effluents into the environment (Liu et al. 2020). Various sectors, including mining, metal polishing, electroplating, textile production, ceramics, and battery manufacturing, discharge heavy metals into ecosystems (Godwin et al. 2019). These metals, being non-biodegradable, persist in the environment, bioaccumulation in organisms, and impose significant ecological and health risks (Poo et al. 2018). Toxic metals pose severe hazards to living organisms and disrupt ecological balance, with most heavy metals recognized for their carcinogenicity and toxicity (Patra et al. 2017).

Heavy metals interact with soil components through processes like adsorption, fixation, and surface precipitation, making remediation of contaminated sites essential (Campillo-Cora et al. 2020). There are several methods for removing inorganic contaminants, (Rathod et al. 2019) particularly heavy metals, from soil. These include phytoremediation, where plants absorb or stabilize contaminants; soil washing, which uses water or chemicals to leach out pollutants; and soil immobilization, where amendments like lime or phosphate bind metals, reducing their mobility. Ex-situ soil treatment involves removing contaminated soil for treatment, including thermal desorption, bioremediation, and chemical extraction (Fig. 1). Electrokinetic remediation applies an electric field to move contaminants through soil for removal or immobilization. Vitrification uses high temperatures to melt soil, trapping contaminants in a solid matrix. Bioremediation relies on microorganisms (Singh et al. 2024) or plants to degrade or stabilize pollutants. Nanoremediation employs nanomaterials, such as nanoparticles and modified nanoparticles, (Rathod et al. 2019) (Rathod et al. 2018) (Rathod et al. 2024) to interact with and neutralize contaminants and adsorb the contaminant. Metal organic frameworks and membranes (Chappa et al. 2018) are also useful for this applications. (Rathod et al. 2023) Lastly, solidification/stabilization involves mixing soil with agents like cement to contain pollutants.

These methods vary in effectiveness, cost, and environmental impact. Phytoremediation and bioremediation are eco-friendly, (Margonwar et al. 2024) while soil washing and electrokinetic remediation can be faster but more expensive. Immobilization techniques are cost-effective but do not remove contaminants. In some cases, combining methods offers a comprehensive approach, optimizing the removal or stabilization of inorganic pollutants in contaminated soils.

Bioremediation, which uses organisms, particularly microbes, to degrade pollutants into less toxic forms, is a viable approach for soil and water remediation (Akpomie and Ejechi 2016). Among remediation technologies, adsorption is widely recommended for heavy metal removal due to its effectiveness, ease of application, low cost, and the availability of diverse adsorbents (Bordoloi et al. 2017). Adsorption not only reduces the bioavailability of contaminants but also limits their environmental spread.

Biochar, a carbon-rich material produced from biomass through pyrolysis, has been identified as a promising sorbent for environmental applications. Pyrolysis and

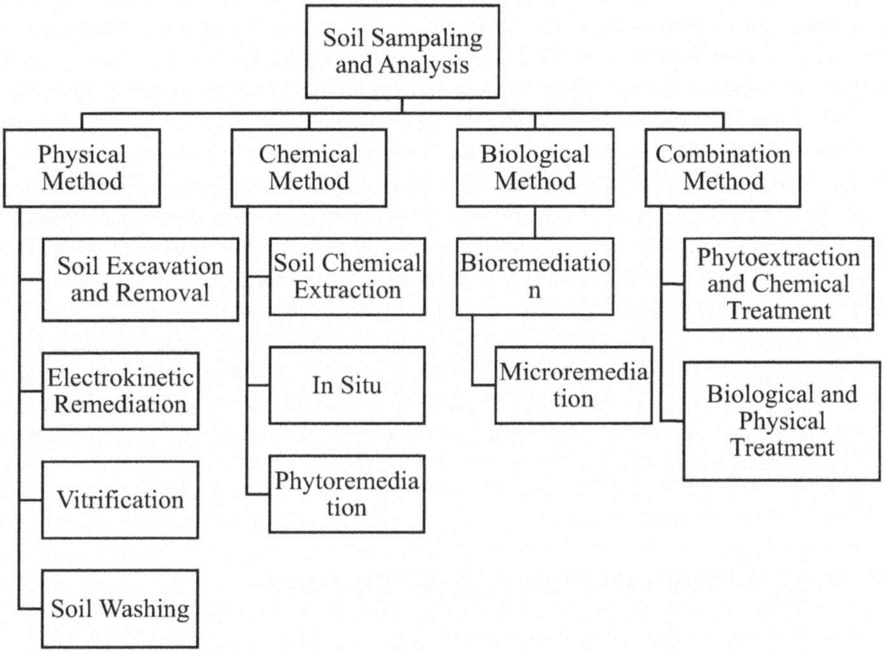

Fig. 1 Flowchart showing the key processes for removing inorganic contaminants from soil

other thermal degradation processes such as combustion, gasification, and liquefaction yield biochar, which is widely studied for its potential in climate change mitigation, waste management, and soil enhancement (Sohi et al. 2010). Pyrolysis, specifically, offers a practical method for converting biomass into biochar through techniques like slow, fast, flash, catalytic, and microwave-assisted pyrolysis (Manya 2012).

Activated carbon is the most commonly used adsorbent due to its high surface area and porosity, often enhanced through heat or chemical treatment (Cao et al. 2011). However, biochar, a cost-effective and versatile alternative, has shown significant potential for contaminant removal. Although biochar production generally requires less energy than activated carbon, it remains a potent option for environmental decontamination and has shown promise in immobilizing contaminants in various environmental matrices (Xu et al. 2012; Lu et al. 2014).

The use of biochar for metal sorption is facilitated by its high surface area, porous structure, and abundance of functional groups that can bind heavy metals (Wu et al. 2012). Its role in carbon sequestration and soil fertility enhancement has also been documented, making it a multifunctional tool for environmental management (Jeffery et al. 2011; Kookana et al. 2011). Biochar is especially effective in metal remediation due to its negative surface charge, which favors the adsorption of cations, including toxic metals like Pb^{2+}, Cu^{2+}, Zn^{2+}, and Cd^{2+} (Beesley and Marmiroli 2011). Methods to enhance biochar's metal-binding capacity, such as treatment with iron oxide, have been explored to improve its efficacy in removing contaminants from aqueous solutions (Chen et al. 2011; Zhang et al. 2013a, b).

Biochar's potential for immobilizing heavy metals through mechanisms like precipitation with phosphate, carbonate, and hydroxide ions is well-supported. Various biochars, including those derived from animal manure, have shown enhanced performance in adsorbing metals like Pb^{2+} due to their phosphorus content, which promotes co-precipitation with metal ions (Xu et al. 2015). The competitive adsorption of metals onto biochar is influenced by the biochar's characteristics, the metals' properties, and environmental factors, necessitating careful selection and optimization of biochar for specific remediation applications.

This chapter focuses on biochar's preparation methods, physicochemical properties, and applications in the remediation of inorganic contaminants. The chapter aims to assess biochar's efficiency in removing heavy metals from soil and water, compare biochars produced from various feedstocks, and discuss the role of biochar in environmental sustainability.

2 Need of Removal of Inorganic Compounds

The need for removing inorganic toxic compounds from soil arises due to the severe environmental and health risks these substances pose shows in Table 1. Inorganic toxic compounds, including heavy metals like lead, cadmium, chromium, and arsenic, persist in the soil due to their non-biodegradable nature. These compounds can accumulate over time, disrupting soil ecosystems, reducing soil fertility, and impairing agricultural productivity. Additionally, toxic metals can leach into groundwater, contaminate water sources, and enter the food chain through crops, leading to bioaccumulation in plants, animals, and humans.

Chronic exposure to heavy metals is linked to serious health problems such as kidney damage, neurological disorders, cancer, and cardiovascular diseases. Children and pregnant women are particularly vulnerable to the toxic effects of these compounds. Soil contamination from inorganic pollutants often stems from industrial activities, mining, agricultural inputs (such as pesticides and fertilizers), and improper waste disposal.

Given these risks, effective removal techniques are essential for restoring soil health, ensuring safe agricultural production, and preventing further contamination of ecosystems. Methods such as phytoremediation, adsorption using biochar or modified clays, soil washing, and stabilization have shown promise in immobilizing or extracting inorganic pollutants, thus safeguarding environmental health and reducing exposure risks to humans and wildlife.

Table 1 Inorganic contaminants in soil and their environmental and health effects

Inorganic contaminant	Source	Environmental effect	Health impact
Lead (Pb)	Industrial emissions, lead-based paints, pesticides	Toxic to plants, can reduce soil fertility, bioaccumulation in the food chain	Neurological damage, developmental delays in children, kidney damage, anemia, gastrointestinal issues
Arsenic (As)	Pesticides, industrial waste, mining activities	Toxic to plants, soil acidification, disrupts microbial communities	Carcinogenic, skin lesions, organ damage (liver, kidneys), neurological effects, cardiovascular issues
Cadmium (Cd)	Industrial processes, sewage sludge, fertilizers	Inhibits plant growth, reduces soil fertility, disrupts soil microbial activity	Kidney damage, skeletal damage, cancer, respiratory issues, reproductive toxicity
Chromium (Cr)	Industrial discharges, leather tanning, pigments	Reduces plant growth, toxic to soil microorganisms, can contaminate groundwater	Carcinogenic, lung and liver damage, skin irritation, kidney and gastrointestinal issues
Mercury (Hg)	Industrial activities, waste disposal, mining	Bioaccumulation in soil and water, affects aquatic ecosystems and soil microorganisms	Neurological damage, kidney damage, developmental effects, respiratory issues, immune system dysfunction
Nickel (Ni)	Mining, industrial activities, fertilizers	Toxic to plants, disrupts soil microorganisms, affects soil structure	Respiratory issues, skin rashes, kidney and liver damage, carcinogenic effects
Zinc (Zn)	Industrial discharges, fertilizers, sewage sludge	High concentrations can inhibit plant growth, toxicity to aquatic organisms, alter soil pH	Gastrointestinal issues, skin irritation, damage to liver and kidneys
Copper (Cu)	Agricultural runoff, industrial discharges	Toxic to soil organisms, affects plant growth, can lead to soil acidification	Gastrointestinal problems, liver damage, kidney toxicity, neurological issues
Aluminum (Al)	Mining, industrial discharges, fertilizers	Toxic to plants, soil acidification, disrupts microbial activity, affects plant nutrient uptake	Toxic to the central nervous system, bone disorders, Alzheimer's disease-like symptoms

(continued)

Table 1 (continued)

Inorganic contaminant	Source	Environmental effect	Health impact
Selenium (Se)	Agricultural runoff, industrial waste	Disrupts plant growth, bioaccumulation in food chains, can alter soil nutrient balance	Toxic to humans and animals, can cause skin, gastrointestinal, and respiratory issues
Fluoride (F)	Agricultural runoff, industrial waste, pesticides	Affects soil and plant health, bioaccumulation in the food chain, can lower soil fertility	Dental and skeletal fluorosis, thyroid dysfunction, gastrointestinal issues, bone deformities

3 Biochar Preparation Techniques and Conditions

The production of biochar (BC) has utilized a variety of feedstocks and production methods. The various steps involved in the preparation of value added biochar are shown in Fig. 2. To prevent explosions in closed furnaces during biochar production, a nitrogen flow rate of 2500 ml/min was maintained to create an inert atmosphere (Poo et al. 2018). In one study, Wang produced BC using slow pyrolysis at 300 °C for three hours in an oxygen-free environment, with a heating rate of 25 °C/min (Wang et al. 2016). Agrafioti et al. (2014) conducted pyrolysis in a muffle furnace at a rate of 17 °C/min. Other researchers employed a bench top quartz tube furnace capable of reaching up to 1200 °C, heating biochar samples at 350 °C, 450 °C, 550 °C, and 650 °C, with a heating rate of 5 °C/min for one hour. Slow pyrolysis at 450–500 °C is preferred for high-quality biochar with optimal yield. A quartz glass vertical tube reactor, heating at 40 °C/min with a nitrogen flow of 100 ml/min, was also used to produce BC (Bordoloi et al. 2017). Nitrogen is commonly introduced to prevent oxygen intrusion during pyrolysis (Sekar et al. 2021).

Pyrolysis, a thermal decomposition reaction, occurs under high temperature and low oxygen conditions. Based on duration and heating rates, it can be classified into slow, fast, or flash pyrolysis (with heating rates up to 1000 °C/s) (Onay and Kockar 2003). Studies indicate that lower temperatures and slower heating rates favor the production of solid products, making slow pyrolysis the preferred method for biochar production (Kambo and Dutta 2015).

Hydrothermal carbonization (HTC) offers an alternative method, using biomass in a water-saturated, high pressure (2–6 MPa), moderate-temperature (up to 350 °C) environment for 5–16 h (Tan et al. 2015; Kambo and Dutta 2015; Berge et al. 2015). HTC generates biochar with effective pollutant adsorption properties, and due to the use of water as a reaction medium, it avoids the formation of toxic by-products (Regmi et al. 2012; Zhang et al. 2013a, b). However, HTC is limited by its requirements for high pressure, expensive equipment, and elevated operational costs, making its large-scale application challenging (Regmi et al. 2012).

Other techniques, such as dehydration, gasification, rapid pyrolysis, and flash pyrolysis, are mainly employed for bio-oil and gas production rather than dense

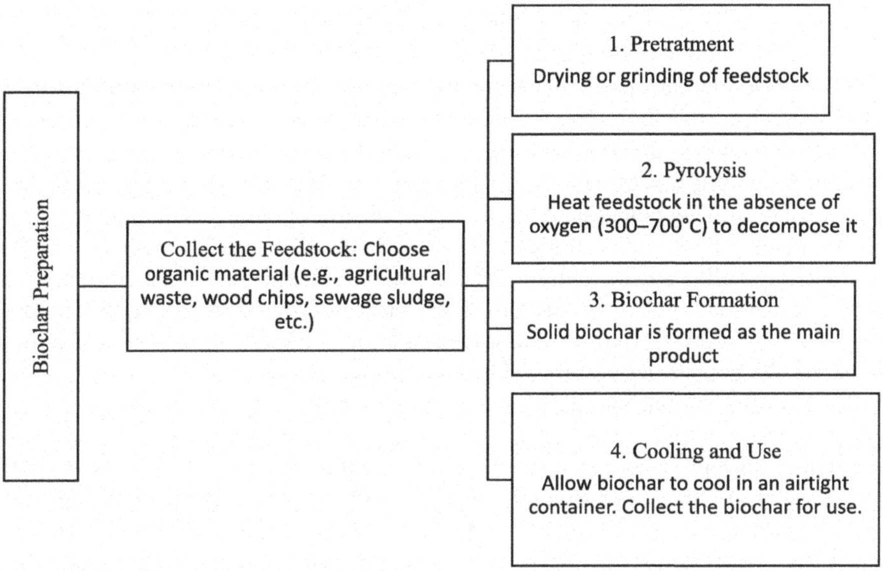

Fig. 2 Flow chart of biochar preparation

biochar, due to high gas-product yields (up to 90%) with increasing temperatures (Tan et al. 2015).

In a controlled study, biochar was produced by pyrolyzing raw biomass in a fixed-bed reactor within an electrothermal oven. Biochar feedstock was placed on a quartz frit in the reactor, heated at a rate of 10 K/min to the target temperature, and maintained for 15 min under ultra-high purity argon (>99.999%) at 300 °C, 500 °C, and 750 °C, representing different pyrolysis conditions.

4 Application of Biochar in Removal of Inorganic Compounds

Carbonaceous materials have been extensively utilized for extended durations to remove both carbonaceous and non-carbonaceous pollutants from the environment (Ahmad et al. 2012; Saeed et al. 2005; Salih et al. 2011). Activated charcoal, which undergoes oxygen activation to increase its porous structure and surface area, is widely recognized as a leading carbonaceous adsorbent. Similarly, biochar is produced through pyrolysis, resulting in a material with moderate to high surface area (Cao et al. 2011). Uncarbonized portions of biochar can interact with soil contaminants, and the presence of oxygen-containing functional groups in biochar can significantly enhance its ability to adsorb soil pollutants (Uchimiya et al. 2011). These characteristics position biochar as an eco-friendly and effective sorbent for both organic and inorganic pollutants in water and soil.

Inorganic pollutants, particularly metals, primarily originate from anthropogenic activities such as mining, metal plating, fertilizer use, sewage, and industrial effluents (Adriano 2001; Ok et al. 2011; Lim et al. 2013). Metals are of particular concern due to their bioavailability and non-biodegradable nature (Adriano 2001). Recently, biochar has shown promising potential in removing metals from both aqueous and terrestrial environments. The mobility of metals in soils can be influenced by biochar, with its alkalinity typically causing a liming effect, immobilizing metals and facilitating the sorption of oxyanions (Almaroai et al. 2013). However, biochars produced at higher temperatures (>600 °C) may lack dissolved carbon, potentially reducing the stability of metals like copper in the soil (Uchimiya et al. 2011).

Biochar also exhibits significant reactivity with pollutants like Cr, aiding in their reduction through electron transfer. Functional groups such as carbonyl, carboxyl, and hydroxyl groups play a key role in facilitating the reduction of Cr(VI) to Cr(III) (Choppala et al. 2012). For example, biochar derived from soybean stalks demonstrated 86.4% removal of Hg(II) from aqueous solutions, showing its high efficacy in metal removal (Kong et al. 2011). Beyond metal removal, biochar also contributes to climate change mitigation by sequestering carbon in soils, thus reducing CO_2 emissions (Lehmann et al. 2008). This long-term stability in soil makes biochar an important tool for reducing atmospheric CO_2 (Cheng et al. 2008; Kuzyakov et al. 2009; Singh et al. 2012). Additionally, biochar application enhances agricultural productivity by increasing soil fertility, improving microbial activity, and reducing nutrient leaching (Verheijen et al. 2010; Lehmann et al. 2011).

Research has shown that biochar can improve soil conditions, such as increasing pH in alkaline soils and boosting the soil's cation exchange capacity and nutrient content (Qian et al. 2015). For instance, adding 5% biochar to soils in the Kubuqi Desert increased sorghum crop production by up to 30% (Laghari et al. 2015). Furthermore, when biochar was combined with compost, it enhanced the hydrogen potential and microbial populations in the soil (Qayyum et al. 2015). Biochar also promotes plant growth in poor-quality soils by enhancing nutrient availability and capturing phosphorus, making it a potential slow-release fertilizer (Yao et al. 2013a). Studies have shown that biochar can significantly reduce phosphorus levels in agricultural surplus, contributing to more sustainable agricultural practices.

In addition to organic contaminants, biochar is effective at adsorbing inorganic pollutants, including toxic metals. The primary adsorption mechanisms for metals include ion exchange, electrostatic interactions, and precipitation, whereas for organic contaminants, biochar's large surface area and porous structure are crucial for adsorption (Ahmad et al. 2014). The potential of biochar to prevent the degradation of organic contaminants in soils and limit their leaching into groundwater sources further supports its role in environmental remediation (Jones et al. 2011).

Phosphorus, as a nutrient, can also be a significant non-carbonaceous pollutant when it contaminates water bodies. Although research on phosphorus sorption by biochar is limited (Yao et al. 2013b), studies suggest that biochar could potentially reduce phosphorus requirements in agriculture, creating a closed-loop system that minimizes surplus phosphorus in farming and enhances crop development. The study by Pan et al. investigates the use of biochar derived from agricultural waste as an adsorbent for simultaneously removing organic and inorganic phosphorus from circulating cooling water. This process addresses both organic pollutants, such as phosphonates, and inorganic forms like phosphate, which are common in industrial wastewater and can lead to eutrophication if not properly managed (Pan et al. 2023).

5 Advanced Biochar Composites for Enhanced Pollutant Removal and Environmental Remediation

Due to its superior properties including non-hazardous nature, abundant functional groups, porous structure, and high adsorption capacity biochar (BC) has emerged as an innovative sorbent, extensively applied to remove pollutants from aqueous waste (Shaheen et al. 2019; Hu et al. 2020). BC's feedstocks, often derived from agricultural waste, cultivated biomass, and animal litter, are promising for environmental remediation as they do not require activation (Liu and Zhang 2011). By incorporating functional elements, magnetic compounds, and nanoparticles, BC composites can be tailored for specific contaminants, significantly enhancing pollutant removal efficiency.

BC-based composites, as reported, show impressive adsorption capacities. For example, MnO_2-BC demonstrated a high sorption capacity (qm) of 904 mg/g, removing 100% of U(VI) from water (Ioannou et al. 2019). Similarly, TiO_2-BC increased the sonocatalytic degradation efficiency of Reactive Blue 69 (RB69) from 60 to 98%, with 92.1% removal efficiency retained after five reuse cycles (Khataee et al. 2017). Thus, advanced BC-based materials, potentially incorporating immobilized microorganisms, should be developed to further improve pollutant eradication. However, the release of minor constituents from BC particles and composites into the environment is inevitable, posing potential risks to animal health (Lu et al. 2020; Lian and Xing 2017; Zhang et al. 2019). For instance, BC derived from rice straw, when added to contaminated soil, released DOC, PO_4^{3-}, Cl^-, and SO_4^{2-}while reducing Pb bioavailability. Other studies found that BC application in loam soils under oxidizing conditions increased soluble arsenic and cobalt levels, raising concerns about food chain contamination and ecological health impacts (Rinklebe et al. 2020; El-Naggar et al. 2019).

Heavy metals, non-biodegradable and prone to bioaccumulation, present substantial risks to human health, including carcinogenic and endocrine effects. BC effectively adsorbs metal ions through mechanisms such as adhesion, ion exchange, electrostatic interaction, complexation, and precipitation (Inyang et al. 2016). The ongoing industrial and economic expansion has exacerbated water contamination, posing serious threats to environmental and human health. Water pollutants, primarily toxic metals (e.g., Cu, Cr, Pb, Ni) and organic contaminants (e.g., dyes), are highly persistent, mobile, and toxic (Islam et al. 2015; Schwarzenbach et al. 2010; Hao et al.

2021). Among cost-effective pollution control techniques, adsorption stands out due to its high efficiency, affordability, and ease of application (Huggins et al. 2016).

Komkiene and Baltrenaite (2016) prepared BC from silver birch and found it effectively adsorbed Zn(II) (128 mg/g) and Cu(II) (107 mg/g), indicating BC's potential for removing toxic metals from water sources. Other studies demonstrated the efficacy of BC pyrolyzed from various biomass types. For instance, BC produced from water hyacinth at 450 °C achieved high Cd(II) removal within one hour. Similarly, BC from jujube branches and animal waste, pyrolyzed at 700 °C, removed 93% of Cr(VI) from soil, reducing leachate concentrations by 59% (Rafique et al. 2019). Various studies have shown BC to be effective in immobilizing heavy metals in soils, with significant reductions in metal concentrations in plant tissues (Xu et al. 2020). BC's Cr(VI) removal mechanisms primarily involve direct reduction, as revealed by XANES analysis showing the reduction of Cr(VI) to Cr(III) post-interaction with BC (Rajapaksha et al. 2018).

BC modification with Mn, Na_2S, and KOH has proven successful in enhancing its adsorption capabilities. For instance, Mn-modified BC improved Pb removal from 6 to 98% at pH 5, while Na_2S/KOH treatments increased surface area and functional groups, leading to a significant increase in Hg(II) adsorption (Tan et al. 2016; Wang et al. 2015). $KMnO_4$-modified BC effectively removed U(VI) through enhanced oxygen-containing functional groups and MnO_2 nanoparticle formation, aiding in sorption through coordination and Lewis acid–base interactions (Li et al. 2019). Other functionalized BC composites, such as nZVI-BC, significantly reduced as mobility in soil through surface reactions and co-precipitation (Fan et al. 2020).

In the study by Liu et al., researchers explored the effectiveness of silicon (Si)-modified biochars in removing cadmium (Cd) from water. This biochar was created by modifying agricultural wastes, such as rice husks and wood, with silicon-bearing precursors like calcium silicate. The Si modification was achieved through ball milling, enhancing the biochar's properties for cadmium adsorption (Liu et al. 2023).

The study by Gao et al. (2023) examines the use of iron-modified biochar for the removal of antimony (Sb(III)) from contaminated water through combined sorption and oxidation mechanisms. Iron-modified biochar was synthesized and tested for its antimony removal efficiency. The study explored how electron-donating capacity and reactive iron species in the biochar influenced the removal process. The biochar showed enhanced removal of Sb(III) due to its high electron-donating ability and efficient interaction with reactive iron species, achieving substantial sorption and oxidation efficiency (Gao et al. 2023).

Ćwieląg-Piasecka et al. (2023) investigates the use of deashed wheat-straw biochar as an effective sorbent for various pesticides. The biochar was created by high-temperature pyrolysis, and a deashing process was applied to enhance its properties by removing inorganic residues, which resulted in increased porosity and a larger surface area. This modified biochar displayed high efficiency in adsorbing hydrophobic pesticides like carbaryl and metolachlor (Ćwieląg-Piasecka et al. 2023).

The study by Chakhtouna et al. (2023) explores a microwave-assisted synthesis of a composite material, MIL–53(Fe)/biochar, derived from date palm rachis. This composite was developed as an effective adsorbent for removing fluoroquinolone antibiotics, specifically ciprofloxacin (CPX) and ofloxacin (OFX), from wastewater, addressing the persistent issue of pharmaceutical contamination in aquatic environments (Chakhtouna et al. 2023).

Katuwal et al. explores the enhanced capture of dissolved inorganic phosphorus (DIP) from water using a biochar modified with gypsum. The modified biochar is created by pyrolyzing woody biomass pre-treated with flue gas desulfurization gypsum, significantly improving its phosphorus removal efficiency (Katuwal et al. 2023).

The study by Yang et al. investigates the use of ionic liquid-assisted biochar production to enhance Cu^{2+} removal efficiency from aqueous solutions. The authors explore how pyrolysis temperature (350 °C, 600 °C, 700 °C) and the presence of $CaCO_3$ affect the biochar's performance. The biochar produced with ionic liquids showed higher adsorption capacities for copper ions compared to untreated biochars. The best-performing biochar was achieved at 600 °C with a $CaCO_3$ content of 10%. The enhanced removal was attributed to better porosity, surface functionalization, and increased metal ion interactions.

Biochar's effectiveness has been validated by various studies employing analytical techniques, such as CEC, SEM, FTIR, TEM, and XRD, which have documented its high sorption capacities for heavy metal remediation in contaminated soils (Karami et al. 2011; Jiang et al. 2012; Kumar et al. 2011). Biochar's ability to lower Pb, Cd, and Cu levels in polluted soils has also been demonstrated (Jiang et al. 2012).

6 Modified Biochar and Their Effect

The surface area, functionality, and porosity of chemically modified biochar can be significantly enhanced through the use of potassium-based chemicals such as KOH and K_2CO_3 (Yu et al. 2016; Dehkhoda et al. 2016). For instance, the Cu(II) sorption capacity of biochar was increased to 50.70 mg/g with KOH activation, highlighting the importance of the modified surface features, particularly enhanced porosity and functionality, in improving the adsorption of Cu(II) (Jin et al. 2016). Similarly, when modified with a 2 mol/L sodium sulfate solution, the surface and porosity of biochar were significantly enhanced, with biochar derived from corncob showing the best performance for nickel (II) adsorption (Hu et al. 2018).

In the case of Pb(II) adsorption, biochar treated with potassium-based chemicals exhibited a significant increase in Pb capture, from 74.41 mg/g to 200 mg/g. This was accompanied by a 42% reduction in ash content, while the presence of carboxyl groups (COOH) and oxygen content in biochar derived from manure increased by 64 to 101% (Wang and Liu 2018). Chemical activation with $ZnCl_2$ under high temperatures also improved Pb(II) adsorption capabilities of biochar compared to untreated biochar (Rwiza et al. 2018). Furthermore, alkaline activation of biochar demonstrated improved sorption of PAH compounds compared to raw biochar and activated charcoal, with a recovery of surface area and pore structure (Anasonye et al. 2018).

Acid-activated biochar, produced using 30% oxalic acid and H_2SO_4, exhibited significantly higher adsorption capacity for SMZ than biochar produced at 700 °C (BB700). The acid-activated biochar absorbed SMZ at 190 mg/g, whereas BB700 absorbed only 140 mg/g, with a 250-fold increase in BET surface area (Vithanage et al. 2015). Similarly, biochar produced from wheat straw and activated with $ZnCl_2$ (ZnBC) and H_3PO_4 (PBC) demonstrated much higher atrazine removal efficiency compared to untreated biochar, with ZnBC and PBC showing five times the contact area and eight times the micropore capacity of pure biochar.

Biochar derived from crayfish shells and treated with zinc chloride (ZnO-CFS600) exhibited enhanced adsorption of trichloroacetic acid (TCAA) due to the improved porosity and formation of ZnO nanoparticles (Long et al. 2017). Magnetic biochar has also gained significant attention, with studies indicating that it cannot only remove pollutants but also isolate them using external magnets (Kundu et al. 2015; Thines et al. 2017). For example, aromatic pollutants such as anisole (70.4 mg/g), guaiac (23 mg/g), and phenol (17 mg/g) exhibited significant adsorption onto magnetic biochar derived from rice straw, with the adsorption performance being linked to the material's graphene sheets, pore volume, heat resistance, and protonation (Li et al. 2017a, b). Magnetic biochar synthesized with Fe_3O_4 and Douglas fir biochar also demonstrated rapid adsorption of lead and cadmium (Karunanayake et al. 2018).

While chemically modified biochar shows significantly improved adsorption capacities for various contaminants, there is a slight reduction in micropore quantity after chemical activation, which could occasionally affect its ability to adsorb certain chemicals (Wu et al. 2016). Additionally, the cost of using modified biochar is higher than that of using pure biochar, and further research is needed to evaluate the cost-effectiveness and durability of modified biochar in pollutant remediation applications.

7 Comparison with Other Substrate

Biochar's efficiency in removing inorganic contaminants varies significantly based on the substrate used for its production. For example, biochar derived from agricultural waste often exhibits high surface area and porosity, making it effective for adsorbing heavy metals like cadmium (Cd^{2+}) and lead (Pb^{2+}). Conversely, biochars from forestry residues may have higher reactivity due to greater mineral content. Adding functional groups or incorporating modifiers like gypsum can further enhance metal ion removal. Pyrolysis temperature also plays a key role, influencing biochar's structural properties and adsorption capacity.

Biochar and nanoparticles (NPs) both show potential in removing inorganic contaminants, but they differ in mechanisms and effectiveness. Biochar mainly removes contaminants through adsorption due to its high surface area and functional groups. It is effective for heavy metals (e.g., Pb^{2+}, Cd^{2+}) in soil. In contrast, NPs, such as zero-valent iron (ZVI) or titanium dioxide, offer enhanced removal via both adsorption and catalytic reduction, often exhibiting faster reaction rates.

Nanoparticles may also be more effective in targeting specific contaminants due to their smaller size and higher reactivity.

Membrane technologies differ significantly in their mechanisms for removing inorganic contaminants. Biochar primarily works through adsorption, benefiting from its large surface area and functional groups, effective for adsorbing metals like Pb^{2+} and Cd^{2+}. Membranes, on the other hand, rely on filtration, including reverse osmosis or nanofiltration, which physically separates contaminants from water. Membranes tend to offer higher removal rates and are more selective, but they can suffer from fouling and require regular cleaning, whereas biochar is simpler to use but may have lower efficiency in some cases.

8 Future Scope of This Shapter

Biochar may effectively remove dangerous heavy metals from earth and aqueous condition because of its special assets, though. It is becoming increasingly important to comprehend the mechanisms of adsorption in order to utilize eagerly accessible biomass feedstocks. The pyrolysis temperature and the feedstock being utilized may change the features of biochar, such as its structure, chemistry, composition, and mode of action. The texture of the biochar produced by different levels of pyrolysis, as well as its properties and dependence on various biomass sources, provide a variety of challenges in terms of their effects on the soil and soil biota.

The chemical interactions between biochar and soil surface also require in-depth study, particularly in relation to micronutrients, toxic ion sorption and ion exchange processes. Future study on biochar must assess a larger range of distinctive traits and entail collaborative efforts in order to close knowledge gaps since studies to yet have concentrated on a small number of criteria. Knowledge must be based on thorough data that is trustworthy and repeated in the field, like the qualities of biochar that relate to the target application. To get the best benefits, it's important to pick the right amount of biochar and application techniques. The most efficient rates, application windows, and methodology should be the focus of future study, taking cropping system and crop growth stage into consideration.

Extrapolating from wastewater treatment outcomes to soil mitigation enhancement is necessary. The essential to this extension will be the shared goals of the maintenance and alteration systems. Before applying biochar on a broad scale, more in situ or field testing need to be carried out over an extended period of time to imitate the reality and ascertain the true impact of biochar.

9 Conclusion

The creation of biochar and the necessity of harmful metal ion sorption from water and soil are discussed in this chapter. In the chapter, the creation of biochar and its uses are also covered. The chapter provides a brief explanation of the methodology used to separate inorganic compounds from soil as well as information on published studies. The results of the separation of inorganic compounds using biochar and other materials with a similar probability were also compared by the author. The biochar and their effectiveness in separating inorganic components were better highlighted throughout this chapter, which will be helpful for future research and advancement in the biochar study.

References

Adriano DC (2001) Trace elements in terrestrial environments. Biogeochemistry, bioavailability and risks of metals. Springer New York

Agrafioti E, Kalderis D, Diamadopoulos E (2014) Arsenic and chromium removal from water using biochars derived from rice husk, organic solid wastes and sewage sludge. J Environ Manag 133:309–314

Ahmad M, Rajapaksha AU, Lim JE, Zhang M, Bolan N, Mohan D, Vithanage M, Lee SS, Ok YS (2014) Biochar as a sorbent for contaminant management in soil and water: a review. Chemosphere 99:19–33

Ahmad M, Usman ARA, Lee SS, Kim SC, Joo JH, Yang JE, Ok YS (2012) Eggshell and coral wastes as low cost sorbents for the removal of Pb^{2+}, Cd^{2+} and Cu^{2+} from aqueous solutions. J Ind Eng Chem 18:198–204

Akpomie OO, Ejechi BO (2016) Bioremediation of soil contaminated with tannery effluent by combined treatment with cow dung and microorganisms isolated from tannery effluent. J Bioremediat Biodegradation. https://doi.org/10.4172/2155-6199.1000354

Almaroai YA, Usman ARA, Ahmad M, Moon DH, Cho JS, Joo YK, Jeon C, Lee SS, Ok YS (2013) Effects of biochar, cow bone, and eggshell on Pb availability to maize in contaminated soil irrigated with saline water. Environ Earth Sci https://doi.org/10.1007/s12665-013- 2533-6

Anasonye F, Tammeorg P, Parshintsev J, Riekkola M, Tuomela M (2018) Role of biochar and fungi on PAH sorption to soil rich in organic matter. Water Air Soil Pollut 229:49. https://doi.org/10.1007/s11270-018-3708-2

Beesley L, Marmiroli M (2011) The immobilisation and retention of soluble arsenic, cadmium and zinc by biochar. Environ Pollut 159(2):474–480

Berge ND, Li L, Flora JRV (2015) Assessing the environmental impact of energy production from hydrochar generated via hydrothermal carbonization of food wastes. Waste Manage 43:203–217

Bordoloi N, Goswami R, Kumar M, Kataki R (2017) Biosorption of Co (II) from aqueous solution using algal biochar: kinetics and isotherm studies. Bioresour Technol Biortech 05:139. https://doi.org/10.1016/j.biortech.2017.05.139

Campillo-Cora C, Conde-Cid M, Arias-Estevez M (2020) Specific adsorption ofheavy metals in soils: individual and competitive experiments. Agronomy 10:1113. https://doi.org/10.3390/agronomy10081113

Cao X, Ma L, Liang Y, Gao B, Harris W (2011) Simultaneous immobilization of lead and atrazine in contaminated soils using dairy-manure biochar. Environ Sci Technol 45:4884–4889. https://doi.org/10.1021/es103752u

Chakhtouna H, Benzeid H, Zari N et al (2023) Microwave-assisted synthesis of MIL–53(Fe)/biochar composite from date palm for ciprofloxacin and ofloxacin antibiotics removal. Sep Purif Technol 308:122850. https://doi.org/10.1016/j.seppur.2022.122850

Chappa S, Rathod PB, Debnath AK et al (2018) Palladium nanoparticles hosted in poly(ethylenimine) and poly(ethylene glycol methacrylate phosphate) anchored membranes for catalyzing uranyl ions reduction and mizoroki-heck coupling reaction. ACS Appl Nano Mater 1:3259–3268. https://doi.org/10.1021/acsanm.8b00474

Chen BL, Chen ZM, Lv SF (2011) A novel magnetic biochar efficiently sorbs organic pollutants and phosphate. Bioresour Technol 102(2):716–723

Cheng CH, Lehmann J, Thies JE, Burton SD (2008) Stability of black carbon in soils across a climatic gradient. J Geophys Res 113:1–10

Choppala GK, Bolan NS, Mallavarapu M, Chen Z, Naidu R (2012) The influence of biochar and black carbon on reduction and bioavailability of chromate in soils. J Environ Qual 41:1–10

Ćwieląg-Piasecka I, Jamroz E, Medyńska-Juraszek A et al (2023) Deashed wheat-straw biochar as a potential superabsorbent for pesticides. Mater (Basel) 16:2185. https://doi.org/10.3390/ma16062185

Dehkhoda AM, Gyenge E, Ellis N (2016) A novel method to tailor the porous structure of KOH-activated biochar and its application in capacitive deionization and energy storage. Biomass Bioenergy 87:107–121

El-Naggar A, Lee MH, Hur J, Lee HY, Igalavithana AD, Shaheen SM, Ryu C, Rinklebe J, Tsang DCW, Ok YS (2019) Biochar-induced metal immobilization and soil biogeochemical process: an integrated mechanistic approach. Sci Total Environ 698:134112

Fan J, Chen X, Xu ZB, Xu XY, Zhao L, Qiu H, Cao XD (2020) Onepot synthesis of nZVI-embedded biochar for remediation of two mining arsenic-contaminated soils: arsenic immobilization associated with iron transformation—science direct. J Hazard Mater 398:122901

Gao Y, Chen H, Fang Z et al (2023) Coupled sorptive and oxidative antimony(III) removal by iron-modified biochar: mechanisms of electron-donating capacity and reactive Fe species. Environ Pollut 337:122637. https://doi.org/10.1016/j.envpol.2023.122637

Godwin PM, Pan Y, Xiao H, Afzal MT (2019) Progress in preparation and application of modified biochar of improving heavy metal ion removal from wastewater. J Bioresources Bioprod 4(1):31–42

Hao MJ, Qiu MQ, Yang H, Hu BW, Wang XX (2021) Recent advances on preparation and environmental applications of MOF-derived carbons in catalysis. Sci Total Environ 760:143333

Hu BW, Ai YJ, Jin J, Hayat T, Alsaedi A, Zhuang L, Wang XK (2020) Efficient elimination of organic and inorganic pollutants by biochar and biochar-based materials. Biochar 2:47–64

Hu X, Xue Y, Liu L, Zeng Y, Long L (2018) Preparation and characterization of Na_2S-modified biochar for nickel removal. Environ Sci Pollut Res 26:1–9

Huggins TM, Haeger A, Bifnger JC, Ren ZYJ (2016) Granular biochar compared with activated carbon for wastewater treatment and resource recovery. Water Res 94:225–232

Inyang MI, Gao B, Yao Y, Xue YW, Zimmerman A, Mosa A, Pullammanappallil P, Cao XD (2016) A review of biochar as a low-cost adsorbent for aqueous heavy metal removal. Crit Rev Environ Sci Technol 46(4):406–433

Ioannou K, Hadjiyiannis P, Liatsou I, Pashalidis I (2019) U(VI) adsorption by biochar fiber–MnO_2 composites. J Radioanal Nucl Chem 320:425–432

Islam MS, Ahmed MK, Raknuzzaman M, Islam MK (2015) Heavy metal pollution in surface water and sediment: a preliminary assessment of an urban river in a developing country. Ecol Ind 48:282–291

Jeffery S, Verheijen F, Velde M, Bastos A (2011) A quantitative review of the effects of biochar application to soils on crop productivity using meta-analysis. Agric Ecosyst Environ 144:175–187. https://doi.org/10.1016/j.agee.2011.08.015

Jiang J, Xu RK, Jiang TY, Li Z (2012) Immobilization of Cu(II), Pb(II) and Cd(II) by the addition of rice straw derived biochar to a simulated polluted ultisol. J Hazard Mater 229:145–150

Jin H, Hanif MU, Capareda S, Chang Z, Huang H, Ai Y (2016) Copper (II) removal potential from aqueous solution by pyrolysis biochar derived from anaerobically digested algaedairy-manure and effect of KOH activation. J Environ Chem Eng 4:365–372

Jones DL, Edwards-Jones G, Murphy DV (2011) Biochar mediated alterations in herbicide breakdown and leaching in soil. Soil Biol Biochem 43(4):804–813

Kambo HS, Dutta A (2015) A comparative review of biochar and hydrochar in terms of production, physico-chemical properties and applications. Renew Sustain Energy Rev 45:359–378

Karami N, Clemente R, Moreno-Jimenez E, Lepp NW, Beesley L (2011) Efficiency of green waste compost and biochar soil amendments for reducing lead and copper mobility and uptake to ryegrass. J Hazard Mater 191:41–48

Karunanayake AG, Todd OA, Crowley M, Ricchetti L Jr, Anderson R, Mohan D, Mlsna T (2018) Lead and cadmium remediation using magnetized and nonmagnetized biochar from douglas fir. Chem Eng J 331:480–491

Katuwal S, Circenis S, Zhao L, Zheng W (2023) Enhancing dissolved inorganic phosphorous capture by gypsum-incorporated biochar: synergic performance and mechanisms. J Environ Qual 52:949–959. https://doi.org/10.1002/jeq2.20505

Khataee A, Kayan B, Gholami P, Kalderis D, Akay S (2017) Sonocatalytic degradation of an anthraquinone dye using TiO_2-biochar nanocomposite. Ultrason Sonochem 39:120–128

Komkiene J, Baltrenaite E (2016) Biochar as adsorbent for removal of heavy metal ions [Cd(II), Copper(II), Lead(II), Zinc(II)] from aqueous phase. Int J Environ Sci Technol 13(2):471–482

Kong H, He J, Gao Y, Wu H, Zhu X (2011) Cosorption of phenanthrene and mercury (II) from aqueous solution by soybean stalkbased biochar. J Agric Food Chem 59:12116–12123

Kookana RS, Sarmah AK, Van-Zwieten L, Krull E, Singh B (2011) Chapter three–biochar application to soil: agronomic and environmental benefits and unintended consequences. In: Donald LS (eds) Advances in agronomy, vol 112. Elsevier: Amsterdam Netherlands, pp 103–143 https://doi.org/10.1016/B978-0-12-385538-1.00003-2.

Kumar S, Loganathan VA, Gupta RB (2011) An Assessment of U(VI) removal from groundwater using biochar produced from hydrothermal carbonization. J Environ Manage 92(10):2504–2512

Kundu A, Gupta BS, Hashim MA, Sahu JN, Mujawar M, Redzwan G (2015) Optimisation of the process variables in production of activated carbon by microwave heating. RSC Adv 5:35899–35935

Kuzyakov Y, Subbotina I, Chen H, Bogomolova I, Xu X (2009) Black carbon decomposition and incorporation into soil microbial biomass estimated by 14C labeling. Soil Biol Biochem 41:210–219

Laghari M, Hu Z, Mirjat MS, Xiao B, Tagar AA, Hu M (2015) Fast pyrolysis biochar from sawdust improves the quality of desert soils and enhances plant growth. J Sci Food Agric (November 2014)

Lehmann J, Rillig MC, Thies J, Masiello CA, Kockaday WC, Crowley D (2011) Biochar effects on soil biota—a review. Soil Biol Biochem 43:1812–1836

Lehmann J, Skjemstad J, Sohi S, Carter J, Barson M, Falloon P, Coleman K, Woodbury P, Krull AE (2008) Australian climate–carbon cycle feedback reduced by soil black carbon. Nature Geosci 1:832–835

Li B, Yang L, Wang CQ, Zhang QP, Liu QC, Li YD, Xiao R (2017a) Adsorption of Cd(II) from aqueous solutions by rape straw biochar derived from different modification processes. Chemosphere 175:332–340

Li H, Mahyoub SAA, Liao W, Xia S, Zhao H, Guo M, Ma P (2017b) Effect of pyrolysis temperature on characteristics and aromatic contaminants adsorption behavior of magnetic biochar derived from pyrolysis oil distillation residue. Bioresour Technol 223:20–26

Li N, Yin ML, Tsang DCW, Yang ST, Liu J, Li X, Song G, Wang J (2019) Mechanisms of U(VI) removal by biochar derived from *Ficus microcarpa* aerial root: A comparison between raw and modified biochar. Sci Total Environ 697:134115

Lian F, Xing BS (2017) Black carbon (biochar) in water/soil environments: molecular structure, sorption, stability, and potential risk. Environ Sci Technol 51(23):13517–13532

Lim JE, Ahmad M, Usman ARA, Lee SS, Jeon WT, Oh SE, Yang JE, Ok YS (2013) Effects of natural and calcined poultry waste on Cd, Pb and As mobility in contaminated soil. Environ Earth Sci 69:11–20

Liu J, Huang Z, Chen Z, Sun J, Gao Y, Wu E (2020) Resource utilization of swine sludge to prepare modified biochar adsorbent for the efficient removal of Pb (II) from water. J Clean Prod 257:120322

Liu L, Yang X, Ahmad S et al (2023) Silicon (Si) modification of biochars from different Si-bearing precursors improves cadmium remediation. Chem Eng J 457:141194. https://doi.org/10.1016/j.cej.2022.141194

Liu Z, Zhang FS (2011) Removal of copper (II) and phenol from aqueous solution using porous carbons derived from hydrothermal chars. Desalination 267(1):101–106

Long L, Xue Y, Zeng Y, Yang K, Lin C (2017) Synthesis, characterization and mechanism analysis of modified crayfish shell biochar possessed ZnO nanoparticles to remove trichloroacetic acid. J Clean Prod 166:1244–1252

Lu K, Yang X, Shen J, Robinson B, Huang H, Liu D, Bolan N, Pei J, Wang H (2014) Effect of bamboo and rice straw biochars on the bioavailability of Cd, Cu, Pb and Zn to Sedum plumbizincicola. Agr Ecosys Environ 191:124–132. https://doi.org/10.1016/j.agee.2014.04.010

Lu L, Yu WT, Wang YF, Zhang K, Zhu XM, Zhang YC, Wu YJ, Ullah H, Xiao X, Chen BL (2020) Application of biochar-based materials in environmental remediation: from multi-level structures to specifc devices. Biochar 2:1–2

Manya JJ (2012) Pyrolysis for biochar purposes: a review to establish current knowledge gaps and research needs. Environ Sci Technol 46(15):7939–7954. https://doi.org/10.1021/es301029g

Margonwar AS, Rathod PB, Singh MP (2024) Microbial populations, function, and impact on environmental changes. In: Microbial metagenomics in effluent treatment plant. Elsevier, pp 141–166

Ok YS, Usman ARA, Lee SS, Abd El-Azeem SAM, Choi B, Hashimoto Y, Yang JE (2011) Effects of rapeseed residue on lead and cadmium availability and uptake by rice plants in heavy metal contaminated paddy soil. Chemosphere 85:677–682

Onay O, Kockar OM (2003) Slow, fast and flash pyrolysis of rapeseed. Renew Energy 28(15):2417–2433

Pan W, Xie H, Zhou Y et al (2023) Simultaneous adsorption removal of organic and inorganic phosphorus from discharged circulating cooling water on biochar derived from agricultural waste. J Clean Prod 383:135496. https://doi.org/10.1016/j.jclepro.2022.135496

Patra JM, Panda SS, Dhal NK (2017) Biochar as a low cost adsorbent for heavy metal removal: a review. Int J Res Biosci 6(1):(1–7)

Poo KM, Son EB, Chang JS, Ren X, Choi YJ, Chae KJ (2018) Biochars derived from wasted marinemacro-algae (Saccharina japonica and Sargassumfusiforme) and their potential for heavy metalremoval in aqueous solution. J Environ Manage 206:364–372

Qayyum MF, Ashraf I, Abid M, Steffens D (2015) Effect of biochar, lime, and compost application on phosphorus adsorption in a Ferralsol. J Plant Nutr Soil Sci 576–581

Qian K, Kumar A, Zhang H, Bellmer D, Huhnke R (2015) Recent advances in utilization of biochar. Renew Sustain Energy Rev Elsevier 42:1055–1064

Rafique MI, Usman ARA, Ahmad M, Sallam A, Al-Wabel M (2019) In situ immobilization of Cr and its availability to maize plants in tannery waste–contaminated soil: effects of biochar feedstock and pyrolysis temperature. J Soils Sediments 20:330–339

Rajapaksha AU, Alam MS, Chen N, Alessi DS, Igalavithana AD, Tsang DCW, Ok YS (2018) Removal of hexavalent chromium in aqueous solutions using biochar: chemical and spectroscopic investigations. Sci Total Environ 625:1567–1573

Rathod PB, Chappa S, Ajish Kumar KS et al (2018) Poly(ethylenimine) functionalized magnetic nanoparticles for sorption of Pb, Cu, and Ni: potential application in catalysis. Sep Sci Technol 00:1–11. https://doi.org/10.1080/01496395.2018.1520731

Rathod PB, Chappa S, Kumar KSA et al (2019) Cadmium(II)-loaded Fe_3O_4 @MPTS nanoparticles: preparation and application as catalyst for C-N coupling reactions. ChemistrySelect 4:11796–11800. https://doi.org/10.1002/slct.201902432

Rathod PB, Kalel RA, Singh Tomar MP et al (2023) Reformed metal-organic frameworks (MOFs) for abstraction of water contaminants—heavy-metal ions. Met Org Fram Wastewater Contam Remov 227–250. https://doi.org/10.1002/9783527841523.ch10

Rathod PB, Singh MP, Taware AS et al (2024) Comprehensive insights into water remediation: chemical, biotechnological, and nanotechnological perspectives. Environ Pollut Bioavailab 36. https://doi.org/10.1080/26395940.2024.2329660

Regmi P, Moscoso JLG, Kumar S (2012) Removal of copper and cadmium from aqueous solution using switchgrass biochar produced via hydrothermal carbonization process. J Environ Manage 109(17):61–69

Rinklebe J, Shaheen SM, El-Naggar A, Wang HL, Laing JD, Alessi DS, Ok YS (2020) Redox-induced mobilization of Ag, Sb, Sn, and Tl in the dissolved, colloidal and solid phase of a biochar-treated and un-treated mining soil. Environ Int 140:105754

Rwiza MJ, Oh S, Kim K, Kim SD (2018) Comparative sorption isotherms and removal studies for Pb(II) by physical and thermochemical modification of low-cost agro-wastes from Tanzania. Chemosphere 195:135–145

Saeed A, Akhter MW, Iqbal M (2005) Removal and recovery of heavy metals from aqueous solution using papaya wood as a new biosorbent. Sep Purif Technol 45:25–31

Salih HH, Patterson CL, Sorial GA, Sinha R, Krishnan R (2011) The fate and transport of the SiO_2 nanoparticles in a granular activated carbon bed and their impact on the removal of VOCs. J Hazard Mater 193:95–101

Schwarzenbach RP, Egli T, Hofstetter TB, Gunten UV, Wehrli B (2010) Global water pollution and human health. Soc Sci Electron Publ 35(1):109–136

Sekar M, Mathimani T, Alagumalai A, Chi NTL, Duc PA, Bhatia SK, Pugazhendhi A (2021) A review on the pyrolysis of algal biomass for biochar and bio-oil—bottlenecks and scope. Fuel 283:119190. https://doi.org/10.1016/j.fuel.2020.119190

Shaheen SM, El-Naggar A, Wang JX, Hassan NEE, Niazi NK, Wang HL, Tsang DCW, Ok YS, Bolan N, Rinklebe J (2019) Biochar as an (Im)mobilizing agent for the potentially toxic elements in contaminated soils-science direct. Biochar Biomass Waste 14:255–274

Singh BP, Cowie AL, Smernik RJ (2012) Biochar carbon stability in a clayey soil as a function of feedstock and pyrolysis temperature. Environ Sci Technol 46:11770–11778

Singh MP, Rathod PB, Margonwar AS (2024) Industrial wastewater remediation by using microbial communities

Sohi SP, Krull E, Lopez-Capel E, Bol R (2010) A review of biochar and its uses and function in soil. Adv Agron 105:47–82

Tan X, Liu Y, Zeng G (2015) Application of biochar for the removal of pollutants from aqueous solutions. Chemosphere 125:70–85

Thines KR, Abdullah EC, Mubarak NM, Ruthiraan M (2017) Synthesis of magnetic biochar from agricultural waste biomass to enhancing route for waste water and polymer application: a review. J Renew Sustain Energy 67:257–276

Uchimiya M, Klasson KT, Wartelle LH, Lima IM (2011) Influence of soil properties on heavy metal sequestration by biochar amendment: 1. copper sorption isotherms and the release of cations. Chemosphere 82:1431–1437

Verheijen F, Jeffery S, Bastos AC, van der Velde M, Diafas I (2010) Biochar application to soils. A critical scientific review of effects on soil properties, processes and functions. Eur Comm Italy

Vithanage M, Rajapaksha AU, Zhang M, Thiele-Bruhn S, Lee SS, Ok YS (2015) Acid-activated biochar increased sulfamethazine retention in soils. Environ Sci Pollut Res Int 22:2175–2186

Wang C, Gu L, Liu X, Zhang X, Cao L, Hu X (2016) Sorption behavior of Cr(VI) on pineapple-peel-derived biochar and the influence of coexisting pyrene. Int Biodeterior Biodegrad 111:78–84

Wang MC, Sheng GD, Qiu YP (2015) A novel manganese-oxide/ biochar composite for efficient removal of lead(II) from aqueous solutions. Int J Environ Sci Technol 12(5):1719–1726

Wang Y, Liu R (2018) H_2O_2 treatment enhanced the heavy metals removal by manure biochar in aqueous solutions. Sci Total Environ 628–629:1139–1148

Wu J, Yi Y, Li Y, Fang Z, Tsang EP (2016) Excellently reactive Ni/Fe bimetallic catalyst supported by biochar for the remediation of decabromodiphenyl contaminated soil: reactivity, mechanism, pathways and reducing secondary risks. J Hazard Mater 320:341–434

Wu W, Yang M, Feng Q, McGrouther K, Wang H, Lu H, Chen Y (2012) Chemical characterization of rice straw-derived biochar for soil amendment. Biomass Bioenerg 47:268–276. https://doi.org/10.1016/j.biombioe.2012.09.034

Xu C, Lin X, Wu Q (2012) Impacts of biochar on availability of heavy metals and nutrient content of contaminated soil under waterlogged conditions[J]. J Soil Water Conserv 26(6):194–199

Xu CB, Zhao JW, Yang WJ, Wei WX, Tan X, Wang J, Lin AJ (2020) Evaluation of biochar pyrolyzed from kitchen waste, corn straw, and peanut hulls on immobilization of Pb and Cd in contaminated soil. Environ Pollut 261:114133

Yao Y, Gao B, Chen J, Yang L (2013a) Engineered biochar reclaiming phosphate from aqueous solutions: mechanisms and potential application as a slow-release fertilizer. Environ Sci Technol 47:8700–8708

Yao Y, Gao B, Chen J, Zhang M, Inyang M, Li Y, Alva A, Yang L (2013b) Engineered carbon (biochar) prepared by direct pyrolysis of Mg-accumulated tomato tissues: characterization and phosphate removal potential. Bioresour Technol Elsevier Ltd 138:8–13

Yu F, Zhou Y, Gao B, Qiao H, Li Y, Wang E, Pang L, Bao C (2016) Effective removal of ionic liquid using modified biochar and its biological effects. J Taiwan Inst Chem Eng 67:318–324

Zhang LK, Mao JF, Chen BL (2019) Reconsideration of heterostructures of biochars: morphology, particle size, elemental position, reactivity and toxicity. Environ Pollut 254:113017

Zhang M, Gao B, Varnoosfaderani S, Hebard A, Yao Y, Inyang M (2013a) Preparation and characterization of a novel magnetic biochar for arsenic removal. Bioresour Technol 130:457–462

Zhang ZB, Cao XH, Liang P (2013b) Adsorption of uranium from aqueous solution using biochar produced by hydrothermal carbonization. J Radioanal Nucl Chem 295(2):1201–1208

Chapter 9
Removal of Anionic Contaminants Using Biochar Application

Gowhar Rashid, J. A. Ruley, Nihad Ashraf Khan, Haleema Ahmad, Mohd Umar Rehmani, Amaan Rais, and Ayaz Bashir

Abstract Biochar is a carbon-rich source which is produced by burning organic matters like agricultural waste, dung, pinewoods, rice, and maize husk. The process of producing biochar is a temperature dependent and an endothermic process. The process takes place in absence of oxygen at a specific temperature governed by slow or fast pyrolysis. The ancient technique of burning agricultural wastes and other materials only led to increase in the greenhouse gases and global warming. With advent of this technology it not only proved to increase the quality of a soil by enhancing the nutrients and retention of minerals but also plays a great role in the waste-water treatment. The surface of a biochar acts as a great adsorbent for several ions, inorganic materials, and other pollutants. Its surface has active sites containing functional groups of carboxylate –COOH and hydroxylate –OH which forms several bonds like hydrogen, covalent, electrostatic, and non-covalent bonds with the anionic contaminants to be removed from the aqueous medium. Biochar act as one of the finest sorbent as compared to an activated charcoal and has highest efficiency. Apart from the function of certain functional groups on the surface of a biochar in adsorption other factors like temperature, dosage of biochar, pH of a solution also equally plays an important role in the work efficiency of a biochar. The nature of biochar has a direct influence on the effectiveness of sorption which needs to be taken care, as biochar produced from swine dung has large mass and shows less efficiency in adsorption as compared to biochar derived from the agricultural products.

G. Rashid (✉)
Department of Clinical Biochemistry, SKIMS, Srinagar, India
e-mail: dr.gowhar63@gmail.com

J. A. Ruley
Department of Agricultural Sciences, University of Juba, Juba, South Sudan

N. A. Khan
Department of Biosciences, Jamia Milia Islamia, New Delhi, India

H. Ahmad · M. U. Rehmani · A. Rais
Department of Biochemistry, Faculty of Life Sciences, Aligarh Muslim University, Aligarh, India

A. Bashir
Department of Biochemistry, University of Kashmir, Srinagar, India

A. Roy et al. (eds.), *Biochar Production Engineering*,
https://doi.org/10.1007/978-3-031-92594-8_9

Keywords Biochar · Pyrolysis · Agricultural products · Waste-water treatment · Adsorption rate · Gasification · Carbonization

1 Introduction

For farmers, managing agricultural residue has always been a difficult issue to reconcile economic growth with environmental sustainability. Burning crop debris on a regular basis has increased the amount of greenhouse gases in the atmosphere and reduced the amount of nutrients in the soil. Agricultural waste, bones, manure, and other organic materials are heated to a precise temperature and with little to no oxygen to create biochar (Sohi 2012). It may be used to enhance the soil quality with carbon for agricultural purposes, as shown in Fig. 1. Burning carbon is released into the sky more gradually with the help of biochar. It enhances soil quality and stores carbon by being applied to agricultural areas. Additionally, the bioenergy created from biochar offers an excellent substitute for fossil fuels (Lehmann 2007). Numerous publications on biochar applications are being published, which suggests that the scientific community is becoming more interested in biochar research. This growing interest in biochar has led to the emergence of diverse fields of engineering study. Numerous research on biochar has emphasized its significance and advantages for reducing greenhouse gas emissions, improving soil quality, increasing agricultural yields, and storing carbon (Lehmann 2007). The use of biochar in removing anionic contaminants from soil, water, and other aqueous solutions is of great interest. The unique characteristic of biochar include large surface area and porous structure. The surface of the biochar has functional groups which helps in efficient adsorption. Due to these characteristics biochar may be used as an effective and efficient adsorbent to remove contaminants from aqueous solutions, similarly as activated charcoal, which is the most widely used and highly effective sorbent for the removal of various pollutants globally. In comparison with activated carbon, biochar is a more affordable adsorbent and has a porous structure. The porosity and nanostructure of biochar promotes aeration and soil water retention capacity while lowering evapotranspiration. By turning invasive plants into biochar, can enhance the control of invasive species and safeguard the environment. In order to have a "win–win" approach, turning biomass into biochar as a sorbent offers better waste management and preserves the environment. Biochar is a viable resource for environmental technology which is widely used to remove water pollution since it has both economic and environmental advantages as a renewable resource. The majority of research has found that biochar has a good capacity to remove contaminants from aqueous solutions, including organic pollutants, heavy metals, and other pollutants which gets adsorbed on its surface. Compared to commercially available activated carbon, several biochars have higher adsorption capacities (Meyer et al. 2011). As a slow-release fertilizer to increase soil fertility, biochar that has been mixed with phosphate PO_4^{-3}, ammonium NH_4^+, and nitrate NO_3^- has also been recommended. This is because after adsorption, it may contain a wealth of crucial nutrients. With the development of technology,

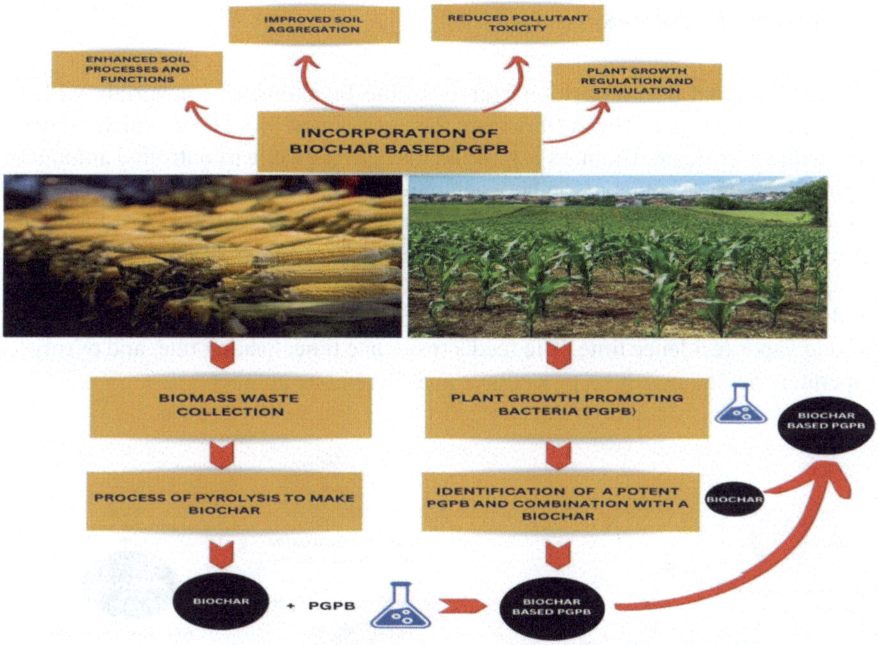

Fig. 1 Predominantly explaining process of formation of biochar from waste materials and plant debris called biomass to enhance the soil quality for a good yield

biochar has become increasingly used in the fields of environment, business, technology, and agriculture (Lima et al. 2010).The main goal of this study is to highlight the uses of biochar for removing anionic pollutants from an aqueous medium. The manufacturing and characteristics of biochar are also briefly summarized in this chapter.

2 Derivation and Properties of the Biochar Sorbent—A Promising Approach Toward Finest Adsorbents

Biochar is a porous, carbon-rich substance that demonstrates a variety of chemical and physical properties with oxygen functional groups and aromatic groups present on its surfaces. Biochar is created using a variety of thermochemical processes from feedstocks. These methods include torrefaction, gasification, hydrothermal carbonization, flash carbonization, slow pyrolysis, and quick pyrolysis. Wood biomass, animal waste, crop leftovers, and solid wastes are a few types of feedstocks. (Brewer et al. 2009).

3 Biomass Pyrolysis

The term "thermal decomposition" refers to the breakdown of materials without the presence of oxygen. This differs from the gasification process, in which biomass reacts with air or steam. Biomass is gasified into syngas while a controlled amount of oxygen is present. A comparison between pyrolysis and carbonization, the principal byproduct of which is charcoal, is made in a number of ancient works of literature. However, pyrolysis now refers to a procedure where the desirable end products are liquids (bio-oils) Fig. 2. The quick pyrolysis method favors the generation of liquids, whereas the slow pyrolysis method depends on the temperature, heating rate, and vapor residence time. The feed's residence time, heating rate, and pyrolysis temperature set the pyrolysis procedures apart from one another.

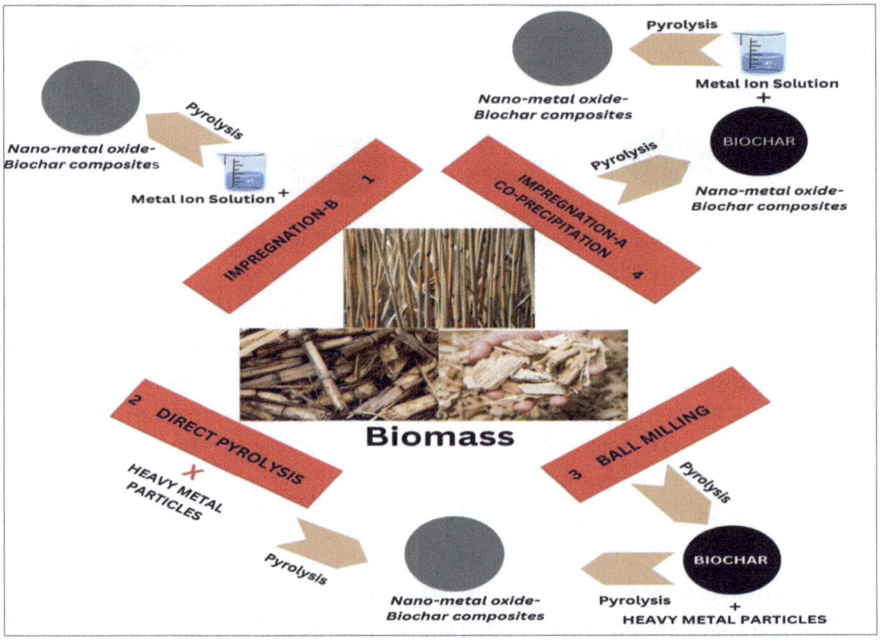

Fig. 2 Importance of pyrolysis in the process of biochar formation. Fast and slow pyrolysis plays a significant role in producing Biochar of different porosity, nature and composition. The different methods; Impregnation-B, Direct pyrolysis, Ball milling and Impregnation-A Co-precipitation are followed by pyrolysis

3.1 Slow Pyrolysis

Charcoal has been made through slow pyrolysis for a number of years. It has been looked into how charcoal is made and its qualities throughout the past 38 millennia (Antal and Grønli 2003). This method gradually raises the temperature of the biomass to about 500 °C in the absence of air, whereas vapor residency could remain for anywhere between 5 and 30 minutes. Vapors do not readily escape during slow pyrolysis, in contrast to fast pyrolysis.

3.2 Fast Pyrolysis

Fast pyrolysis requires high temperature by heating tiny biomass particles (1–2 mm) to 400–500 °C and vapor residence durations of 1 s (maximum 5 s). Dry feedstock (less than 10% moisture) is also required in fast pyrolysis (Lima et al. 2010). Fundamentally different is rapid pyrolysis.

3.3 Gasification

A gas fuel created by the gasification process can be used to provide a lot of power or direct heat. It is known as gasification to partially burn substances. The temperature, particle size, residence time, pressure, gas composition, and, if a catalyst is used, the catalyst's identity, as well as the product mix, are all adjusted during this process (gas, liquid, and solid). Switchgrass and maize stover were quickly pyrolyzed and gasified to produce biochar, which Brewer et al. described. Compared to gasification or rapid pyrolysis, slow pyrolysis generated more aromatic char. The diameters of the fused aromatic ring compounds in the fast and slow pyrolysis chars (7–8 rings per compound) were comparable, but the gasification char was denser (17 rings per compound) (Libra et al. 2011).

4 Hydrothermal Carbonization (HTC)

It is crucial to distinguish between hydrochars and biochars since they differ from one another. A char-water-slurry is created when biomass is hydrothermally carbonized (HTC) at a high temperature and pressure in water (Libra et al. 2011). From the original biomass, the chemical and physical characteristics have considerably changed. Additionally, the solid char is simple to detach. In contrast, different pressures, temperatures, and carbonization times are used to create biochar through gasification or pyrolysis (slow or rapid). Agricultural waste was converted into hydrochar, which was used to purify soil and water. Regarding their chemical composition, biochars made through pyrolysis or gasification were contrasted with hydrochars from HTC. Hydrochar is less stable than biochar (alkyl moieties predominate) (and dominated by aromatics). Hydrochars are excluded from the "European Biochar Certificate" (EBC) due to their unusual chemical characteristics (Wiedner et al. 2013).

5 Torrefaction and Rectification

Between 230 and 250 °C, "rectification" takes place in wood. A reddish-brown or chocolate hue develops in the wood. This product is thought to be immune to biological attack. It occurs between 250 and 280 °C at moderate heating rates, producing a brown or black material with little strength. Energy yields range from 77 to 90%, whereas torrefied wood yields range from 67 to 84%. Furthermore, torrefaction promotes the commercial use of wood for energy generation by lowering the cost of transportation, increasing biomass energy density, significantly reducing weight, and enhancing hydrophobic character. Torrefaction and rectification normally do not result in adsorbent chars since only a small portion of the biomass is broken down to prevent rot and cause minor water loss.

6 Applications

The porous and large surface area of biochar has been a promising tool for the high efficiency removal of toxic materials from the water for the wastewater treatment. There has been several reports published on biochar adsorption properties for several contaminants such as heavy metals, organic and inorganic pollutants. Following below are the applications of biochar.

6.1 Adsorption Isotherms

Water contamination by industry, antrhopogenic activities, toxic metals, and other sources has become a growing worry for the sustainability of the resource available globally. Aluminum (Al), cadmium (Cd), arsenic (As), nickel, chromium (Cr), copper (Cu), lead (Pb), mercury (Hg), uranium (Ur), and zinc are the most prevalent pollutants (Zn). Adsorption isotherm, which characterizes the interaction between the adsorbing substance and the adsorbent, is a crucial component in maximizing the utilization of adsorbents. Several empirical ideas describe how heavy metals interact on the surface of the widely used biochar. These theories are Langmiur theroy, Freundlich theory, Langmiur-Freundlich, and Temkin equations.

6.2 Adsorption Kinetics

The interaction between the adsorption kinetics, which affects the adsorption mechanism, and the physical and chemical properties of biochar can be used to illustrate these aspects in great detail. It gives a clearer picture of the method that biochar uses to remove various types of contaminants. The three most popular models of adsorption kinetics are the intra-particle diffusion model, the pseudo-first-order kinetic model, and the pseudo-second-order kinetic model. Among these models the most widely used models for studying the adsorption of contaminants on the biochar are pseudo-first-order kinetics and pseudo-second-order kinetics.

6.3 Presence of Anionic Contaminants in Water/Wastewater

Due to anthropogenic activities like industry, agriculture, etc., that took place throughout the previous century, water contamination is currently one of the biggest global challenges. It has been stated that human activities emit more than 10 million tons of trash annually, particularly in the form of chemicals (Ali and Gupta 2006). The majority of anionic pollutants are anthropogenic which are easily dissolved and ionized in water with a negative charge. Some of these toxins, like fluoride, reach water systems from geological contexts. Anions are necessary for organisms to continue growing, but their concentrations must be kept within a particular range to prevent them from becoming toxic or hazardous (like CN) or waste products from home use (like ClO^- and anionic surfactants) (Fang et al. 2017). Even at low concentrations, anionic pollutants that are released along with household and industrial waste can be harmful to organisms and the environment. For example, an excessive amount of fluoride in the drinking water supply has detrimental impacts on human health both immediately and over time (Steenbergen et al. 2011). Ingesting perchlorate could lead to neurological problems or a decline in IQ (Ginsberg et al. 2007). Because of the production of carcinogenic amines during its metabolic conversion and breakdown, the anionic dye Congo Red is dangerous to humans (Sponza and Işik 2005). Cr(VI), which occurs as an anion, is very damaging to living beings because of its oxidizing properties and ability to puncture biological membranes (Mearns et al. 1976). Cancer and other fatal diseases could develop as a result (Wiedner et al. 2013; Ali and Gupta 2006). Several nations and organizations have established stringent regulations to stop any potential negative effects brought on by anion exposure. The World Health Organization (WHO) suggested a level of F^{-1} in drinking water of 1.5 mg L^{-1} for reassurance (World Health Organization 2006), at which point dental and skeletal fluorosis may manifest (Harrison 2005). As P and N are necessary elements for organisms, certain anions, such as PO_4^{3-} and NO_3^-, can be eliminated by employing biological processes in conventional wastewater treatment systems. Even at incredibly low concentrations, the majority of anions can infiltrate the food chain, therefore endangering the safety of humans and the environment. This is because they are difficult to clean and remove from water. Researchers and engineers have employed physical, chemical, and biological procedures to successfully remove anions from water.

Recently, researchers have focused more on biochar, a stable carbon-rich byproduct produced by carbonizing biomass in an oxygen-restricted system. As it has a high sorption capacity and is cheaper to produce, biochar has been recognized as an important material for environmental applications with several functions (Lehmann and Joseph 2015). It is a great option for adsorbing cation organics and heavy metals because of its negatively charged surface. Additionally, it is widely claimed that biochar can adsorb or partition non-ionic organics (Ahmad et al. 2014). Historically, the anionic pollutants' sorption in water using biochar was viewed as unpromising because of the potential static repulsion between the anionic biochar surface and anions. Recent findings on the anionic sorption of pollutants using biochar or modified/engineered biochar, however, have been more encouraging.

6.4 Sorption Properties of Water

The ability of biochar to sorb anions, like nutrients, heavy metals, etc., on its surface has been investigated recently. The paragraphs that follow discuss these.

6.4.1 Negatively Charged Nutrients in Water

Increased levels of naturally occurring anionic nutrients like phosphorus (P) and nitrogen (N) are a major factor in the eutrophication of water bodies such as rivers, lakes, and even seas. Conventionally, excess nutrients have been removed from water bodies via natural biophysical or biochemical processes since they are economical and environmentally friendly. However, these processes take a long time, especially in water or waste that contains a lot of anionic nutrients. These impurities must be eliminated using procedures such as ion exchange, reverse osmosis treatment, adsorption, and other physical and chemical processes for P and N removal (Chatterjee and Woo 2009). The most prevalent anionic forms of phosphorus (PO_4^{3-}) and nitrate (NO^{3-}) in water are biochars made from various biomasses have recently been studied in order to test their ability to adsorb these contaminants found in water. A few of these studies are discussed in the sections given below.

Phosphate (PO_4^{3-})

Biochar derived from macroalgae roots when pyrolyzed at different temperatures can adsorb P between 12.97 and 18.21 mg g^{-1}, at the pyrolysis temperature range of 400° to 800 °C. However, phosphate sorption by raw macroalgae root and the biochar thus produced at 200 °C was not observed. The phosphate sorption of biochar derived from orange peel, pyrolyzed at 250° and 400 °C, was observed and found to have very weak phosphate sorption (Chen et al. 2011). However, after modifications using iron-Fe, the orange peel biochar's adsorption capabilities significantly got improved. This improvement was due to the presence of nano-sized magnetite particles in the orange peel biochar that enhances the phosphate adsorption (Chen et al. 2011). Similarly, it was reported that an $AlCl_3$-modified biochar nanocomposite (biochar/AlOOH) could adsorb phosphate effectively (Zhang and Gao 2013).

Nitrate (NO_3^-)

The sorption properties of wheat straw, mustard straw, and bamboo stalks were assessed while they were utilized to make biochar at a regulated temperature. According to reports, bamboo charcoal has a Qm of 1.25 mg g^{-1}, which is 15% greater than commercial AC (CAC, 1.09 mg g^{-1}) (Mizuta et al. 2004). After analysis, it was discovered that the Qm of wheat straw biochar, mustard straw biochar, and CAC were, respectively, 1.10, 1.30, and 1.22 mg g^{-1} (Mishra and Patel 2009). A significant increase in nitrate sorption can also result from the modification of biochar using metal ions. For instance, oak sawdust biochar demonstrated noticeably improved nitrate removal following treatment with lanthanum (La) (Wang et al. 2015). Biochar modification was also performed using $MgCl_2$and $ZnCl_2$, and their sorption of nitrate was reported to be several folds higher than that of unmodified biochar.

6.5 Heavy Metals in Water

Heavy metals in their anionic state are frequently found in nature and include chromium and arsenic. As a result, they cannot be removed from water or trash using the same method as other heavy metals, which are cations. The following excerpts address the active research being done on the sorption activity of these anionic heavy metals using biochar.

6.5.1 Hexavalent Chromium

Hexavalent dichromate [$Cr_2O_7^{2-}$ Cr (VI)] is the anionic form of chromium. Biochar was made using ramie residue (Zhou et al. 2016), sugar beet tailings (Dong et al. 2011), rice straw (Hsu et al. 2009), coconut fiber and shells (Mohan et al. 2005), rubber wood sawdust (Karthikeyan et al. 2005), and waste tires (Hamadi et al. 2001). The maximum Cr(VI) sorption capacity was found in ramie residue biochar produced at 300 °C, with a capacity of 82.32 mg g^{-1} (Yoon et al. 2017), whereas sugar beet tailing biochar had a capacity of 123 mg g^{-1}. (Niazi et al. 2018a). At 30°, 40°, and

50 °C, Q_m for rubber wood sawdust biochar was 44.05, 59.17, and 65.78 mg g^{-1}, respectively (Hamadi et al. 2001).

6.5.2 Arsenic

Arsenic has two anionic forms that are commonly present in water which are arsenate [As(V)] and arsenite [As (III)]. Removal of these using various biomass-derived biochars has been reviewed. For instance, sludge from a paper mill was effectively used to produce biochar in a CO_2 environment, and it was found to have the maximum arsenate sorption capacity of 22.76 mg g^{-1} (Yoon et al. 2017). As the coagulant utilized throughout the paper mill wastewater treatment process was $FeSO_4$, the biochar hence produced had magnetic properties making easy to separate it after undergoing the sorption process (Yoon et al. 2017). The sorption capabilities of Japanese oak wood and perilla leaf derived for arsenate and arsenite were also investigated, and it was discovered that Q_m for biochar from the Japanese oak wood was 3.16 for arsenate and 3.89 mg g^{-1} for arsenate whereas Q_m for biochar from perilla leaf was 7.21 for arsenate and 11.01 mg g^{-1} for arsenite (Niazi et al. 2018a, b).

6.6 *The Anionic Contaminants Present in Water*

It has also been examined if biochars and/or modified biochars can sorb additional anionic pollutants such as fluoride (F^-), perchlorate (ClO_4^-), and bromate (BrO_3^-) (Chatterjee and Woo 2009). Although fluoride is a necessary element, high levels of fluoride can be harmful to human health (Halder et al. 2016). Surface and groundwater are used to spread fluoride, which is mostly created by human activity and natural processes. One of the most difficult ways for eliminating fluoride from water is sorption. It was discovered that the biochar made from Cocos nucifera shells had a removal rate of 83.04% for a synthetic solution with a fluoride content of 10 mg L^{-1} compared to a removal rate of 79.3% for a genuine sample with 7 mg L^{-1} of fluoride (Halder et al. 2016). The US EPA has begun to regulate perchlorate (Kounaves et al. 2010), a recently identified pollutant because it interferes with the thyroid gland's capacity to control hormones and adsorb iodine in humans (Fang et al. 2014). It has been reported that high-temperature (700 °C) biochar made from wood can successfully adsorb perchlorate, with a Q_m of 10.55 mg g^{-1} (Fang et al. 2014). Because many colors used in the textile business are poisonous, mutagenic, or carcinogenic, there have been significant issues with the environment and human health (Yan et al. 2015). Some of these dyes include azo compounds, which are challenging to degrade using conventional techniques for wastewater treatment. As an illustration, the capacity of sorption of Congo Red using vermicompost-derived biochar ranges from 11.63 to 31.28 mg g^{-1} at various pyrolysis temperatures, demonstrating a considerable increase in sorption with production temperature (Yan et al. 2015). In a study about the Congo Red sorption by Fe_3O_4-modified biochar made from cellulose, it was reported that the biochar's Q_m was 66.09 mg g^{-1}, which was in line with earlier studies in the literature (Zhu et al. 2011).

7 Anionic Contaminant Sorption On Biochar

7.1 Filling of Pores on the Surface of Biochar

Macro-, meso-, and micro-pores are created on the biochar's surface as a consequence of the pyrolysis of the organic parts of the biomass. Generally, the surface area of biochar rises as the temperature of pyrolysis goes up, and its sorption capacity grows as its surface area grows (Fig. 3). It was found that raising the pyrolysis temperature from 200° to 400 °C significantly increased the Brunauer–Emmett–Teller (BET) surface area and total pore volume of biochar made from the roots of marine macroalgae, which in turn increased the sorption of phosphate. (Jung et al. 2015). Pore block caused the surface area and pore volume to substantially reduce when the pyrolytic temperature increased to 600° and 800 °C Phosphate sorption on the biochar accordingly declined as a result. The phosphate adsorption onto the surface of biochar in this study had a linear connection to the pore volume and BET surface area, thus demonstrating that the pore structure of the biochar plays a substantial role in phosphate sorption. Metal-based biochar modification can improve sorption capacity and pore structure. Experiments into the MgO-biochar composite's adsorption of anions (PO_4^{3-} and NO_3^-) indicated that a nanoporous structure had developed on the modified biochar composite, greatly expanding its surface area and adsorption efficiency. This also makes biochar more porous (Zhang et al. 2012).

7.2 Hydrogen Bonding

Hydrogen bonding forms the main mechanism through which anionic contaminants, such as perchlorate adsorb on the surface of a biochar (Fang et al. 2014) (Fig. 4). The hydrogen of these oxygen-containing groups were ionized with a further rise in pH in the solution, and the biochar's anionic surface adsorbed almost no perchlorate. The fundamental process is the formation of hydrogen bonds between oxygen atoms of perchlorate and the hydrogen of oxygen-containing functional groups that are present on the surfaces of the biochar.

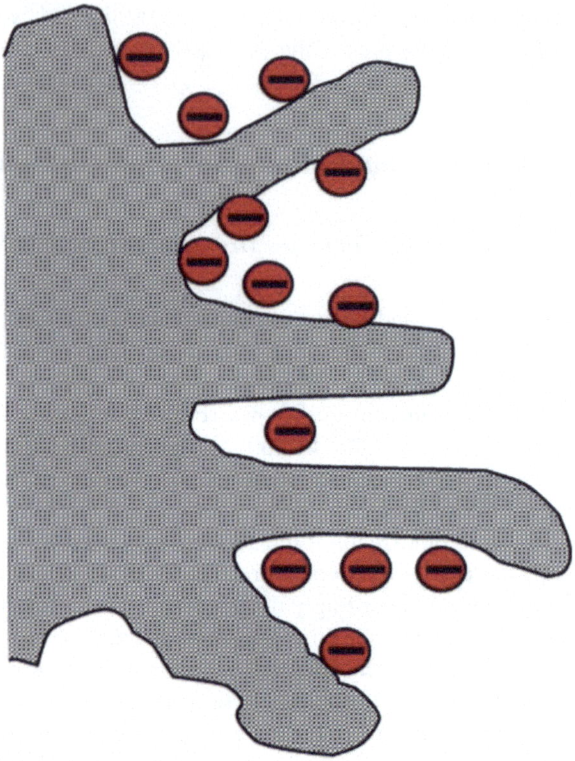

Fig. 3 Anionic pollutants' capacity to fill biochar's micro- or meso-pores

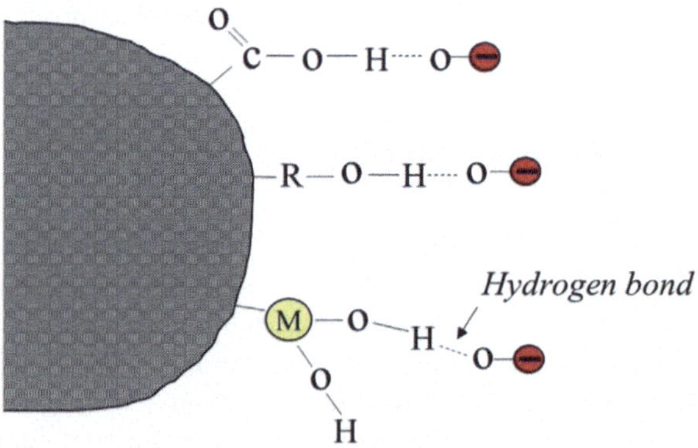

Fig. 4 H-bond formation occurs between an H-containing functional group present on the surface of biochar and an O-containing group found in the anionic pollutants

7.3 Complex Formation/Precipitation on Biochar Surface

As an illustration, Halder et al. (2016) used X-ray diffraction (XRD) to show the formation of fluoride-metal complex on the surface of biochar (Fig. 5). They also found that when fluoride was adsorbed by the biochar, the minerals potassium iron fluoride ($KFeF_3$) and K_2MgF_4 were created. The presence of reactive components in biomass, such as potassium, iron, and magnesium, facilitated the formation of chemical compounds with F-. When $ZnCl_2$-modified biochar made from pellets of coconut was tested for its ability to bind with NO_3^-, the findings of Bhatnagar et al. (2008) revealed that even though the modified carbon's surface area was less than that of carbon with no modifications, the ZnO formation in the macro- and meso-pores of carbon made the sorption of NO_3^- much more effective. As a result, NO_3^- was better sorbed on the biochar surface with modifications and an active site for adsorption (Bhatnagar et al. 2008). Biochar, derived from orange peels, was pyrolyzed at temperatures of 250° and 400 °C exhibited low phosphate sorption (Chen et al. 2011). The phosphate adsorption by magnetic orange peel biochar, however, considerably increased following Fe alteration. Biochar synthesized from orange peels contains nano-sized magnetite particles that help phosphate adsorption (Chen et al. 2011).

Fig. 5 Anionic pollutants on the surface of biochar complex with metals or metallic oxides

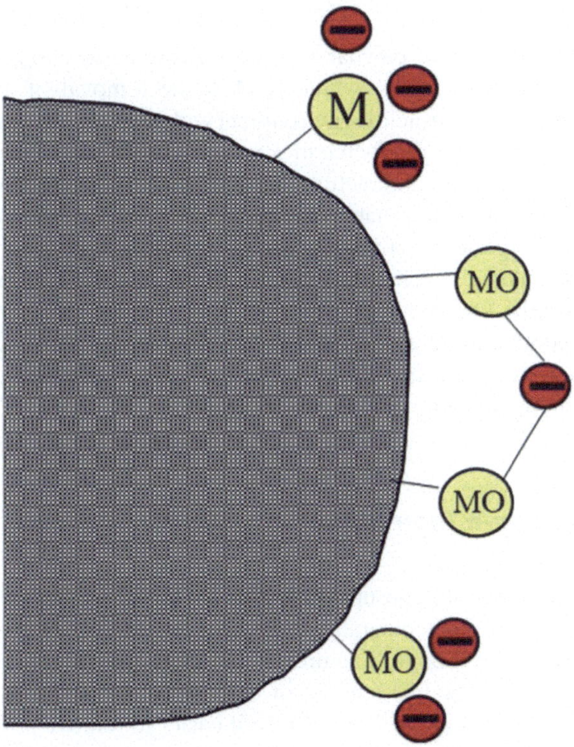

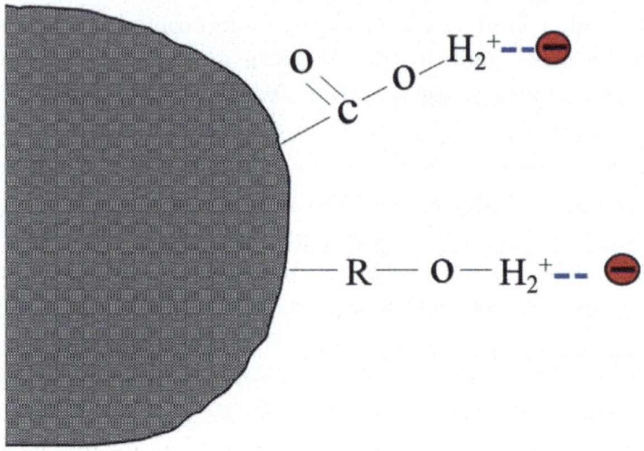

Fig. 6 Polar interactions between protonated biochar surface and various anionic contaminants

7.4 Polar Interactions

The surface of biochar is widely assumed to carry a negative charge, especially in alkali solutions, which renders the removal of anionic impurities unfeasible. However, in an acidic environment with high H^+ levels, the oxygen-containing groups on the surface of biochar pick up a proton. The resulting positive charge causes polar or electrostatic interactions between the protonated biochar surface and the anions in solutions, which enables the anions to be sorbed by the biochar. According to reports, the sorption of perchlorate and protonation of the functional groups on the surface of biochar are much higher in acidic environments compared to alkaline environments. Similar findings in the sorption of the anionic dye Congo Red by biochar were also reported by Zhu et al. (2011) and Yang et al. (2016). Lower pH levels showed more sorption, and the biochar surface had a positive charge and electrostatic properties interaction with anionic dye (Fig. 6).

7.5 π-π Interaction

The functional groups with H and O, i.e., polar functional groups, reduced when the pyrolysis temperature rose, and the number of aromatic structures increased as well. The aromatic nature of pollutants can act as an electron acceptor, whereas the graphene-like surface of charcoal can operate as an electron donor. Sorbents firmly bind impurities to make it easier to remove pollutants that contain aromatics by utilizing this electron donor–acceptor interaction. The mole ratio of hydrogen to carbon atoms (H/C ratio) decreased as the pyrolysis temperature was elevated, according to Yang et al. (2016), indicating that the biochar's aromaticity had increased. As a result of the aromatic structure of the biochar surface, it is possible to develop an EDA interaction with aromatic groups containing Congo Red, boosting

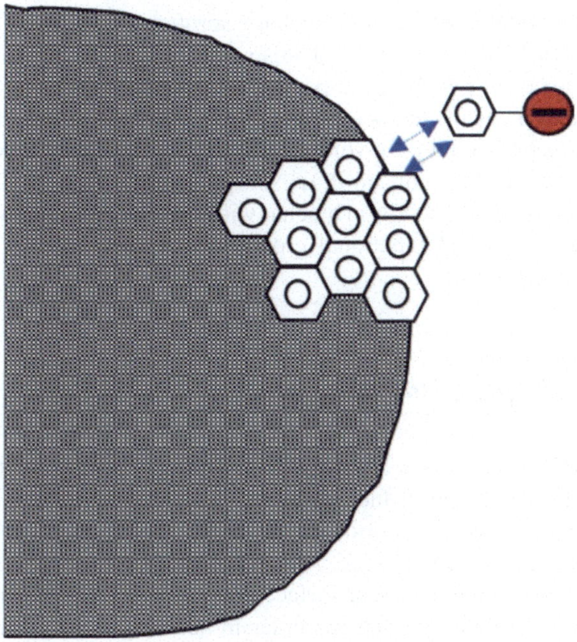

Fig. 7 π-π EDA interaction involving an aromatic group and the graphene-like structure of biochar surface

the sorption capacity (Fig. 7). However, with the sorption of Methylene Blue in the same investigation, no comparable result was seen (Yang et al. 2016).

8 Factors Affecting the Anionic Contaminants' Sorption

8.1 Temperature of Pyrolysis

Polysaccharides like cellulose, hemicellulose, and lignin comprise the bulk of the biomass utilized to create biochar, with trace amounts of lipids and proteins also involved. For hemicellulose and cellulose temperature rate for decomposition is between 250–350 °C and 325–400 °C, respectively. Lignin is the more stable component and disintegrates between 300° and 550 °C (Jung et al. 2016). The organic component content of various biomass varies, resulting in diverse biochar properties. However, the overall trends in the properties of biochar are virtually comparable as the pyrolysis temperature rises: decreased polar functional groups (H-, O-, and N-containing groups) and increasing aromaticity. Using sugar beet bagasse biochar treated with $ZnCl_2$, a study on the sorption of NO_3^- found that biochar produced at higher temperatures (700 °C) has the highest sorption capacity due to its larger pore volume and surface area than biochar produced at lower temperatures (Demiral and Gunduzoglu 2010). As the pyrolytic temperature is raised from 300° to 700 °C, the vermicompost-derived biochar's sorption capability of Congo Red improved noticeably. The important mechanism was thought to be the enhanced aromaticity due to increased pyrolytic temperature. However, the inverse behavior has been recorded. By raising the temperature of pyrolysis, it was shown that waste-marine macroalgae biochar's total pore volume and surface area decreased because blocking the pores considerably reduced PO_4^{3-} sorption. According to Zhou et al., pyrolysis at lower temperatures results in biochar with greater Cr(V) sorption (2016) caused by carboxyl and hydroxyl functional groups which enhanced Cr(V) sorption at low pyrolytic temperatures.

8.2 pH of the Solution

HF weakly ionized created by strongly acidic solutions, this prevented F-from sticking to the surface of the biochar. Around pH 6.25, the highest level of sorption was observed. Fluoride and hydroxyl ions fought for sorption sites on the charcoal surface as the pH rose. At lower pH environment, the functional groups that have oxygen on the biochar surface get protonated. This generated interactions of biochar with perchlorate that are polar in nature and increasing the solution's pH lowered perchlorate sorption. With increasing pH, the surface of biochar became neutral, and hydrogen bonding significantly aided perchlorate sorption. Electrostatic repulsion between the anions and the negatively charged surface of the biochar prevented practically all sorption when the pH increased. A lower pH increases the sorption capacity due to electrostatic interaction between the protonated biochar surface and NO_3^-, which was also demonstrated to have an impact on the NO_3^- sorption by biochar generated from maize stover. Strongly acidic solutions, however, produced weakly ionized HF, which prevented F-from adhering to the surface of the biochar. The highest level of sorption was noticed around pH 6.25. As the pH increased, fluoride and hydroxyl ions competed for sorption sites on the biochar surface.

8.3 Coexisting Ions

The equilibrium of the adsorption capacity gets greatly influenced by co-existed ions. Contaminants do not exists independently due to the complex nature of wastewater or natural water posing an interaction effect on adsorption efficiency. The studies of adsorption application of biochar provides a better insight regarding the adsorption mechanism of contaminants in the environment. When assessing the sorption capacity of biochars, coexisting ions, particularly those with similar characteristics, should be taken into account. Pakade et al. (2019) identified that in the presence of ions that co-exists, quaternized AC's ability to remove Cr(VI) decreased, with the major explanation being a competition of carbon surfaces with positive charges with other anions such as SO_4^{2-}, PO_4^{3-}, NO_3^-, and $Cr_2O_7^{2-}$. The reduction of Cr(VI) adsorption was also significantly influenced by the size of the anions [65]. Additionally, it was demonstrated that the presence of strong anions like PO_4^{3-} and SO_4^{2-} in an aqueous solution had a substantial impact on the NO_3-sorption by biochar generated from maize stover.

8.4 Temperature

A sorbate's sorption on a sorbent is a process that is related to energy and includes both endothermic and exothermic reactions. As a result, changes in environmental temperature may have a considerable impact on sorption performance (Jung et al. 2015). Through an endothermic process, the pollutants are adsorbed onto the surface of a biochar, and as the temperature rises, so does the process' adsorption capacity. The maximal phosphate adsorption capacities were determined to be at 335, 480, and 727 mg g-, respectively. This research examined the phosphate's adsorption by biochar/Mg–Al constructed nanocomposites at the temperature range of 10–30 °C to identify temperature's influence on anion's sorption by biochar. This implies that phosphate adsorption on biochar is an endothermic process. The endothermic nature of the process of Cr(VI) adsorption by worn-out tire biochar and sawdust biochar possessing ΔG negative and ΔH positive values was established by the thermodynamic characteristics found by Hamadi et al. (2001). The studies reported favorable adsorption of lead ions on the biochar surface derived from pinewood and rice husk with the absolute increase in the values of ΔG due to increasing temperature.

8.5 Dosage of Adsorbent

Dosage adsorbent also influences the adsorption efficiency of a biochar in which appropriate dosage of a biochar derived from organic material adsorbs contaminants for its effective removal from the environment. It was seen that the increasing concentration of the adsorbent (biochar) the adsorption rate increased significantly. This is due to the fact that increased concentration of adsorbent provided a greater surface area for the adsorbate to get adsorbed and smoothly removed from the heterogeneous mixture. The increase in the surface area can also be attributed to the increased in the number of the functional groups and adsorbent lines on the biochar surface. Whereas a subsequent decrease in the adsorption efficiency was seen in the swine-derived biochar due to increase in the mass. This suggests that even the proper selection of a biochar for various applications and procedure is a crucial step which needs to be properly taken care of.

9 Conclusions and Perspectives

Heavy metal, anionic, and non-ionic pollutants can all be removed from wastewater using biochar as a suitable sorbent. Electrostatic interactions, hydrogen bonds, the pore-filling effect, surface complexation or precipitation, and EDA interactions are the principal mechanisms by which biochar adsorbs anionic pollutants. Environmental factors such the H^+ level, surrounding ions, temperature, and the biochar's pyrolysis temperature have a significant impact on the adsorption process. Comparatively to the adsorption of contaminants of an organic and inorganic (heavy metal) type, biochar is only moderately effective in removing the majority of anionic pollutants. In most cases, biochars can only adsorb up to 20 mg/g of anionic pollutants. Therefore, new methods are required to boost biochar's ability to absorb anionic pollutants. The biochars may also have N and P, in anionic forms like NO_3^- and PO_4^{3-}, as they are extremely common in biomass and may be formed during pyrolysis. Therefore, the concentrations of the anions present in the wastewater must be found and determined prior to the application of biochar for treating the wastewater.

References

Ahmad M, Rajapaksha AU, Lim JE, Zhang M, Bolan N, Mohan D et al (2014) Biochar as a sorbent for contaminant management in soil and water: a review. Chemosphere 99:19–33

Ali I, Gupta VK (2006) Advances in water treatment by adsorption technology. Nat Protoc 1(6):2661–2667

Antal MJ, Grønli M (2003) The art, science, and technology of charcoal production. Ind Eng Chem Res 42:1619–1640

Bhatnagar A, Ji M, Choi YH, Jung W, Lee SH, Kim SJ et al (2008) Removal of nitrate from water by adsorption onto zinc chloride treated activated carbon. Sep Sci Technol 43(4):886–907

Brewer CE, Schmidt-Rohr K, Satrio JA, Brown RC (2009) Characterization of biochar from fast pyrolysis and gasification systems. Environ Prog Sustain Energy 28(3):386–396

Chatterjee S, Woo SH (2009) The removal of nitrate from aqueous solutions by chitosan hydrogel beads. J Hazard Mater 164(2–3):1012–1018

Chen BL, Chen ZM, Lv SF (2011) A novel magnetic biochar efficiently sorbs organic pollutants and phosphate. Bioresour Technol 102(2):716–723

Demiral H, Gunduzoglu G (2010) Removal of nitrate from aqueous solutions by activated carbon prepared from sugar beet bagasse. Bioresour Technol 101(6):1675–1680

Dong XL, Ma LNQ, Li YC (2011) Characteristics and mechanisms of hexavalent chromium removal by bio-char from sugar beet tailing. J Hazard Mater 190(1–3):909–915

Fang QL, Chen BL, Lin YJ, Guan YT (2014) Aromatic and hydrophobic surfaces of wood-derived biochar enhance perchlorate adsorption via hydrogen bonding to oxygen-containing organic groups. Environ Sci Technol 48(1):279–288

Fang C, Dharmarajan R, Megharaj M, Naidu R (2017) Gold nanoparticle-based optical sensors for selected anionic contaminants. TrAC. Trends Anal Chem 86:143–154

Ginsberg GL, Hattis DB, Zoeller RT, Rice DC (2007) Evaluation of the U.S. EPA/OSWER preliminary remediation goal for perchlorate in groundwater: focus on exposure to nursing infants. Environ Health Perspect 115(3):361–369

Halder G, Khan AA, Dhawane S (2016) Fluoride sorption onto a steam-activated biochar derived from Cocos nucifera shell. Clean: Soil, Air, Water 44(2)

Hamadi NK, Chen XD, Farid M, Lu MGQ (2001) Adsorption kinetics for the removal of chromium(VI) from aqueous solution by adsorbents derived from used tyres and sawdust. Chem Eng J 84:95–105

Harrison PTC (2005) Fluoride in water: a UK perspective. J Fluor Chem 126:1448–1456

Hsu NH, Wang SL, Liao YH, Huang ST, Tzou YM, Huang YM (2009) Removal of hexavalent chromium from acidic aqueous solutions using rice straw-derived carbon. J Hazard Mater 171(1–3):1066–1070

Jung KW, Hwang MJ, Jeong TU, Ahn KH (2015) A novel approach for preparation of modified-biochar derived from marine macroalgae: dual purpose electro-modification for improvement of surface area and metal impregnation. Bioresour Technol 191:342–345

Jung KW, Kim K, Jeong TU, Ahn KH (2016) Influence of pyrolysis temperature on characteristics and phosphate adsorption capability of biochar derived from waste-marine macroalgae (Undaria pinnatifida roots). Bioresour Technol 200:1024–1028

Karthikeyan T, Rajgopal S, Miranda LR (2005) Chromium(VI) adsorption from aqueous solution by Hevea Brasilinesis sawdust activated carbon. J Hazard Mater 124(1–3):192–199

Kounaves SP, Stroble ST, Anderson RM, Moore Q, Catling DC, Douglas S et al (2010) Discovery of natural perchlorate in the antarctic dry valleys and its global implications. Environ Sci Technol 44(7):2360–2364

Lehmann J (2007) A handful of carbon. Nature 447:143–144

Lehmann J, Joseph S (2015) Biochar for environmental management: science and technology, 1st edn. Earthscan, London

Libra JA, Ro KS, Kammann C, Funke A, Berge ND, Neubauer Y, Titirici MM, Fühner C, Bens O, Kern J, Emmerich KH (2011) Hydrothermal carbonization of biomass residuals: a comparative review of the chemistry, processes and applications of wet and dry pyrolysis. Biofuels 2:71–106

Lima IM, Boateng AA, Klasson KT (2010) Physicochemical and adsorptive properties of fast-pyrolysis bio-chars and their steam activated counterparts. J Chem Technol Biotechnol 85(11):1515–1521

Mearns AJ, Oshida PS, Sherwood MJ, Young R, Reish DJ (1976) Chromium effects on coastal organisms. J Water Pollut Control Fed 48(8):1929–1939

Meyer S, Glaser B, Quicker P (2011) Technical, economical, and climate-related aspects of biochar production technologies: a literature review. Environ Sci Technol 45:9473–9483

Mishra PC, Patel RK (2009) Use of agricultural waste for the removal of nitrate-nitrogen from aqueous medium. J Environ Manage 90(1):519–522

Mizuta K, Matsumoto T, Hatate Y, Nishihara K, Nakanishi T (2004) Removal of nitrate-nitrogen from drinking water using bamboo powder charcoal. Bioresour Technol 95(3):255–257

Mohan DS, Singh KP, Singh VK (2005) Removal of hexavalent chromium from aqueous solution using low cost activated carbons derived from agricultural waste materials and activated carbon fabric cloth. Ind Eng Chem Res 44(4):1027–1042

Niazi NK, Bibi I, Shahid M, Ok YS, Burton ED, Wang H et al (2018a) Arsenic removal by perilla leaf biochar in aqueous solutions and groundwater: an integrated spectroscopic and microscopic examination. Environ Pollut 232:31–41

Niazi NK, Bibi I, Shahid M, Ok YS, Shaheen SM, Rinklebe J et al (2018b) Arsenic removal by Japanese oak wood biochar in aqueous solutions and well water: investigating arsenic fate using integrated spectroscopic and microscopic techniques. Sci Total Environ 621:1642–1651

Pakade VE, Tavengwa NT, Madikizela LM (2019) Recent advances in hexavalent chromium removal from aqueous solutions by adsorptive methods. RSC Adv 9(45):26142–26164

Sohi SP (2012) Carbon storage with benefits. Science 338:1034–1035

Sponza DT, Işik M (2005) Toxicity and intermediates of C.I. Direct Red 28 dye through sequential anaerobic/aerobic treatment. Process Biochem (20):2735—2744

Steenbergen F, Haimanot RT, Sideil A (2011) High fluoride, modest fluorosis: Investigation in drinking water supply in Halaba (SNNPR, Ethiopia). J Water Resour Protect 3:120–126

Wang ZH, Guo HY, Shen F, Yang G, Zhang YZ, Zeng YM et al (2015) Biochar produced from oak saw-dust by Lanthanum (La)-involved pyrolysis for adsorption of ammonium, nitrate and phosphate. Chemosphere 119:646–653

Wiedner K, Rumpel C, Steiner C, Pozzi A, Maas R, Glaser B (2013) Chemical evaluation of chars produced by thermochemical conversion (gasification, pyrolysis and hydrothermal carbonization) of agro-industrial biomass on a commercial scale. Biomass Bioenergy 59:264–278

World Health Organization (2006) Guidelines for drinking water quality, 3rd edn. World Health Organization

Yan LL, Kong L, Qu Z, Lo L, Shen GQ (2015) Magnetic biochar decorated with ZnS nanocrytals for Pb (II) removal. ACS Sustain Chem Eng 3(1):125–132

Yang G, Wu L, Xian QM, Shen F, Wu J, Zhang YZ (2016) Removal of Congo red and methylene blue from aqueous solutions by vermicompost-derived biochars. PLoS ONE 11(5):e0154562. Available from: https://doi.org/10.1371/journal.pone.0154562

Yoon K, Cho DW, Tsang DCW, Bolan N, Rinklebe J, Song H (2017) Fabrication of engineered biochar from paper mill sludge and its application into removal of arsenic and cadmium in acidic water. Bioresour Technol 246:69–75

Zhang M, Gao B (2013) Removal of arsenic, methylene blue, and phosphate by biochar/AlOOH nanocomposite. Chem Eng J 226:286–292

Zhang M, Gao B, Yao Y, Xue Y, Inyang M (2012) Synthesis of porous MgO-biochar nanocomposites for removal of phosphate and nitrate from aqueous solutions. Chem Eng J 210:26–32

Zhou L, Liu YG, Liu SB, Yin YC, Zeng GM, Tan XF et al (2016) Investigation of the adsorption-reduction mechanisms of hexavalent chromium by ramie biochars of different pyrolytic temperatures. Bioresour Technol 218:351–359

Zhu HY, Fu YQ, Jiang R, Jiang JH, Xiao L, Zeng GM et al (2011) Adsorption removal of Congo red onto magnetic cellulose/Fe_3O_4/activated carbon composite: equilibrium, kinetic and thermodynamic studies. Chem Eng J 173(2):494–502

Chapter 10
Decontamination of Inorganic and Organic Compounds from Wastewater Using Biochar

Vishwa Nayani, Anoushka Prabhu, Jaya Lakkakula, and Nilesh Shirish Wagh

Abstract Adsorption is the most preferred method for removing contaminants from water bodies out of all the available water treatment technologies. Numerous studies have been conducted recently to look at using biochar produced by pyrolysis to eliminate pollutants from solutions that are aqueous in nature. Given the abundance of feedstock, relatively inexpensiveness, and advantageous physical surface properties, biochar has considerable potential to effectively combat water contaminants. Biochar is a heterogenous carbon-containing char produced by thermally heating significant amounts of totally degraded organic waste biomass like crop residues, forest biomass, algae, and animal manures. Yet, there is a dearth of knowledge on the functions of various sorption mechanisms for various contaminants, and the latest developments in biochar modification to attain an effective adsorption capacity that is essential for the forthcoming prospects of biochar. This chapter aims to explain the several kinds of biochar that can be used to efficiently remove heavy metals, ions, organic compounds, and dyes from wastewater as well as the ideal conditions needed for the optimal removal of the contaminants. The objective is to give a detailed summary in the use of biochar as a cost-effective and sustainable adsorbent to get rid of harmful substances from aqueous environments and to draw attention to the need for further research in this intriguing area of bioremediation that could be used in wastewater treatment projects in the future.

V. Nayani · A. Prabhu · J. Lakkakula (✉) · N. S. Wagh
Amity Institute of Biotechnology, Amity University Maharashtra, Mumbai-Pune Expressway, Panvel, Mumbai 410206, Maharashtra, India
e-mail: spencerjaya@gmail.com

J. Lakkakula
Centre for Computational Biology and Translational Research, Amity Institute of Biotechnology, Amity University Maharashtra, Mumbai-Pune Expressway, Panvel, Mumbai 410206, Maharashtra, India

N. S. Wagh
Centre for Drug Discovery and Development, Amity Institute of Biotechnology, Amity University Maharashtra, Mumbai-Pune Expressway, Panvel, Mumbai 410206, Maharashtra, India

A. Roy et al. (eds.), *Biochar Production Engineering*,
https://doi.org/10.1007/978-3-031-92594-8_10

Keywords Biochar · Dyes · Heavy metals · Methylene blue · Organic pollutants

1 Introduction

Water is a source that all living matter depends on and is necessary for all socioeconomic development, and energy generation. 70% of Earth's surface is covered by water, but just 3% of it is freshwater, and only a tiny percentage of that sustains all land-based life. Eighty percent of the world's wastewater, which can range from highly toxic industrial discharges to human waste, is not treated (Shirvanimoghaddam et al. 2021). It is challenging, expensive, and nearly impossible to eliminate pollutants from the water after it has become contaminated. Freshwater's fitness for many human purposes, including drinking, bathing, and agriculture, depends on the types and concentrations of contaminants present (Liu et al. 2022). In a world where industrialization, mining, and the use of fertilizers and pesticides are rapidly increasing and growing at a global scale, especially in developing countries, it is inevitable that the concentration of a number of pollutants in wastewater will rise (Nguyen et al. 2021). Most of these contaminants which include heavy metals and inorganic ions, organic compounds (dyes, phenols, and polyaromatic hydrocarbons), and pharmaceutical waste such as antibiotics and various types of drugs are non-biodegradable due to their stable chemical and structural properties and hence persist in the water for an extended period, resulting in bioaccumulation (Nguyen et al. 2021). These pollutants are carcinogenic and growth inhibitory in nature. They also form stable intermediates which are toxic in nature and interact with other coexisting pollutants (Shoaib et al. 2024). One of the most urgent environmental problems is caused by heavy metals, which are generally dangerous and pose significant danger to human health and the environment (Ge et al. 2024). Various heavy metals are also known to interfere with the ion balance of various aquatic species and disrupt the optimum physical and chemical condition, posing a grave risk to human health as a result, aquatic species, as well as the bionetwork in general, making the need for a solution urgent (Nyamunda et al. 2019).

Several methods such as solvent extraction, chemical precipitation, ion exchange, and the activated carbon method have been implemented to eliminate these contaminants, however, the downside of these methods is that they are inefficient in the removal of the contaminant as well as require a high amount of energy or reagents and therefore becoming an expensive treatment method that discourages the industries from implementing them. For example, solvent extraction requires a huge amount of organic extractants, and the treatment also faces hydrodynamic constraints (Tian et al. 2022). Similarly, in the coagulation/flocculation method there is the uptake of pollutants followed by separation of product formed requires utilization of non-reusable chemical agents thereby generating huge volumes of sludge. Chemical precipitation which has the same characteristics as that of the coagulation technique is inefficient in the removal of metal ions and an additional oxidation step is required if the metals are too complex in nature. The ion-exchange method which is a non-destructive process is also not economically feasible due to the rapid saturation and clogging of reactors (Dalahmeh et al. 2020). Thus, there was a need for environment-friendly and inexpensive technology to reduce the bioavailability of these contaminants in wastewater. Therefore, a novel bio-adsorption technique was introduced, particularly the biochar method (Zoghlami et al. 2021).

Biochar is essentially a solid black stable carbonaceous material that is produced by the pyrolysis of waste materials like wood, manure, sludge, plant wastes, etc., with little to no air at high temperatures (300–800 °C) (Mansouri et al. 2021). Owing to its high specific surface area, porous structure, high carbon content, and water insolubility, it is one of the most effective methods for the adsorption of different effluents and pollutants in wastewater. Biochar exhibits a strong attraction for heavy metals due to its highly porous structure and a variety of functional groups such as hydroxyl, carboxyl, and phenolic groups. Additionally, they can also be easily modified and enhanced with various nanoparticles to add active functional groups on the surface (Song et al. 2021). They are highly effective in the adsorption of the most common effluents found in wastewater such as organic dyes, metal ions, antibiotics, phenols, polyaromatic hydrocarbons, and various inorganic ions. Recycling biomass, they reduce pollution and the overall cost of production (Shafiq et al. 2021). Furthermore, they do not harm the environment and are environmentally benign in both their manufacturing as well as their use. All these advantages make biochar an ideal and promising green technology used for wastewater treatment, and numerous studies have been conducted recently to examine the use of biochar for wastewater treatment (Yang et al. 2021).

In this chapter, we aimed to highlight the characteristics, preparation method, and efficiency of biochar for the adsorption of different contaminants from wastewater as discussed in Table 1. The sorption mechanisms, thermodynamics, kinetics, and ideal physical conditions are considered. We will also discuss the current drawbacks of this treatment method and future research opportunities to overcome them.

1.1 Removal of Organic Dyes Using Biochar

1.1.1 Removal of Methylene Blue and Methylene Orange

Methylene blue is one of the major effluents of the textile industry discharged in wastewater. It is non-biodegradable and carcinogenic in nature hence, its removal from wastewater is essential. Nabilah Suhaimi et al. prepared the biochar using *Gigantochloa* bamboo, which was washed, sun dried, oven dried at 80 °C and then underwent through pyrolysis at 500 °C. It was then grounded with pestle-mortar and passed through 150 μm mesh sieves. BET imaging showed the specific surface area to be 4.1519 m^2/g, total volume 0.015165 cm^3/g and a high average pore diameter of 14.61015 nm. SEM spectroscopy revealed the surface morphology to be distorted with irregular micro-structures. XRD technique suggested that before adsorption, it was crystalline due to KCl, but after adsorption, it was more amorphous in nature due

Table 1 Details of biochar, its characteristics, and ability to remove contaminants

Sr. No.	Types of biochar	Details of biochar	Characteristics of biochar	Contaminant removed	Technique used	References
1	Plant based	*Gigantochloa* Bamboo	Sorption capacity: 86.6 mg/g pH: 7–74.3% removal At 2 there is 80.6% removal	Methylene blue	Pyrolyzed at 500 °C	(Suhaimi et al. 2022)
2		Maple leaf (*Acer palmatum*)	Temp: 30 °C Adsorption capacity: 9.5 mg/g Surface area: 2–191 m^2/g pH: 9–10 for cationic dyes and 4–6 for anionic dyes, chosen experimental pH 7	Methylene blue and congo red dyes	Pyrolyzed at 550 °C with 10 °C per min under nitrogenous atmosphere for 2 h	(Choi et al. 2020)
3		Triarrhena stalks loaded with TiO_2 nanoparticles	Surface area: 161.5 m^2/g Sorption capacity: approximately 62.5 mg/g	Congo red	Pyrolysis at 300 °C	(Yu et al. 2018)
4		*Caulerpa scalpelliformis* marine seaweed	Sorption capacity: 0.1905 mmol/g pH: near neutral	Remazol brilliant orange 3R	Muffle furnace at temperature of 350 °C	(Ravindiran et al. 2021)

(continued)

Table 1 (continued)

Sr. No.	Types of biochar	Details of biochar	Characteristics of biochar	Contaminant removed	Technique used	References
5		Green seaweed (*Ulva lactuca*)	Temp: 30 °C Sorption capacity: 0.301, 0.292, 0.265, and 0.224 mmol/g for Remazol brilliant orange 3R, Remazol brilliant blue R, Remazol brilliant violet 5R, Remazol black B Removal efficiency: above 77.5% and the decolorization rate was high with more than 95% of total dye decolorization completed within 240 min pH: 2	Remazol dye	Pyrolyzed at 300 °C for 2 h in the presence of N_2 for 10 min at 100 mL/min	(Gokulan et al. 2019)
6		Chili seeds	Temp: 25 °C Sorption apacity: 26.13 mg g^{-1} at 0.01 mol/L ionic strength Surface area: 0.52 m^2 g^{-1} pH: 7	Ibuprofen	Pyrolysis at 600 °C	(Ocampo-Perez et al. 2019)
7		Coconut Husk and Kaolinite clay (KCB)	Temp: 30 °C Sorption capacity: CIP uptake by KCB, KCB-A and KCB-B are 71.1, 139.9 and 228.7 mg g^{-1}, respectively TET uptake e 118, 116.6 and 231.9 mg g^{-1} for KCB, KCB-A and KCB-B, respectively	Ciprofloxacin (CIP) Tetracycline (TET)	Heating under microwave irradiation at 540 W for 10 min then activated with either a basic or acidic medium	(Egbedina et al. 2021)

(continued)

Table 1 (continued)

Sr. No.	Types of biochar	Details of biochar	Characteristics of biochar	Contaminant removed	Technique used	References
8		Nano-manganese dioxide modified biochar from corn stalk	Temp: 25 °C Sorption capacity: 0.0236 mmol/g 0.0178 mmol/g (DNP adsorption) Surface area: 60.97 m^2/g pH: 7	Oxytetracycline (OTC), Di-*n*-butyl phthalate (DNP)	Pyrolyzed at 600 °C and then modified with co-precipitation method	(Gao et al. 2018)
9		Sewage sludge and Bamboo waste	Temp: 30 °C Surface area: 47.48 Sorption capacity: 62.48 pH: 6	Ciprofloxacin	Pyrolysis at 700 °C	(Li et al. 2020)
10		*Ulva prolifera*	Temp: 200–600 °C Sorption capacity: 21.20 mg/g Surface area: 0.0189 cm^3/g pH: 7	Ofloxacin	Pyrolysis at 200–600 °C	(Yang et al. 2021)
11		Sunflower stalk derived biochar enhanced with thermal activation of persulfate (BC)	Temp: 40–70 °C. Max at 70 °C Surface area: 324 m^2/g pH: 2.58	*p*-Nitrophenol (PNP) oxidation rate	Pyrolysis at 650 °C	(Sun et al. 2019)
12		Magnetic biochar derived from Palm kernel	Temp: 25–55 °C Sorption capacity: 10.84 mg/g Surface area: ?? Adsorption dosage: 0.6 g pH: 8	Phenols	24 h of drying at 80 °C in a Memmert oven	(Hairuddin et al. 2019)

(continued)

Table 1 (continued)

Sr. No.	Types of biochar	Details of biochar	Characteristics of biochar	Contaminant removed	Technique used	References
13		Biomass obtained from olive mill	Temp: 23 °C Adsorption capacity: 73.9 mg/g Surface area: 28.03–31.95 m^2/g pH: 5, keeping the pH close to natural pH	Phenols	Pyrolysis temperature 550 °C	(El Hanandeh et al. 2020)
14		Nano-manganese dioxide modified biochar from corn stalk	Temp: 25 °C Sorption capacity: 0.0236 mmol/g (OTC adsorption) 0.0178 mmol/g (DNP adsorption) Surface area: 60.97 m^2/g pH: 7	Di-*n*-butyl phthalate (DNP) and oxytetracycline (OTC)	Pyrolyzed at 600 °C and then modified with co-precipitation method	(Gao et al. 2018)
15		*Arnudo donax* loaded with copper ions (Cu–BC)	Temp: 20–40 °C Sorption capacity: 94.09% removal pH: near neutral	Polyaromatic hydrocarbons (PAH), Phenanthrene	Pyrolysis at 450 °C then followed by dipping the biochar in $Cu(NO_3)_2$ which is again pyrolyzed at 300 °C	(Shen et al. 2020)
16		Fibers from date palm and wood waste (mixture of two woods)	Temp: 30 °C Surface area: 123.63 m^2/g pH: neutral 7	*Geobacter sulfurreducens* strain bacteria	Pyrolyzed at 450 °C	(Van Limbergen et al. 2022)
17		Birch, aspen, and alder wood	Temp: 17–25 °C Adsorption: 73% removal of DOC pH: 7–8	Dissolved organic compound	Pyrolysis at 500 °C	(Kholoma et al. 2020)

(continued)

Table 1 (continued)

Sr. No.	Types of biochar	Details of biochar	Characteristics of biochar	Contaminant removed	Technique used	References
18		Wilt-infected banana straw modified with KOH	Temp: 25 °C Surface area: 1297.05 m^2/g pH: 5.5 Adsorption capacity: 150.09 mg/g	Cd(II)	Pyrolysis at 800 °C for 2 h	(Gao et al. 2024)
19		Brown seaweed derived modified with iron oxide	Temp: room temperature Surface area: 168.29 m^2/g pH: 6 Adsorption capacity: 1.64 mg/g	Arsenic	Pyrolysis at 400 °C	(Devrajani et al. 2024)
20		*Atropa belladona* L.	Temp: 25 °C Surface area: 812 m^2/g pH: 7.4 Adsorption capacity: 252 mg/g	Norfloxacin	Pyrolysis at 900 °C under N_2 flow	(Li et al. 2024a, b)
21		Gac fruit (*Momordica cochinchinensis* Spreng.)	Temp: 30 °C Surface area: 697 m^2/g Adsorption capacity: 909 mg/g	Crystal violet dye	Pyrolysis at 550 °C	(Nguyen et al. 2024)
22		Bamboo	Temp: Surface area: 85.5 m^2/kg Sorption capacity: 93.46 mg/g pH: 7	Cd(II)	Pyrolysis at 900 °C under N_2 flow using vacuum annealing furnace	(Huang et al. 2022)
23		Corn stalk biochar and Klebsiella grimontii by sodium alginate encapsulation	Temp: 40 °C pH: 7 Absorption capacity: 278.69 mg/g and 71.75 mg/g for Pb(II) and Cd(II), respectively	Pb(II) Cd(II)	Pyrolysis at 500 °C	(Li et al. 2024a, b)

(continued)

Table 1 (continued)

Sr. No.	Types of biochar	Details of biochar	Characteristics of biochar	Contaminant removed	Technique used	References
24		Peat, Olive husk, Volcanic rock	Sorption: VR/BO 99.3% Peat/BO 99.1% Water holding capacity: VR/BO 0.5 g/cm^3 peat/BO 0.4 g/cm^3 pH: 7.9 peat and olive husk 8.7 volcanic rock and olive husk	Chromium	Slow pyrolysis at 450 °C	(Piscitelli et al. 2018)
25		Cotton stalks modified with ZnO nanocomposites	Temp: 25 + − 1.5 °C Sorption capacity: 107.53 mg/g Adsorbent dosage: 2 g/L pH: 4	Chromium Cr(VI)	Pyrolysis at 400 °C followed by precipitation with 0.1 M $ZnSO_4$ followed by Calcination at 350 °C	(Tariq et al. 2020)
26		Rice straw (RS) and Anaerobically digested residues (DRS)	Temp: 15–45 °C Sorption capacity: RS (276.3) and DRS (90.5) mg/g	Pb(II)	Pyrolyzed at 500 °C	(Li et al. 2018)
27		Perilla leaf derived biochar	Temp: 20 °C Sorption capacity: As(III): 11.01 mg/g As(V): 7.21 mg/g (Langmuir model) Surface area: 473.4 m^2/g pH: 7–9	As(III) As(V)	Pyrolysis at 700 °C	(Niazi et al. 2018)
28		Green seaweed (*Ulva reticulate*)	Temp: room temperature Adsorption capacity: 7.67 mg/g pH: 4	Arsenic (As)	Pyrolysis at 300 °C for 120 min	(Senthilkumar et al. 2020)

(continued)

Table 1 (continued)

Sr. No.	Types of biochar	Details of biochar	Characteristics of biochar	Contaminant removed	Technique used	References
29		Coffee grounds soaked in NaOH	Sorption capacity: max is 51.52 mg/g Surface area: 46.318 m^2/g pH: 7	Ammonium ions NH_4^+	Soaked with 5MNaOH followed by pyrolysis at 500 °C	(Nguyen et al. 2021)
30		Sorghum straw	Temp: 45 °C Adsorption capacity: 7.09 mg/g Total pore volume: 0.036 cm^3/g pH: 6–10	Ammonium	Pyrolysis at 300 °C under the condition of introducing N_2 and then kept at the target carbonization temperature for 30 min	(Xu et al. 2022a, b)
31		Tomato green waste	Temp: room temperature Sorption capacity: 169.5 mg/g Surface area: 18 m^2/g pH: 13	Cr(III) ions	Pyrolysis at 800 °C	(Mokrzycki et al. 2021)
32		Sawdust and bark	Temp: room temperature Adsorption capacity: 19.998 mg/g Surface area: 5.70 m^2/g pH: 7	Cadmium ions	It was pyrolyzed at 600 °C at a thermal rate of 5 °C/min and kept at a constant temperature for 2 h	(Zhu et al. 2020)

(continued)

Table 1 (continued)

Sr. No.	Types of biochar	Details of biochar	Characteristics of biochar	Contaminant removed	Technique used	References
33	Agricultural waste	Rice husk with acid modification at 8 M H_2SO_4	Temp: 30–50 °C Sorption capacity: Almost 77% of 750 ppb Chromium (VI) adsorbed by 20 mg of AMB within 2 h at 30 °C, removal 86% with the temperature being 50 °C and adsorption capacity of 963 µg g^{-1} pH: 7 or near neutral	Cr(VI)	Pyrolysis at 520 °C	(Sarkar et al. 2019)
34		Rice straw	Temp: 298 K Adsorption capacity: 165.88 mg/g Surface area: 26.434 m^2/g pH: 4–5	Lead Pb(II)	Pyrolyzed at 300 °C for 4 h and desiccated at 70 C	(Tang et al. 2022)
35		Egg shell	Temp: room temperature Sorption capacity: 6.3 mg/g pH: 4.5	Arsenic As(V)	Heating at 450 °C for 2 h	(Akram et al. 2022)
36		Sugarcane bagasse	Temp: 90 °C Adsorption rate: nearly 100% of Cr(VI) pH: 5–8	Chromium Cr(VI)	Pyrolyzed at 350 °C for 30 min with a 10 C/min rate of heating	(Peng et al. 2021)
37		Rice straw	Temp: 25 °C Adsorption capacity: 171.54 mg/g Surface area: 20.50–212.65 m^2/g pH: 5	Phosphorus	Pyrolysis at 450 °C with a heating rate of 2 °C /min under N_2 atmosphere and maintained for 2 h	(Xu et al. 2022a, b)

(continued)

Table 1 (continued)

Sr. No.	Types of biochar	Details of biochar	Characteristics of biochar	Contaminant removed	Technique used	References
38		Chicken manure	Temp: 303 K Adsorption capacity: 60.69 mg/g for non magnetic and 41.07 mg/g for magnetic Surface area: 25.56 m^2/g in chicken manure biochar to 5.44 m^2/g in Magnetic-Chicken Manure Biochar pH: 11–12	Cadmium Cd^{2+}	Pyrolyzed under nitrogenous atmosphere at 600 °C for about 4 h	(Huang et al. 2020)
39		Rice husk biochar modified with oyster shell	Surface area: up to 46.19 m^2/g Maximum sorption capacity: 196.2 mg/g pH range: pH 3–11	Phosphorus (P)	Pyrolysis under nitrogen atmosphere; variation in pyrolysis temperature (700–900 °C)	(Xu et al. 2023)
40	Food processing industry waste	Walnut shells	Temp: room temperature Adsorption capacity: 2.592 mg/g for As and 1.43 mg/g Surface area: 112.84 m^2/g pH: 5.5–6	Arsenic (As) and Phosphorus (P)	Pyrolyzed at 500 °C for 60 min (thermal rate of 30 °C/min)	(Frišták et al. 2021)
41	Forestry biomass	Lignin content of forestry biomass and which are graphitic carbon derived (pitch pine sawdust) (*Pinus rigida*)	Temp: 25 °C Sorption capacity: 159.7 mg/g Surface area: 457.4m^2/g pH: 6.5	Diclofenac sodium	Heated in tube furnace at 900 °C in presence of atmospheric N_2	(Tam et al. 2020)

(continued)

Table 1 (continued)

Sr. No.	Types of biochar	Details of biochar	Characteristics of biochar	Contaminant removed	Technique used	References
42	Magnetic material + biochar	Magnetite nanoparticles biochar	Temp: 40 °C Surface area: 319 m^2/g pH: low	Chromate	Pyrolysis at 60 °C. 2 g of $FeCl_3 \cdot 6H_2O$ and 1 g of $FeCl_2 \cdot 4H_2O$ were dissolved in 200 mL water that had bubbled with N_2 gas for 30 min	(Alam et al. 2020)
43	Algae derived	Sodium alginate fiber from marine brown algae	Temp: 303 K Adsorption capacity: 104.9 mg/g Surface area: 52.08 m^2/g pH: 4.6	Hexavalent chromium (Cr(VI))	Pyrolysis at 900 °C for 1 h (with a heating rate of 5 °C per min)	(Sun et al. 2021)
44	Sewage waste	Sewage sludge	Sorption capacity: Toluene (79.1%) and Xylene (86.6%) removal Surface area: 52.210 m^2/g pH: 10	Petrochemical industrial waste	Pyrolyzed at 550 °C	(Kumi et al. 2022)
45		Sludge	Temp: room temperature Sorption capacity: 112.40 mg/g Surface area: 241.42 m^2/g pH: 6	Uranium	Pyrolyzed at 350–700 °C for 40–50 min under N_2 atmosphere	(Yu et al. 2022)
46	Animal waste	Giant mud crab (*Scylla serrata* species)	Sorption capacity: 3139 mg/g Surface area: 59.73 m^2/g Pore size: 31.3 nm	Methyl violet dye	Pyrolysis at 500 °C	(Mohd Faizal et al. 2024)

to the washing away of KCl. The MB is adsorbed maximum at acidic pH 2 (80.6%) as predicted; otherwise, there is not much difference in neutral and higher pH (74.3%). At initial concentration of 200 ppm of MB, the most efficient adsorption dosage of biochar was 0.03 g and the equilibrium was achieved in 120 min of contact time. The adsorption isotherm was best explained in Langmuir model (monolayer formation) with q_m 86.6 mg/g. The adsorption mechanics involved electrostatic interaction, hydrogen bonding, and hydrophobic-hydrophobic interaction of functional groups OH of carboxyl group and C=O of carbonyl group, and kinetics was best explained by pseudo-second-order model (Suhaimi et al. 2022)

Li Liu prepared biochar with cornstalk, which was washed with deionized water, dried in oven, crushed, and pyrolyzed at 500 °C in muffled furnace. It was then washed and dried in oven at 80 °C. The biochar was activated with KOH and H_3PO_4 and again pyrolyzed at 700 °C. It was then washed and dried in oven to remove impurities. BET method showed the specific surface area to be 474 m^2/g for KOH modified BC(KOH–BC) and 3 m^2/g for H_3PO_4 modified BC (H_3PO_4–BC). Pore volume was 0.24 cm^3/g and 0.0025 cm^3/g, respectively. The SEM image observation is given in Fig. 1. Maximum pH adsorption was achieved at 11 with 87% for KOH–BC and 86% for H_3PO_4–BC, although at pH 5 nearly 100% adsorption is achieved for KOH–BC. Adsorption at MB = 200 mg/L for KOH–BC is 372.3 mg/g and for H_3PO_4–BC is 105 mg/g. Both followed pseudo-second-order reaction model. KOH–BC adsorption isotherm was best explained through Langmuir model, and H_3PO_4–BC adsorption isotherm was best explained through Freundlich model. Adsorption mechanism involved physical interaction, ion exchange, hydrogen bonding, electrostatic interaction, π–π interaction, and pore filling. For KOH–BC functional group involved in adsorption was oxygen containing functional group and for H_3PO_4–BC was –OH, –COOH, P–OH, and amino phosphonic acid functional group (Liu et al. 2019)

Methylene orange is a very common effluent from textiles, laboratories, and paper industries. It has low biodegradability and genotoxic in nature. Zhang et al prepared biochar using sewage sludge which was dried at 60 °C for 12 h and then pyrolyzed at 800 for 12 h. It was further modified by adding 1.364 g $SrCO_3$ and then adding 1N HCl to dissolve $SrCO_3$ followed by addition of 1N NaOH to adjust pH at 10. 75% of the total quality of $SrCO_3$ and $FeCl_3.6H_2O$ was added in the mixture which was dried at 60 °C for 12 h and then calcined at 800 °C for 2 h with the protection of N_2 gas to obtain modified biochar MSBC. FTIR results showed presence of active functional groups such as C–O–C and O–H involved in the adsorption process. Adsorption was maximum at pH 5 (145.7 mg/g) and at MO concentration of 600 mg/L (149.31 mg/g). The adsorbent dosage was kept at 2 g/L and effective contact time was kept at 40 min to reach the equilibrium (Adsorption obtained under these conditions was 149.18 mg/g). The adsorption isotherm was best explained by Langmuir model due to uniform distribution of active sites on adsorbents surface and the kinetics followed pseudo-second-order reaction (Zhang et al. 2017)

Phuong Thu Le et al synthesized biochar using rice husk and coconut coir which were washed and then dried for 5 h at a temperature of 105 °C. After this, they were kept in a stainless-steel pot and covered with commercial charcoal to avoid formation of ash and coke during carbonization. Next, the biochar from coconut coir (BC1)

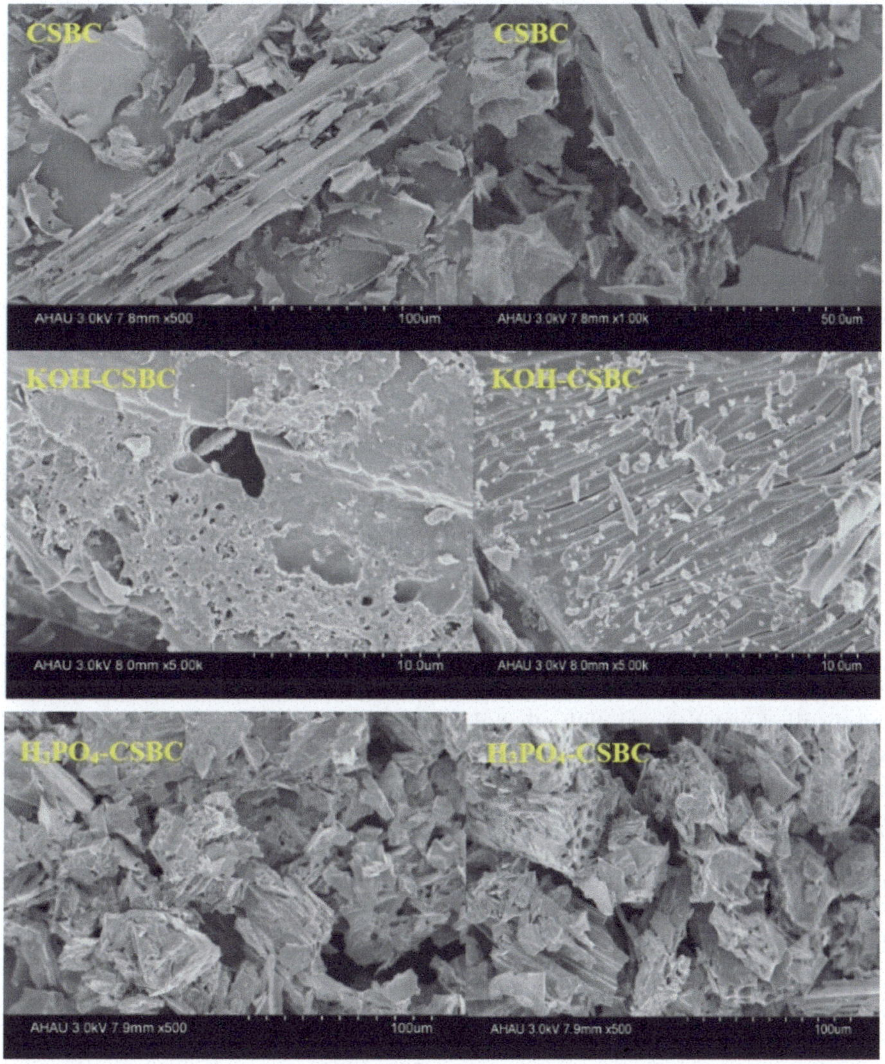

Fig. 1 Scanning electron microscope (SEM) micrographs of corn stalk biochar (CSBC), KOH-CSBC and H_3PO_4-CSBC. Reprinted with permission from reference (Liu et al. 2019)

and rice husk (BC2) were activated using NaOH 25% to increase the surface area and ion-exchange capacity. Thermogravimetric analysis and differential gravimetric analysis showed that greatest mass loss occurred at 250–300 °C which suggested that all organic bound compounds and water have been decomposed. After carbonization, the density decreases by 20% for BC1 and 40% for BC2. SEM imaging showed that BC1 had varying pore size, whereas for BC2 the structure is intact with uniform pores. The surface area of BC1 was found to be 329.7 m^2/g and pore volume 0.17 cm^3/g.

For BC2, it was 364 m^2/g and 0.26 cm^3/g; hence, BC2 is more efficient than BC1. XRD showed that the BCs are of amorphous carbon and has graphene structure. A buffer solution of NaCl with NaOH or HCl was prepared to determine PZC. For BC1, it was 9, and for BC2, it was 8. Methylene blue (MB) was adsorbed maximum at pH 6, and this is because the surface of adsorbent becomes negative at high pH values so there is a strong electrostatic force interaction with the cationic groups of methylene blue and adsorbent. Methylene orange (MO) was maximum adsorbed at pH 2, as at higher pH values the adsorbent surface slowly becomes neutral and then ultimately positive which reduces the efficiency. For BC1, MB followed Freundlich model and MO followed Langmuir model. For BC2, the MO adsorption followed both the model, whereas MB followed the Langmuir model (Le et al. 2021)

1.1.2 Removal of Congo Red (CR)

Printing and textile industry with their rapid development leads to large amount accumulation of dyes in wastewater contributing to water and soil pollution. In order to tackle this problem, biochars are being used to efficiently treat wastewater. Congo red is the most widely used Azo dye due to its low cost, but it is highly carcinogenic due to the presence of benzidine. Peng Yu et al prepared the biochar by the pyrolysis of Triarrhea stalks at 300 °C. This was followed by preparing a solution of anhydrous ethanol and 4 *n*-butyl titanates slowly dripped in another solution of acetic acid and distilled water mixed with ethanol at pH 3. To this obtained solution biochar was added and stirred at a water bath of 80 °C continuously for 1h followed by drying it in oven at 105 °C for 12 h and finally pyrolyzed at 300 °C TiO_2–BC. BET method showed the specific surface area to be 161.5 m^2/g and mean pore diameter of 60 nm. SEM analysis showed that surface had rod shaped grooves with various TiO_2 nanoparticles. The adsorption efficiency was found to be 83.59% with maximum value being 61.67 mg/g at mass of 0.015 g. The adsorption mechanism involved was the weak interaction of the organic functional group of TiO_2–BC (C–O, C–H, and CO_3^{2-}) with organic molecule of CR and MB dye, grain size of TiO_2 at saturation time (60 min), and physical adsorption at the enlarged pores (Yu et al. 2018).

Yong-Keun Choi et al prepared biochar using Maple leaves in which the leaves were desiccated at 60 °C, compressed to 500–2360 μm particle dimension, and pyrolyzed at 350, 550, and 750 °C under N_2 atmosphere. The produced biochar (BC) was washed several times for 3 days to remove impurities. BET analysis showed the specific surface area to be increasing from 2 to 191 m^2/g as temperature increased. The XRD patterns showed that Biochar550 and Biochar750 had high $CaCO_3$ crystallinity and Biochar350 showed high $MgCO_3$ crystallinity. The pH_{pzc} of Biochar350, Biochar550, and Biochar750 was 6.3, 9.2, and 9.3 correspondingly with higher pH due to concentrated alkali salts. Out of three pH experimented (4, 7, 10), both MB and CR showed good adsorption quality at pH 7 for Biochar550 and Biochar750 at temperature 35–40 °C. FTIR spectra for Biochar550 showed participation of various functional groups such as carbonates, C–O and C–H, and ion-exchange mechanism for adsorption. The adsorption efficiency for both dyes at single reaction MB is 25%, and for CR 97% alternatively, adsorption effectiveness for each dye in the dye mixture for MB is 68% and CR is 74% suggesting a competition for adsorption between the dyes (Choi et al. 2020).

1.1.3 Removal of Crystal Violet (CV)

Crystal Violet dyes is constituted of stable functional groups which makes the molecule stable and difficult to decompose, which is why biosdsorbants like biochars are implemented to treat wastewater. Kyi et al prepared biochar using Palm Kernel shell pyrolyzed at 350 °C temperature in a rotary kiln for 20 min and then crushed and dried in oven for 24 h at 105 °C. The concentration of dye was 400 mg/L. The adsorbent dosage was 0.5 g. Although 30.9 g/L of BC is required to remove 80% of Crystal Violet but improved concentration results in the formation of concentration gradient between solute and solution, and the solute concentration on BC surface. The average pore diameter was found to be 4.1 nm. It showed maximum adsorption efficiency at pH 4–6 at temperature 25 °C. As the pH of solution increases, BC surface becomes more negatively charged hence favoring the adsorption of cation via the electrostatic attraction mechanism. The maximum adsorption efficiency was 24.45 mg/g. The kinetics satisfied the pseudo-second-order model and adsorption isotherm best suited in accordance with the Langmuir model. Presence of cations in the solution however does decrease the CV adsorption because of competitive adsorption (Kyi et al. 2020).

Moharm et al prepared biochar using bagasse samples from sugarcane which were converted into fine powder dried in oven at 70 °C for 3 days and then pyrolyzed at 350 °C for 30 min at a rate of 10 °C/min, and the resulting BC was treated with NaOH at 1% now referred as T-BC1. SEM was used for surface morphology which showed smooth and even surfaces for TBC and indicated in the presence of nanosheets which was also confirmed by XRD spectroscopy. FTIR showed the active oxygenated functional groups which interact with the MB and CV dyes. BET technique was utilized to determine specific surface area which was 8.2175 m^2/g and pore volume 6.3102 cm^3/g. The equilibrium was achieved at 15 min for MB and 45 min for Crystal Violet (with both dye concentration being 30 mg/L) at 25 °C optimum temperature and pH 7 with adsorbent dosage 20 mg. On increasing the pH from acidic to neutral, the active sites become more ionized and hence are suitable for efficient adsorption. The adsorption kinetic was best explained through pseudo-second-order reaction. The adsorption capacity was 73.55 mg/g and 68.8008 mg/g for MB and CV, respectively. Langmuir model explains the adsorption isotherm suggesting the process occurs on a homogeneous active site as monolayer for both dyes (Moharm et al. 2022).

Nguyen et al. focused on two major objectives which were optimization of carotenoids extraction from gac fruit aril and evaluation of the potential of gac fruit peel as a precursor for biochar production. Optimal extraction occurred at 70 °C, and higher temperatures increased carotenoid yields until degradation began. The carbon content in biochar increased with higher pyrolysis temperatures, reaching approximately 62.98% at 550 °C. The highest absorption intensity took place at 120 min. After that, the yield started decreasing. Second-order kinetics better explained the extraction yield and antioxidant activity as compared with first-order kinetics. It is thus evidenced that the extraction process involves solute-solvent interactions. Biochar prepared at a temperature of 550 °C has the greatest adsorption efficiency toward Crystal Violet dye; accordingly, it might be applied for environmental remediation purposes as an effective adsorbent (Nguyen et al. 2024).

1.1.4 Removal of l Brilliant Orange 3R (RBO_3R) and Remazol Brilliant Blue R Dye (RBBR)

RBBR is extensively used in various industries and more than 700,000 metric tons of dye is released into wastewater. Ravindiran et al. prepared biochar by using a marine seaweed *Caulerpa scalpelliformis* which was washed and shredded to 7.5 mm proceeded by drying it in oven at 103 °C for 24 h. It was then pyrolyzed at 350 °C in a muffled furnace. Batch adsorption study showed that the biochar had good adsorption potential even at low concentration with maximum removal efficiency of 76.2%. Real time experiment results showed maximum adsorption efficiency of 72.69% with 0.3 L/h flow rate of wastewater. A decreased and slow flowrate allowed more interaction time of BC with RBO_3R, and hence this optimum flow rate of 0.3 L/h was selected for real time experiment study. The sorption capacity was 0.087 mmol/g at BC depth of 25 cm (Ravindiran et al. 2021).

Gokulan et al. prepared the biochar using *Ulva lactuca* a marine green alga which were washed extensively, dried and crushed to 1 mm average size which pyrolysis was done at 300 °C in an electric muffle furnace for 2 h. SEM analysis showed BC surface to be rough with varying pore sizes although after dye adsorption the surface became relatively smooth due to the removal of impurities and other ions from biochar. The dye concentration was between 0.05–1 mmol/L and BC adsorption capacity was favored at acidic pH 1.75–2. The reason being alga derived BC have high sulfate, carboxyl, and amine functional groups which become protonated making an overall positively charged surface which interact (electrostatic interaction) better with the negatively charged dye. Due to the limited bonding sites for interaction, low dosage was found to be more efficient and hence kept at 2 g/L. The adsorption isotherm was described accurately in Langmuir model and the kinetics in pseudo-second-order reaction. Overall, 77.5% discoloration efficiency was obtained making it a suitable treatment method for RBBR removal from wastewater (Gokulan et al. 2019).

1.2 *Removal of Antibiotics Using Biochar*

Ibuprofen is a very common anti-inflammatory and anti-pyretic drug and hence is accumulated in the wastewater. To remove this drug, bio-adsorbent technique biochar is used. Ocampo-Perez et al produced biochar using chili seeds which were washed and dried at 80 °C for approximately 24 h in an oven. It is then pyrolyzed at 450, 550, and 600 °C for 2 h with heating rate of 10 °C/min. The surface nature of the BC was found to be nearly neutral. Raman spectra showed the presence of polyaromatic hydrocarbons and graphitic material on surface. BC450 had C=O groups and BC550 and BC600 had aryl ring and aromatic structures and also presence of lactonic sites. The degree of amorphous carbon also increased with increasing pyrolyzed temperature. SEM analysis revealed incomplete oval geometry with homogeneous nonporous morphology as shown in Fig. 2. The adsorption capacity was 11.5, 14.4 and 19.1 mg/g for BC450, BC550, and BC600, respectively. The adsorption mechanism was hydrophobic interactions induced by Van der waal's forces and π-acceptor cloud interactions with IBP and graphenic planes of BC. At pH 7, there is a significant electrostatic interaction with the negative IBP and the positively charged BC surface. The adsorption isotherm was best explained in Langmuir isotherm model (Ocampo-Perez et al. 2019).

Ciprofloxacin and Tetracycline are commonly used antibiotics which are stable and can't be easily degraded. They decrease the nitrifying activities and include heavy metals to increase the bacteriocidal property making it more difficult to biodegrade through conventional method. Egbedina et al prepared biochar using coconut husks

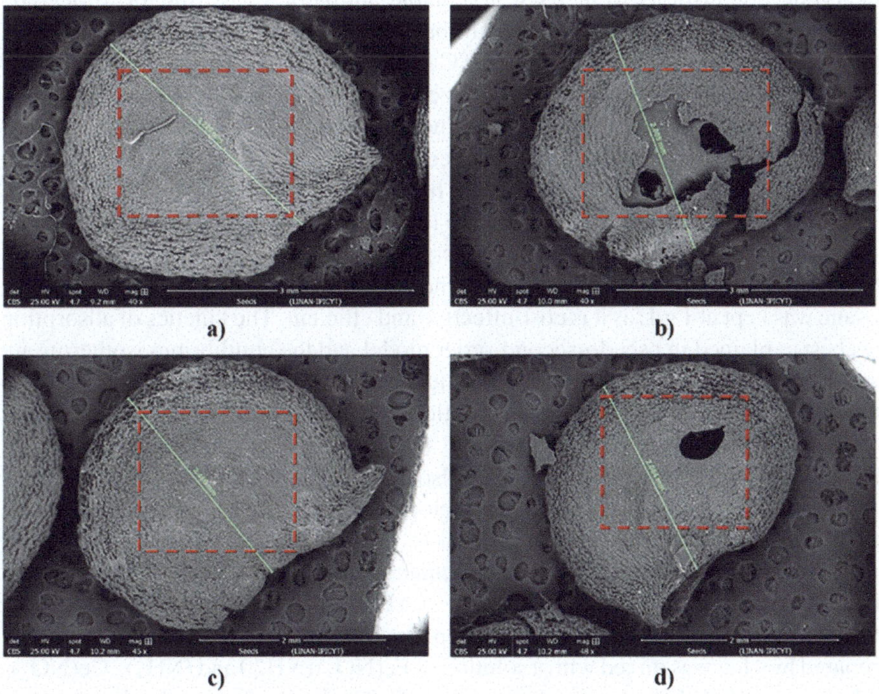

Fig. 2 SEM micrographs of samples **a** CS, **b** C450, **c** C550 and **d** C600. Reprinted with permission from reference (Ocampo-Perez et al. 2019)

which was washed with distilled water, sun dried, milled. Biomass along with Kaolinite clay and $ZnCl_2$ (1:1:2) and finally 0.1 M NaOH was added to this mixture and dried in oven for 24 h at 150 °C. It was brought to pH 7 and again dried at 105°C. The prepared KCB was then activated in two ways—one with HCl acid (KCB-A) and other with KOH alkali (KCB-B). SEM images showed irregularly shaped ellipsoidal pores with surface having sharp edges showing good crystallinity. Activation leads to generation of more exposed pores as compared to normal KCB. Elimination of TET was maximum at pH 10 and both adsorbents and TET are negatively charged. Hence, instead of electrostatic repulsion there is hydrophobic interactions. CIP is better adsorbed in acidic medium pH 4 by KCB-A. At this pH KCB-A surface is positively charged and so is CIP. Hence, in this case as well instead of electrostatic repulsion there is hydrophobic interaction. Other interactions involved in adsorption are π–π interactions, London dispersion forces and Van der Waal's forces. The isotherm equilibrium was best explained with Langmuir model. The CIP uptake for KCB, KCB-A, KCB-B is 71.1, 139.9 and 228.7 mg/g and for TET it is 118, 116.6 and 231.8 mg/g, respectively. The pseudo-first-order kinetic model is best explained for high concentrations of pollutants and hence can be successfully removed (Egbedina et al. 2021).

Oxytetracycline when released in the aquatic system is found to significantly affect the microbial growth pattern at genes level and disrupt the growth of certain eukaryotic organism as well. As it is a stable antibiotic, it is difficult to degrade naturally. Gao et al prepared corn stalk-based biochar in which corn stalks were washed, dehydrated in an oven at 80 °C for 24 h and then crushed and filtered through 0.5 mm size mesh. It was pyrolyzed at 600 °C in muffled furnace and then again dried at 80 °C overnight. The obtained biochar was modified to $nMnO_2$–BC using co-precipitation method. The obtained biochar was then again washed, vacuum filtered, and dried in oven at 80 °C for 8 h. SEM imaged revealed that the surface had namo-MnO_2 adhered to it in a sphere like morphology. These nanospheres had nanosheets projecting from the center perpendicular to the surface. FTIR spectra shows that the $nMnO_2$ loading increases the O containing in the BC. The adsorbent dosage was kept at 1 g/L as it is cost-effective and efficient. The kinetics of adsorption was best explained in pseudo-second-order model and the equilibrium isotherms are explained in both Freundlich and Langmuir model. The interaction is seemed to have π–π electron donor acceptor interaction and oxidation of MnO_2. Involvement of functional groups such as C≡N C–O–C and C–N and their interactions with OTC and CIP also plays an important role in adsorption mechanism (Gao et al. 2018)

Ampicillin is a non-biodegradable antibiotic and accumulate in wastewater which causes bioaccumulation which is harmful to the aquatic life. Gómez-Vilchis et al prepared the biochar using *Punica grantanum* peels which were washed at 120 °C and sun dried for 3 days. It was then treated with 0.05 M NaOH solution for 20 min and then washed with deionized water. This step was followed by pyrolysis at 700 °C. The prepared biochar was mixed with a solution of $Fe(NO_3)_3.9H20$ and $(NH_4)_2.Ce(NO_3)_6$ and again pyrolyzed at the same temperature. Fe/Ce–BC is characterized with rough surface having hexagonal cavities, forming a honeycomb like structure. The pore diameter was about 17–20 μm in diameter. Specific surface area was found to be

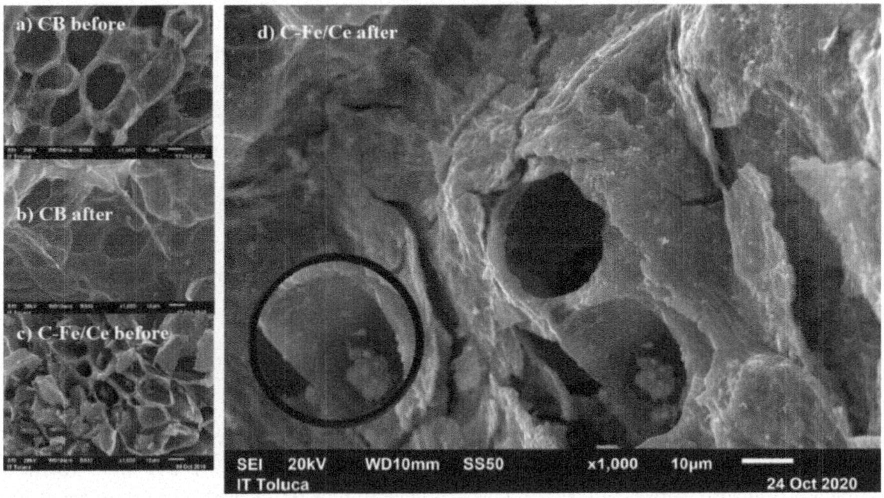

Fig. 3 Micrograph of CB **a** before sorption, **b** after sorption; C–Fe/Ce **c** before sorption, **d** after sorption, **e** CB pore size distribution after sorption, **f** C–Fe/Ce pore size distribution, and **g** Fe/Ce particle size distribution. Reprinted with permission from reference (Gómez-Vilchis et al. 2022)

249.44 m^2/g, pore volume 0.21 cm^3/g and pore size of 3.44 nm (BET method) as shown in Fig. 3. FTIR analysis suggested the electrostatic interaction between AMP anionic groups and positively charged surface of Fe/Ce–BC. Thus, at lower pH values the biochar can efficiently remove the drug. Oxygen atoms in esters and hydroxyl functional group also interact with AMP functional groups via hydrogen bonding. The adsorption kinetics and isotherms were best explained by Elovich and Langmuir model correspondingly. One of the significant properties of this biochar is that it can be reused upto 5 times although the efficiency does decrease with each reapplication (Gómez-Vilchis et al. 2022).

Ofloxacin is a broad-spectrum fluoroquinolone derivate antibiotic which affects the growth of non-targeted prokaryotes and induces inhibitory growth factors. Yang et al prepared biochar using *Ulva prolifera* a green tide forming macroalgae pyrolyzed at temperature from 200 °C. Two sample of biochar was prepared BC200 °C and washed BCW200 °C. BET method showed that the surface area of BCW200 is greater than BC200. Washed biochar were found to be highly carbonized and hydrophobic in nature due to low O/C ratio. The water washing treatment also decreased the aromaticity and O-containing functional groups. The specific surface area for BC200 and BCW200 is 4.65 m^2/g and 4.88 m^2/g, average pore volume is 4.27 nm and 15.52 nm, respectively. Maximum adsorption capacity for both biochars was found to be between pH 5–7. In very acidic solution the sorbent and sorbate both are positively charged resulting in electrostatic repulsion. In neutral medium there is less electrostatic repulsion and adsorption mechanism involved were, π–π interactions, hydrophobic interactions and hydrogen bonding. In alkaline medium both the biochars as well as OFL become negatively charged resulting in strong electrostatic repulsion. The kinetics is best described in pseudo-first-order model. The adsorption isotherm follows the Freundlich model. Thermodynamic parameters demonstrated that the adsorption for both the biochar samples was spontaneous and endothermic. Hence, BCW is an effective low-cost treatment method for removal of Ofloxacin drug from wastewater although optimal pyrolysis temperature needs to be selected (Yang et al. 2021).

Li et al prepared biochar by preparing a mixture of sludge and bamboo waste (4:1) and then pyrolyzed at 700 °C with ramp rate 10 °C per min under nitrogenous atmosphere for 30 min. Then sludge-based biochar (SLB) and bamboo-based biochar (BBC) were ready. SEM images for SLB showed the surface being comprised of several blocks with different shapes. The blocks and slit pores were comparatively smaller for BBC hence, BBC (47 m^2/g) has higher specific surface area than SLB (26 m^2/g). The maximum adsorption capacity was 46.20 and 62.8 mg/g for SLB and BBC, respectively. XPS analysis showed π–π interaction between aromatic structure of the biochar and CIP. Functional groups –COOH, –NH, –F of CIP also formed hydrogen bonding with C=O, C–O–C, C–N, C=N, C–OH of both biochars. Furthermore, electrostatic force of attraction also exists between piperzine rings of CIP molecule with H^+ of biochar surface in neutral to faintly acidic solution which also contributed to adsorption mechanism through electrostatic interactions. The pseudo-second-order and Freundlich model best explained the adsorption mechanism (Li et al. 2020)

Balofloxacin is third generation Fluoroquinolones antibiotic. Photo-degradation of this antibiotic results in formation of toxic intermediates in the wastewater which are harmful for the aquatic ecosystem. El-Azazy et al synthesized biochar with the help of spent coffee grounds which were dried in oven at 80 °C for 3 days followed by pyrolysis at 550 °C. It was then modified TiO_2 nanoparticles using modified hydro-thermal method. The presence of TiO_2 was proved using FTIR analysis. BET method showed the specific surface area to be 50.54 m^2/g. The average pore radius was 24.4 nm which is less than the unmodified biochar. SEM imaging showed pyramidal shaped TiO_2 nanoparticles. The adsorption capacity of modified biochar is 196.73 mg/g. The kinetics were explained in the pseudo-second-order model and the adsorption isotherms were best explained in Fruendlich model. Moreover, the reusability of the modified biochar is up to 5 times with retained efficiency of 66.32%. The BALX-loaded BC can be unloaded using 0.1 M HNO_3 with desorption efficiency of 72.14%. Hence TiO_2–BC is a suitable and highly economical treatment method for removal of Balofloxacin from wastewater (El-Azazy et al. 2021)

Li et al conducted research on the application of Chinese herbal medicine residue derived biochar for adsorption of Norfloxacin, another commonly used antibiotic. The biochar was produced from *Atropa belladona* L. residues. Pyrolyzed at 1173 K for 4 h under the nitrogen for optimizing its surface area and structure. The calculated adsorption capacity of ABLB4 for NOR reached a maximum value of 252.0 mg/g at 298 K, and this value suggests that removal of NOR from water would be very efficient. Langmuir model fits well for the adsorption data and supports monolayer coverage of NOR on ABLB4. According to the RSM, besides these optimal conditions, using different models, it was confirmed that the study could eventually have a maximum removal rate of 97.4% under such conditions in its optimal ranges: (303 K, 67.1 mg/L, pH of 7.4) (Li et al. 2024a, b)

1.3 *Removal of Organic Compounds Using Biochar*

The dichlorofluorescein (DCF) content in wastewater is found to be as high as 2.51 µg/L. This contaminant is partially adsorbed in the ecosystem and affects the food chain and the marine environment in general. Hence, an efficient removal technique is required. Tam et al prepared biochar using pitch pine sawdust which was mixed in H_2SO_4 (10%w/v) to obtain lignin of the biomass. This solution was adjusted to 4% using deionized water and kept at 120 °C for 1h. It was then dried at 60 °C, and the solid sample obtained was dispersed in a solution of 0.1 M Fe_2SO_4 and dried overnight at 100 °C. The obtained sample was then pyrolyzed at 900 °C in atmospheric N_2. The treated sample (LPGC) is now ready for adsorption. SEM images revealed LPGC surface possessing high rough surface and parallel stripped scratches. Iron particles were loaded on the surface. TEM image showed the presence of graphitic carbon on LPGC. The BET specific surface area for LPGC was 457.4 m^2/g. The adsorption mechanism was mainly electrostatic interaction π–π interaction between the aromatic rings of DCF and LPGC and hydrogen bonding. The adsorption was maximum at pH 2. The adsorption efficiency also increased in presence of other cations and certain background electrolytes. The adsorption capacity was 159.7 mg/g at 298 K and pH 6.5. The kinetics followed was pseudo-second-order model adsorption isotherm was best was explained by Langmuir model. Hence, LPGC is an efficient technique in elimination of DCF from wastewater (Tam et al. 2020).

Sun et al prepared biochar using sunflower stalk biomass for the adsorption of *p*-Nitrophenol (PNP) which were washed and grinded into fine powder of 2.0 mm. It was pyrolyzed in ceramic pot at temperature 500, 650 or 1000 °C for 2 h and the obtained BC was called BC500, BC650 and BC1000. To remove the influence of metals and minerals all biochars samples were subjected to with 200rpm at 40 °C for 24 h in mixture of the 1M HCl and 10% v/v HF.at a 1:5 solid/liquid. The final biochar obtained was used for adsorption BC650. The specific surface area 324 m^2/g. FTIR spectra revealed functional group presence such as COOH, COOC, quinones, and conjugated ketones. The oxidation of PNP in persulfate system was 0.6062 L/mol S. The PNP oxidation was a radical reaction in the PS /BC system. The defective site on biochar weakens the PS O–O bond and eventually cleaved it to form sulfate radicals. When performed under real wastewater the oxidation efficiency did decrease due to presence of high number of organic compounds. But despite of this Biochar did significantly enhance the PNP oxidation and even low percentage of PNP was removed within 30 min. Hence BC with activation of PS could be an efficient and environment-friendly method for removal of PNP (Sun et al. 2019).

Phenols are one of the most common contributing factors to nerve coordination damage, kidney impairment, amentia, cancer and contributes to Iron deficiency.

Hence, their removal from aquatic ecosystem is important. Hairuddin et al. prepared biochar using palm kernel which were soaked for one day and dried in oven at 80 °C for 24 h and pyrolyzed at a temperature of 800 °C. FESEM imaging revealed enlarged pores on BC surface. The increased magnetizing nature leads to high crystallinity and because of this super-magnetic behavior they can be widely used for treating wastewater. The maximum adsorption occurs at pH 8 (90.7%). The reason being at higher pH it has negative active sites, and the phenols exist positively charged. The adsorption isotherm is explained both by Freundlich and Langmuir model and the adsorption kinetics follows the pseudo-second-order reaction model. The adsorption capacity was 10.84mg/g which is approximately 93.399% removal of phenol at pH 8 with adsorption dosage 0.6g and contact time 60 min. Magnetic based biochar hence are the new emerging technique to remove certain contaminants from wastewater in an efficient and environmentally safe way (Hairuddin et al. 2019).

Hanandeh et al prepared biochar using olive mill biomass (jift) for phenol removal. The jift was sieved through 2 mm sieve and pyrolyzed at 550 °C for 90 min to obtain pristine biochar. This was modified soaking the pristine biochar in 0.1 M $FeCl_3$ solution and then dried in oven for 24 h at 105 °C this gives post treated biochar. Another way is to soak the biomass in 0.1 M $FeCl_3$ solution and then undergoing pyrolysis at 350 °C, this is called the pre-treated biochar. The BET surface area for BC_{pre} is 31.95 m^2/g and that of BC_{post} is 29.44 m^2/g. SEM showed that the $FeCl_3$ has been effectively loaded onto the surface of the biochar in the ridges, channels and pores of the biochars. One of the most important adsorption mechanisms is H-bonding, π–π interaction and pore filling. BC550 has very less hydroxyl groups as their presence would lead to increased affinity toward water rather than phenol and this chemical characteristic increases the efficiency of phenol adsorption. It is also noticed that pre-treated biochar has higher adsorption as compared to the post treated biochar. Although for economic reasons post treated biochar is recommended without sacrificing much of the adsorption efficiency. Both samples of biochar followed Langmuir isotherm model and kinetics was best explained by pseudo-second-order model (El Hanandeh et al. 2020).

Polycyclic aromatic hydrocarbons (PAH) are carcinogenic contaminants which are harmful to the ecosystem. Shen et al prepared biochar using *Arundo donax* a tall perennial cane which were smashed and mixed with H_3PO_4 (85%wt) solution for 24 h. It was then heated in muffled furnace at 500 °C under pure N_2 condition for 1 h. It was then dried at 105 °C for 12 h. The obtained biochar was modified with copper using impregnation method and then undergoes pyrolysis once more at 300 °C for 4h. SEM imaging revealed that loading of copper reduced the BET surface area of the BC probably due to the Cu filling up the pores as observed in Fig. 4. However, the mesoporous pores level increased. XPS spectra showed that phenanthrene removal was due to copper ions participation in adsorption. The C=O and decrease in O_2-ions indicates that the combination of copper with phenanthrene. The phenanthrene removal efficiency for Cu–BC is 94.09%. Certain presence of bacterial community like *Pseudomonas, Gordonia* sp., *Bacillus*, and *Desulfovibrio* in the constructed wasteland enhanced the adsorption efficiency of Phenanthrene and degradation capability of PAH. However, in the CW system the substrate is prone to get saturated due to the slow and gradual accumulation of suspended solids, biomass, and plant residue. Therefore, more research is being performed in terms of constructed wetlands (Shen et al. 2020).

Pesticides are one of the most common contaminants of the wastewater. They are toxic in nature and impose health hazards to humans and cause imbalance in the aquatic ecosystem. Limbergen et al prepared biochar from fibers of date palm (*Phoenix dactylifera*) (PDF), coffee bean husks (COF) and wood leftovers (AB). The materials were pyrolyzed at four distinct temperature 350, 450, 500, and 550 °C for 30 min. The BET specific surface area for 450 °C temperature for PDF450, COF450, and AB450 was 32.9736, 2.1162, and 78.2368 m^2/g, respectively. This biochar was activated using *G. sulfurreducens*. SEM images revealed dense coverage of bacteria on the surface. The main adsorption mechanism was electrostatic interaction, hydrophobic effects, H-bonding, and pore filling. Presence of bacteria and Biochar together increased the removal of pesticides and other micro-pollutants by

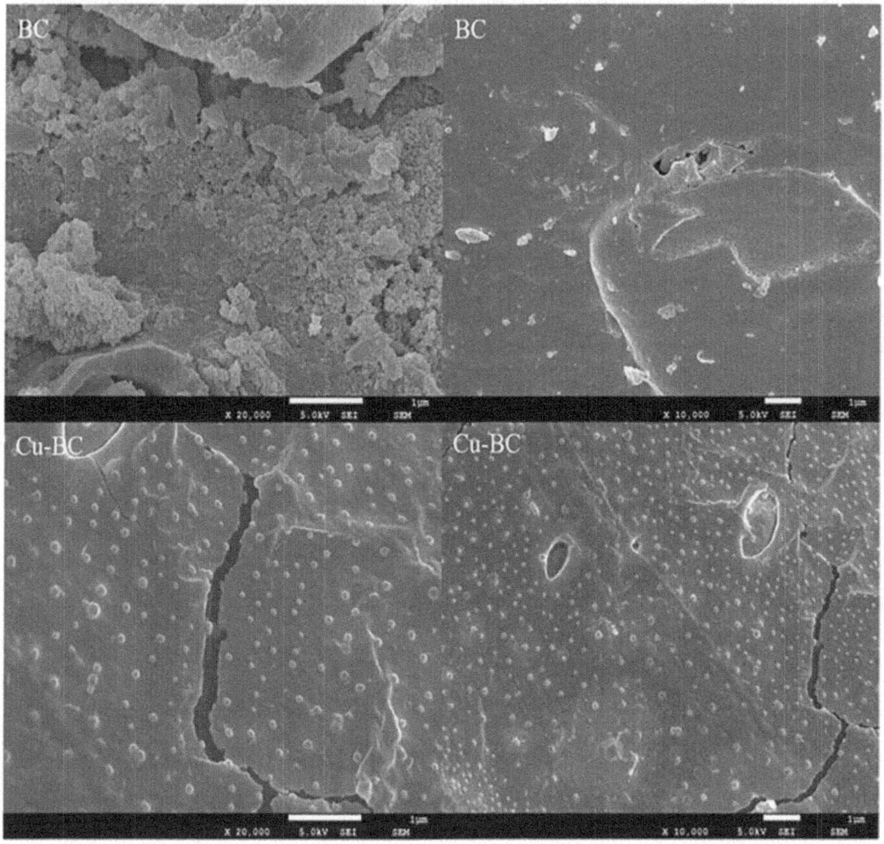

Fig. 4 SEM images of BC and Cu–BC. Reprinted with permission from reference (Shen et al. 2020)

50% more than just bacteria alone. AB wood biochar and coffee husk biochar showed better results as compared to PDF biochar. Furthermore, bacteria isolated from these pesticides can moreover lead to the discovery of novel degrading strains that can be implemented for new modified biochar remediation experiments (Van Limbergen et al. 2022).

Petrochemical industrial wastes (PIW) consist of mainly aromatic hydrocarbon and organic solvents which degrade slowly forming derivatives which are toxic in nature. Kumi et al prepared biochar using sewage sludge which was dried in oven at 100 °C for 24 h. Then it was subjected to pyrolysis at temperature 300, 350, 400, 450, 500, and 550 °C. BET specific surface area for the biochars BC300, BC350, BC400, BC450, BC500, and BC550 were 0.227, 0.282, 1.266, 22.680, 27.240, and 52.210 m^2/g, respectively. The XRD pattern showed two main crystalline minerals namely the quartz and calcite. SEM imaging showed irregular, uneven pores which with increasing temperature became more well-defined vacant pores and improved porosity. The maximum adsorption was seen in pH 10. At low pH there is high amount of H^+ causing electrostatic repulsion. The adsorption mechanism was prominently hydrophobic interaction and π–π bonding interaction (aromatic ring of TX and carboxylic oxygen atom of biochar). The adsorption isotherm followed was Langmuir model and pseudo-second-order model in terms of kinetics. Moreover, the biochar can be reused for up to five adsorption-desorption cycles. The overall removal of Toluene is 79.1% and Xylene 86.6% at adsorption dosage of 2 g/L. within 60 min contact time. Hence, it is safe to conclude that sludge derived biochar is an efficient technology to remove Toluene and Xylene (Kumi et al. 2022).

Kholoma et al prepared biochar from birch, aspen, and alder wood chips for the removal of Dissolved Organic Carbon (DOC). These biomasses underwent pyrolysis condition at 500 °C through wagon retort process. SaC porosity was 0.43 while that of GCC was 0.50. Biochar enhanced with sand filter (SaC) had improved efficiency in removal of dissolved organic compounds. It could remove 58% of the DOC in wastewater. Gas concrete enhanced biochar (GCC) however performed slightly better with 61% removal of DOC. The adsorption mechanism was likely to be through pore filling, ion exchange and hydrophobic interaction (after colonized by fungi). On turbidity, all the filters performed efficiently with removing 51–61% DOC. Therefore, it can be concluded that biochar enhanced with filter is a potential treatment for removal of DOC and other contaminants such as phosphates and with further improvements; it can be used as an upgraded septic system at high environmental standards (Kholoma et al. 2020).

2 Removal of Inorganic Compounds Using Biochar

2.1 Removal of Heavy Metals Using Biochar

2.1.1 Removal of Lead

Iron-manganese oxide-modified biochar is prepared by rice straw being fused with a blended solution of potassium permanganate and ferric nitrate. The biochar's absorption properties were studied by testing the absorption of lead, an extremely harmful element commonly found in water bodies near industries. Research has demonstrated that adding metal or metal oxides to biochar can considerably increase its specific surface area, microporous pore size, load quantity of functional groups, such as hydroxyl and carboxyl groups, and adsorption effectiveness. To create biochar, the powdered straw was pyrolyzed at 300 °C for 4 h in an anaerobic environment. To recover the original biochar made of straw, the biochar temperature was then brought down to 25 °C, cleaned with deionized water, desiccated at 70 °C, and the strongest adsorption capacity for Lead(II) was found in the molar ratio Fe:Mn=2:5. In addition to having more adsorption sites and pore architectures, FM–BC had a specific surface area that was 18.20 times more than BC's. The highest Iron Manganese biochar Pb(II) adsorption capacity was 165.88 mg/g, optimized at 298 K, at a pH of 4–5 and the surface area recorded was 26.434 m^2/g. The adsorption mechanism comprised of chemical precipitation, complexation, and ion exchange, Fe_2O_3, and Fe(III) were transformed into FeOOH, while Mn_2O_3 was transformed into MnO_2. FTIR analysis was employed, and adsorption properties were better described by the pseudo-second-order kinetic model. The Langmuir and Freundlich equations were used in thermodynamics to evaluate the experimental data's isothermal adsorption and investigate the adsorption. Thus, Iron Manganese biochar is an effective biomass adsorption material (Tang et al. 2022).

In most nations that harvest rice, rice straw is used to meet the nutritional needs of grazing animals. However, rice straw's poor nutritive value, phenolic traits, and elevated amounts of lignin and silica make it difficult for grazing animals to digest, making it an ideal candidate for exploring ways of sustainably utilizing it. The elimination of lead from an aqueous solution using dried rice straw and rice straw residue that had been anaerobically digested were both examined. By precipitating phosphate silicates and complexing with carboxylate groups, anaerobically digested rice straw was primarily used for lead adsorption. The highest efficiency was shown by the biochar produced from both types at a pyrolysis temperature of 500 °C under the safeguard of 99.9% pure N_2 flowing at a rate of 0.1 m^3 per h. 10 °C/min was the set temperature for the tube furnace's heating rate. Based on the Langmuir model, maximal sorption capacity for biochar made from digested rice straw residue was determined at 90.5 mg/g and for rice straw at 276.3 mg/g, between 15 and 45 °C and at low pH values. The thermal degradation of cellulose, hydroxyl groups, and aliphatic alkyl groups caused the band intensities at 3400–3430 cm^1 (–OH) and

2850–2950 cm^{-1} (aliphatic) to decline when the pyrolysis temperature was raised, according to the FTIR spectra (Li et al. 2018).

Li et al. introduced a novel method in which immobilized bacterial microsphere, CYB-SA, was constructed from corn stalk biochar pyrolyzed at 500 °C in muffle furnace for 2 h with Klebsiella grimontii, by encapsulation in sodium alginate followed by freeze drying. The ability of the removal of Pb(II) and Cd(II) ions from monometallic solutions at a dosage level of 2.0 g/L and pH of 7 was evaluated as 99.14% and 83.35% and compared to biochar alone for a 10.77% and 18.58% removal increase, respectively. At 40 °C, the calculated maximum adsorption capacities for Pb(II) and Cd(II) using Langmuir isotherm model were 278.69 mg/g and 71.75 mg/g, respectively (Li et al. 2024a, b).

2.1.2 Removal of Cadmium

This study was initiated to develop an inexpensive, environmentally friendly absorbent bamboo biochar as well as chitosan-modified bamboo biochar. Results indicate that bamboo biochar treated with chitosan supplied more active adsorption sites (such as $-NH_2$, –COOH, –OH, and C=O) on the surface to improve the removal rate of Cadmium in polluted wastewater. The results of the adsorption isotherm indicate that surface adsorption, electrostatic adsorption, and ion exchanges are among the potential adsorption mechanisms. This is the first investigation to demonstrate the production of bamboo biochar modified with chitosan without the use of NaOH, $Fe_2(SO_4)_3$, or $FeSO_47H_2O$. Using a vacuum annealing furnace, bamboo particles were pyrolyzed at 900 °C for 4 h while being supplied with N_2 at a flow rate of 100 mL/min. To research the absorption process, FTIR spectroscopy was utilized. The greatest observed sorption capacity of chitosan-modified bamboo biochar was 93.46 mg/g, and the maximum surface area was 85.5 m^2/kg, both of which were optimized at pH 7 and 120 min of contact time. Therefore, chitosan-modified bamboo biochar can be viewed as a workable, optimistic, and highly value-added strategy for detoxifying Cd(II) contaminated water (Huang et al. 2022).

Gao et al. explores the application of biochar prepared from fusarium wilt-infected banana straw in remediation of water cadmium (Cd(II)) pollution. The biochar was prepared by pyrolysis followed by KOH modification at various concentrations to enhance the adsorption capacity. The most optimum was found to be 800 °C for 2 h. Adsorption kinetics model indicated pseudo-second-order phenomenon, and isotherm data could be best fitted into Freundlich model, which represents multilayer adsorption. The adsorption quantity of the modified biochar was 109.8 mg/g (Gao et al. 2024).

2.1.3 Removal of Chromium

The use of environmentally safe and cost-effective, acid-modified (8 M H_2SO_4) rice husk biochar for the elimination of Chromium from water was made at a pH of

7 that makes water safe to consume. The acidic alteration of the original biochar significantly increased the carbon matter by 25.29% and active sites by ten folds in the acid-modified biochar, which led to a Brunauer, Emmett and Teller technique surface area of 142.78, 2.57 m^2/g. Moreover, the pure biochar's surface was altered, resulting in a case of a rise in C=C, C=O, C–O, phenolic, and alcoholic hydroxide bonds, which helped the biochar absorb Cr. The examination of the approximately adsorption site energy distribution function revealed that the elimination was 86% at a temperature of 50 °C, the highest adsorption capacity being 4536 μg g^{-1}, and the Freundlich isotherm model describes the adsorption process most accurately. The biochar was synthesized by pyrolysis at 520 °C; however, modification of the biochar using H_2SO_4 increased its carbon content percentage and increased the number of active sites by ten times on the surface of biochar per gram. From the thermodynamic analyses, a spontaneous and endothermic physical adsorption process was also revealed (Sarkar et al. 2019).

Piscitelli et al. demonstrated the use of olive husk biochar pyrolyzed at 450 °C, along with peat and volcanic rock. Peat olive biochar had a 99.1% sorption capacity, whereas volcanic rock olive biochar had a 99.3% sorption capacity. Water holding capacity was found to be 0.5 g/cm^3 for volcanic rock olive biochar and 0.4 g/cm^3 for peat olive biochar. For the biochar with peat and volcanic rock additions, the pH was kept at 7.9 and 8.7, respectively. One-way analysis of variance (ANOVA) and SPSS software were both used for the statistical analysis of all data. The least significant difference test (LSD test) was used to compare means with a *p*-value less than 0.05 (Piscitelli et al. 2018).

Due to its harmful effects on living things, the elimination of Cr(VI) from aqueous solutions has drawn much attention. Cotton stalk biochar and ZnO nanoparticles were combined to create a new nanocomposite for the batch-scale removal of chromium from contaminated water. The biochar is synthesized by pyrolyzing it at 400 °C, precipitating it with 0.1 M $ZnSO_4$, and then calcining it at 350 °C. The successful impregnation of ZnO particles on cotton stalk biochar with improved surface features were demonstrated by FTIR, scanning electron microscopy, and point of zero charge characterization. Maximum cotton stalk biochar and ZnO modified biochar adsorption was 54.95 and 107.53 mg Cr/g, respectively with the adsorbent dosage being 2 g/L. This was optimized at a pH of 4, at 10–40 °C, and an interaction time of 180 min. According to the findings, doping biochar with metallic nanoparticles was very successful (99.6% at 50 mg/L) and had a high rate of reusability (91%) after five cycles of adsorption/desorption. This suggests that this method is suitable for the decontamination of waters that have been contaminated with Cr(VI) (Tariq et al. 2020).

2.1.4 Removal of Arsenic

Arsenic (As) contamination of water sources has reached dangerous levels and is now a great concern on a global scale. The removal of As from contaminated water using affordable, environmentally friendly sorbents like biochar is currently thought to be a promising option among different treatment methods. The elimination of arsenate As (V) and arsenite As(III) by biochars generated from perilla leaf produced at 300 and 700 °C in aqueous settings was closely examined in aqueous settings. Compared with pyrolysis at 300, 700 °C pyrolyzed biochar provided a superior surface area of 473.4 m^2/g. Predictions of sorption mechanisms were made using a variety of sorption isotherm models, FTIR spectroscopy, microscopic elemental mapping, and X-ray absorption near edge structure spectroscopy. According to the Langmuir model, the sorption capacity was found to be 7.21 mg/g for As(V) and 11.01 mg/g for As(III) at a pH of 7–9 and temperature of 20 °C. Notably, the sorption records and spectroscopic findings revealed that the two main mechanisms of As removal by BC700 were surface complexation/precipitation and pore filling, preferentially for As(III) (Niazi et al. 2018).

Similarly, the use of eggshell biochar to detoxify As-contaminated water was investigated. The studies indicated that the Arsenic (V) removal was 96% with the concentration of 0.6 mg/L, a sorbent dose of 0.9 g/L, and a contact period of 2 h at pH of 4.5, where highest sorption of 6.3 mg/g was noted. In a muffle furnace that has been warmed to 450 °C for 2 h, eggshell biochar is synthesized. The surface morphological characteristics and elemental configuration of biochar were analyzed using a field emission scanning electron microscope, the functional groups of biochar were recognized using FTIR spectroscopy, and the surface characteristics of eggshell biochar with and without As (V) were obtained using X-ray photoelectron spectroscopy. With the aid of well-known kinetic models such as the pseudo-first-order and pseudo-second-order models, the kinetic data of As(V) adsorption on eggshell were further emphasized (Akram et al. 2022).

Green seaweed (*Ulva reticulata*) biochar is one such absorbent that was investigated for the removal of arsenic. About 100 g of grounded algal samples were placed in an electric muffle furnace to prepare the biochar. The electrical muffle furnace was first purged with nitrogen gas at 0.1 L/min for 10 min, and the algal sample was then pyrolyzed for 120 min at 300 °C. The three-parameter models Redlich-Peterson and Toth correctly predicted the experimental isotherm with good precision and high correlation coefficient values among the many model equations investigated to describe the arsenic(V) isotherm. At pH 4, the maximum adsorption capacity of 7.67 mg/g was attained. When compared to the pseudo-second-order model, the pseudo-first-order model clearly illustrated the data and captured the fast sorption kinetics (Senthilkumar et al. 2020).

Devrajani et al. developed iron oxide-impregnated biochar derived from the brown seaweed Sargassum polycystum. It was prepared through pyrolysis at 400 °C followed by impregnating with iron oxide for promoting the adsorptive qualities of the material. maximum arsenic removal efficiency was 96.7% at the optimum pH of 6 while maintaining an adsorbent dosage of 100 mg with an initial arsenic concentration of 0.25 mg/L after contact for 90 min. Kinetics of adsorption fitted into the pseudo-second-order model, indicating a chemisorption mechanism predominantly present in the adsorption process, though the Langmuir isotherm model was shown to best describe the given system. Column studies saturated at 200 min. Results obtained are promising for field applications in scaling up for arsenic removal in large quantities from contaminated water sources (Devrajani et al. 2024)

2.1.5 Removal of Uranium

Environmental engineering's key topics right now include dealing with excess sludge treatment and uranium recovery from wastewater. To extract uranium from water, sludge-based biochar modified with acetic acid (to increase the porosity and add (−COOH)) groups was created by pyrolyzing it at 350–700 °C for 40–50 min in a nitrogen environment. Starting conditions included pH 6.0, U(VI) = 10.0 mg/L (initial concentration), adsorbent dosage = 0.30 g/L, and sorption duration = 5.0 min. This allowed for an ideal U(VI) removal rate of 97.8%. The positive outcome was attributed to the acetic acid modification treatment's dual actions of enlarging pores and boosting –COOH functional groups. Physical and chemical sorption, specifically monodentate coordination binding between –COO– and uranium, are both essential components of the U(VI) sorption process by modified biochar. A maximum sorption capacity of 112.40 mg/g and a surface area of 241.42 m^2/g were measured. The pseudo-second-order and first-order reactions were included. To suit the U(VI) sorption data, Langmuir, and Freundlich sorption isotherm models were utilized (Yu et al. 2022).

2.2 Removal of Miscellaneous Inorganic Compounds

2.2.1 Removal of Ammonium Ions

Nguyen et al. prepared biochar from used coffee grounds was tested for its ability to bind ammonium ions from water. At pH 7, preliminary NH_4^+ concentration of 40 mg/L, and coffee grounds biochar dose of 2 g/L, the best conditions for NH_4^+ adsorption by biochar were discovered. With an R^2 of 0.976, the NH_4^+ adsorption data were highly consistent with the Langmuir isotherm. The biochar is made by pyrolyzing at 500 °C, and further measurements showed the surface area to be 46.318 m^2/g. The highest adsorption capability was determined to be 51.52 mg/g, and the pseudo-second-order model fit the adsorption data accurately. A pseudo-second-order model described how NH_4^+ adsorption of ammonium ions occurred. The results suggest that the Freundlich model was less in agreement with the adsorption data than the Langmuir model (Nguyen et al. 2021).

Sorghum straw was also used to generate biochar, which was then employed as an adsorbent for ammonium adsorption, in order to address the issue of resource use and remove ammonium from water. The biochar was achieved by pyrolyzing it at 300 °C while injecting N_2, and it was then maintained at the desired carbonization temperature for 30 min. At 45 °C, the maximal ammonium adsorption capacity of sorghum straw biochar was 7.09 mg/g, or 7.60 times that of sorghum straw alone. The optimal pH was recorded to be 6–10. After three adsorption-regeneration cycles, the regeneration test showed that sorghum straw biochar had good regeneration capability and that the ammonium adsorption was primarily chemical in nature. The adsorption procedure was more consistent with the Freundlich isotherm adsorption model and the pseudo-second-order kinetic model (Xu et al. 2022a, b).

2.2.2 Removal of Chromium Ions

The International Agency for Research on Cancer has placed Cr(VI), which is chromium pollution, in Group 1 as a major environmental issue. Fish, plants, earthworms, and other living things are all adversely affected by Cr(VI). The risk that chromium poses to living things has made its removal from aqueous solutions a hot topic. Recently, magnetite (Fe_3O_4) nanoparticles and biochar have both drawn a lot of attention, in part because of their potential applications in the purification of water. As a result, the synergetic effect of magnetite biochar was created to test for better outcomes of adsorption of chromate. At 60 °C, it was pyrolyzed. 200 mL of water were dissolved in 2 g of $FeCl_3$ and 1 g of $FeCl_2$ after being bubbled with N_2 gas for 30 min. Adsorption and intraparticle diffusion were used as part of the mechanism for the removal of Cr(VI), which was then reduced to Cr(III), confirmed by Synchrotron-based X-ray absorption spectroscopy. When the pH is low, magnetite and biochar can successfully remove Cr(VI) from aqueous solutions, but when the pH is high, their individual efficacies decline. Furthermore, the magnetite biochar's Cr(VI) adsorption kinetics are quick, reaching equilibrium in just 1 h. The adsorption of Cr(VI) was well fit by both pseudo-first-order and pseudo-second-order models. The highest surface area recorded was 319 m^2/g (Alam et al. 2020).

For Cr(III) ion removal, leftover tomato leaves and stems were used to create biochar. Because the Cr(III) ions precipitate out of the mineral matter in the biochar, this process is critical for removing the ions from aqueous solutions. Although the sorption capacity was reduced four times, it was discovered that biochar could be employed again in the sorption process followed by desorption using 0.1 M HCl. Pyrolysis was used to form the biochar at a temperature of 800 °C and a thermal rate of 10 °C per min. At pH 13, the highest sorption capacity measured was 169.5 mg/g of an aqueous solution having 300 mg/L Chromium(III), and the surface area was 18 m^2/g. The produced biochar was thoroughly examined using FTIR, scanning electron microscopy, pH point of zero charges, ultimate and proximate analysis, and mercury porosimetry. Several prominent models were used to analyze the kinetics of sorption; however, the pseudo-second-order model had the best correlation (R^2 with experimental data) at or above 0.98 (Mokrzycki et al. 2021).

In an experiment, biochar was used to reduce Cr(VI) in a very effective manner. Bagasse from sugarcane was pyrolyzed at 350 °C for 30 min at a heating rate of 10 C/min. The findings revealed that at the quantity of biochar atm(C)/m(Cr) = 3.0, reaction temperature of 90 °C, the reaction time of 60 min, and concentration of H_2SO_4 of 20 g/L, about 100% of Cr(VI) was removed at pH 5–8. The apparent activation energy was determined to be 40.24 kJ/mol based on the reduction kinetics analysis, which showed that the reduction of Chromium (VI) fit well with the pseudo-first-order model. All the experimental factors had a favorable impact on the lowering of Cr, according to response surface methodology (VI). The following was the sequence in which each factor had an impact on the reduction process: (dosage of biochar (m (C)/m(Cr)) > (concentration of H_2SO_4) > (reaction temperature) > (reaction time). The concentration of H_2SO_4 and the dosage of biochar especially had the biggest effects on the reduction process (Peng et al. 2021).

A lanthanum chemical molecule called lanthanum oxide (La_2O_3) has a useful anionic complexing ability but is unstable at low pH levels. A metal chemical compound can be supported by biochar fibers because of their vast surface area and variety of uses. SEM, XRD, and XPS were used to analyze the La_2O_3-modified biochar fiber. An adsorption experiment has shown that the biochar had a good capacity to adsorb chromates, with a saturation adsorption capacity of about 104.9 mg/g. Following pseudo-second-order surface assimilation mechanics, which demonstrated that the Cr(VI) adsorption was characterized by single-phase and polyphase adsorption, primarily chemical adsorption, the information suggested that the adsorption is within phase with the Langmuir and Freundlich models. The endothermic and spontaneous nature of the adsorption process was demonstrated by the thermodynamic characteristics. According to the mechanistic analysis, electrostatic contact, ligand exchange, or complexation, all have a significant impact in the adsorption of Chromium (VI). After pyrolyzing marine brown alginate fiber at 900 °C for 1 h (with a heating rate of 5 C/min), lanthanum oxide was introduced to the biochar. At an ideal temperature of 303 K and a pH of 4.6, the modified biochar had a highest adsorption capacity of 104.9 mg/g and a surface area of 52.08 m^2/g (Sun et al. 2021).

2.2.3 Removal of Phosphorus

Anthropogenic sources of elevated or excessive phosphorus concentrations in water bodies can pose serious risks to both natural ecosystems and food production for humans. Using a solvent-free ball milling technique, MgO-modified biochars were developed in this study and used to remove phosphorus and reduce eutrophication. Rice straw that had undergone pyrolysis at 450 °C at a thermal rate of 2 °C per min under an N_2 atmosphere and maintained for 2 h, forms the required biochar. Phosphorus could be completely removed from livestock effluents comprising of 39.51 mg/L phosphorus by the synthesized 2MgO/BC-450-0.5, which had phosphorus adsorption capabilities of 171.54 mg/g at 25 °C and pH of 5, and also exhibited excellent recycling stability. The pseudo-second-order model ($R^2 = 0.999$) and Langmuir models ($R^2 = 0.982$) both might accurately represent the adoption process, according to the examined kinetics and isotherms. The adsorption of phosphorus on the Magnesium Oxide-modified biochar adsorbent was endothermic and spontaneous. Studies on the impact of pH, FTIR spectra, and XPS spectra revealed that the protonation process, electrostatic attraction, and precipitation process are all involved in the adsorption of phosphorus was spontaneous and endothermic. The surface area obtained was ranging from 20.50 to 212.65 m^2/g (Xu et al. 2022a, b).

Biochar made from Mg (100 mL of 8% $MgCl_2$) modified walnut shells were produced by pyrolysis at 500 °C for 60 min (thermal rate of 30 °C per min), and its capacity to extract Phosphorus from aqueous solutions was evaluated to reduce losses of important Phosphorus when necessary. Mg-based composites (MgO, MgOHCl) are to blame for the 20-fold increase in sorption potential capacity for phosphorus and it gives a way to create sorbents for anionic chemical forms that are stable and effective. Within 360 min, stability is established in this quick procedure. The Sips model (combined Langmuir-Freundlich) has the maximum efficiency when compared to empirical traditional Langmuir and Freundlich models, according to the comparison of coefficients of determination. Both pseudo-first-order and second-order reactions were used. At a pH of 5.5–6, the largest surface area measured was 112.84 m^2/g and the maximum adsorption capacity was 1.43 mg/g (Frišták et al. 2021).

2.2.4 Cadmium Ions

Thiourea a modified biochar was made using poplar bark and sawdust as raw materials. The outcomes showed that the alteration enhanced the biochar's surface structure (surface area of 5.70 m^2/g). The maximum calculated adsorption capacity of 600 °C pyrolyzed sawdust and bark biochar and 600 °C pyrolyzed thiourea-modified poplar biochar was 19.998 mg/g and 9.631 mg/g, correspondingly. The Langmuir adsorption isotherm model and pseudo-second-order kinetic model were better explained with the adsorption behavior. The recommended dosage for the biochar and modified biochar reactions was 4 g/L and 5 g/L, correspondingly and the ideal reaction pH was 7. Due to the surface functional groups, most of the adsorption behavior was chemical. These outcomes revealed that the thiourea-modified biochar has great adsorption ability and superior acid tolerance, that is, encouraging for Cd-polluted water remediation (Zhu et al. 2020).

Using biochar made from chicken manure, the adsorption of Cadmium ions by magnetic and nonmagnetic biochar was tested. Both biochar adsorption processes were better explained by the spontaneous and endothermic Freundlich isotherm model and pseudo-second-order kinetic equation. The highest capacities were revealed to be 60.69 and 41.07 mg/g at the baseline Cadmium (II) content, 180 mg/L for chicken manure biochar and magnetic chicken manure biochar, and a reduction in turbidness of the adsorption-treated solution from 244.3 to 11.3 NTU after 0.5 min of magnetic separation. These showed that, despite being successfully isolated from the treated solutions, the magnetic biochar exhibited a lower capacity for Cd^{2+} adsorption than chicken manure biochar. Additionally, SEM-EDS, XRD, and FTIR tests were performed on both the biochar before and after adsorption. The main adsorption mechanisms were precipitation, ion exchange, complexation, and C-coordination. Ion exchange and precipitation prevailed in the adsorption process by chicken manure biochar, but precipitation was the preferred mechanism in magnetic chicken manure biochar. The surface area measured ranged from 25.56 m^2/g in Chicken Manure Biochar to 5.44 m^2/g in the magnetic biochar, with an optimum condition of pH 11–12 and temperature 303K, and was pyrolyzed at 600 °C for 4 h under an Nitrogenous environment (F. Huang et al. 2020)

3 Conclusion

Although biochar is an efficient method for contaminant adsorption from water, it is mostly done at the laboratory level. Hence focusing only on one or few specific contaminants (Krishnasamy et al. 2022). Natural wastewater is much more complicated and includes various types of heavy metals, organic compounds, and suspended particulate matter coexisting with each other. Due to this, there is conflict for the sorption sites on biochar, which lowers the effectiveness of biochar (Giri et al. 2020). There are some studies on the effect of multi-metal adsorption on biochar, Yang et al. synthesized biochar using Manganese-rich pokeweed and investigated its efficiency on various heavy metal ions which included Cd, Pb, and Cu with Cu being the most effectively adsorbed metal ion and overall showed positive results. On the other hand, the experiment performed by Sheng et al. on the removal of polyaromatic hydrocarbons from constructed wastelands (CW) using copper-loaded biochar as mentioned earlier, saw a significant decline in the efficiency of biochar in the CW system due to the gradual accumulation of the suspended solids, biological film, plant wastes, etc. Therefore, the biochar can be used efficiently only up to a certain saturation level before its adsorption capacity starts to deteriorate (Rong et al. 2024). Although very few studies focus on field application, it cannot be denied that it is one of the most economical methods for water remediation. Some of the biochar used can be reused up to 4–5 times without compromising with the adsorption efficiency (Dalahmeh et al. 2020). Future research could further concentrate on improving the functional group for the adsorption mechanism and the various characteristics of biochar (Iáñez-Rodríguez et al. 2020). To fully explore biochar's potential for wastewater remediation, it is crucial to make it effective in natural wastewater (Kochito et al. 2024).

References

Akram A, Muzammal S, Shakoor MB, Ahmad SR, Jilani A, Iqbal J, Al-Sehemi AG, Kalam A, Aboushoushah SFO (2022) Synthesis and application of egg shell biochar for As(V) removal from aqueous solutions. Catalysts 12(4). https://doi.org/10.3390/catal12040431

Alam MS, Bishop B, Chen N, Safari S, Warter V, Byrne JM, Warchola T, Kappler A, Konhauser KO, Alessi DS (2020) Reusable magnetite nanoparticles–biochar composites for the efficient removal of chromate from water. Sci Rep 10(1). https://doi.org/10.1038/s41598-020-75924-7

Choi YK, Gurav R, Kim HJ, Yang YH, Bhatia SK (2020) Evaluation for simultaneous removal of anionic and cationic dyes onto maple leaf-derived biochar using response surface methodology. Appl Sci (Switz) 10(9). https://doi.org/10.3390/app10092982

Dalahmeh SS, Stenström Y, Jebrane M, Hylander LD, Daniel G, Heinmaa I (2020) Efficiency of iron- and calcium-impregnated biochar in adsorbing phosphate from wastewater in onsite wastewater treatment systems. Front Environ Sci 8. https://doi.org/10.3389/fenvs.2020.538539

Devrajani SK, Ahmed Z, Qambrani NA, Kanwal S, Sundaram UM, Mubarak NM (2024) Mechanism of arsenic removal using brown seaweed derived impregnated with iron oxide biochar for batch and column studies. Sci Rep 14(1). https://doi.org/10.1038/s41598-024-69117-9

Egbedina AO, Adebowale KO, Olu-Owolabi BI, Unuabonah EI, Adesina MO (2021) Green synthesis of ZnO coated hybrid biochar for the synchronous removal of ciprofloxacin and tetracycline in wastewater. RSC Adv 11(30):18483–18492. https://doi.org/10.1039/d1ra01130h

El-Azazy M, El-Shafie AS, Morsy H (2021) Biochar of spent coffee grounds as per se and impregnated with TiO_2: promising waste-derived adsorbents for balofloxacin. Molecules 26(8). https://doi.org/10.3390/molecules26082295

El Hanandeh A, Albalasmeh A, Gharaibeh M, Alajlouni M (2020) Modification of biochar prepared from olive oil processing waste to enhance phenol removal from synthetic and olive mill wastewater. In: Separation science and technology (Philadelphia), pp 1–13. https://doi.org/10.1080/01496395.2020.1794897

Frišták V, Pipíška M, Turčan V, Bell SM, Laughinghouse HD, Ďuriška L, Soja G (2021) Preparation and characterization of novel magnesium composite/walnut shells-derived biochar for as and p sorption from aqueous solutions. Agriculture (Switz) 11(8). https://doi.org/10.3390/agriculture11080714

Gao M, Zhang Y, Gong X, Song Z, Guo Z (2018) Removal mechanism of di-n-butyl phthalate and oxytetracycline from aqueous solutions by nano-manganese dioxide modified biochar. Environ Sci Pollut Res 25(8):7796–7807. https://doi.org/10.1007/s11356-017-1089-5

Gao C, Lan Y, Zhan Y, Li Y, Jiang J, Li Y, Zhang L, Fan X (2024) Preparation of porous biochar from fusarium wilt-infected banana straw for remediation of cadmium pollution in water bodies. Sci Rep 14(1). https://doi.org/10.1038/s41598-024-63954-4

Ge S, Zhao S, Wang L, Zhao Z, Wang S, Tian C (2024) Exploring adsorption capacity and mechanisms involved in cadmium removal from aqueous solutions by biochar derived from euhalophyte. Sci Rep 14(1). https://doi.org/10.1038/s41598-023-50525-2

Giri BS, Gun S, Pandey S, Trivedi A, Kapoor RT, Singh RP, Abdeldayem OM, Rene ER, Yadav S, Chaturvedi P, Sharma N, Singh RS (2020) Reusability of brilliant green dye contaminated wastewater using corncob biochar and *Brevibacillus parabrevis*: hybrid treatment and kinetic studies. Bioengineered 11(1):743–758. https://doi.org/10.1080/21655979.2020.1788353

Gokulan R, Prabhu GG, Jegan J (2019) Remediation of complex remazol effluent using biochar derived from green seaweed biomass. Int J Phytorem 21(12):1179–1189. https://doi.org/10.1080/15226514.2019.1612845

Gómez-Vilchis JC, García-Rosales G, Longoria-Gándara LC, Pérez-Gómez EO, Castilleros DT (2022) Obtention of biochar-Fe/Ce using *Punica granatum* with high adsorption of ampicillin capacity. Heliyon 8(1). https://doi.org/10.1016/j.heliyon.2022.e08841

Hairuddin MN, Mubarak NM, Khalid M, Abdullah EC, Walvekar R, Karri RR (2019) Magnetic palm kernel biochar potential route for phenol removal from wastewater. Environ Sci Pollut Res 26(34):35183–35197. https://doi.org/10.1007/s11356-019-06524-w

Huang F, Zhang L, Wu RR, Zhang SM, Xiao RB (2020) Adsorption behavior and relative distribution of Cd^{2+} adsorption mechanisms by the magnetic and nonmagnetic biochars derived from chicken manure. Int J Environ Res Publ Health 17(5). https://doi.org/10.3390/ijerph17051602

Huang A, Bai W, Yang S, Wang Z, Wu N, Zhang Y, Ji N, Li D (2022) Adsorption characteristics of chitosan-modified bamboo biochar in Cd(II) contaminated water. J Chem. https://doi.org/10.1155/2022/6303252

Kholoma E, Renman A, Renman G (2020) Filter media-packed bed reactor fortification with biochar to enhance wastewater quality. Appl Sci (Switz) 10(3). https://doi.org/10.3390/app10030790

Kochito J, Gure A, Abdissa N, Beyene TT, Femi OE (2024) MnOx-*Coffea arabica* Husk and *Catha edulis* Leftover biochar nanocomposites for removal of methylene blue from wastewater. Sci World J. https://doi.org/10.1155/2024/7585145

Krishnasamy S, Saiatchyuth BA, Ravindiran G, Subramaniyan B, Ramalingam M, Sai Vamsi JUB, Ramesh B, Razack NA (2022) Effective removal of reactive yellow 145 (RY145) using biochar derived from groundnut shell. Adv Mater Sci Eng. https://doi.org/10.1155/2022/8715669

Kumi AG, Ibrahim MG, Fujii M, Nasr M (2022) Petrochemical wastewater treatment by eggshell modified biochar as adsorbent: atechno-economic and sustainable approach. Adsorp Sci Technol. https://doi.org/10.1155/2022/2323836

Kyi PP, Quansah JO, Lee CG, Moon JK, Park SJ (2020) The removal of crystal violet from textile wastewater using palm kernel shell-derived biochar. Appl Sci (Switz) 10(7). https://doi.org/10.3390/app10072251

Le PT, Bui HT, Le DN, Nguyen TH, Pham LA, Nguyen HN, Nguyen QS, Nguyen TP, Bich NT, Duong TT, Herrmann M, Ouillon S, Le TPQ (2021) Preparation and characterization of biochar derived from agricultural by-products for dye removal. Adsorp Sci Technol. https://doi.org/10.1155/2021/9161904

Li J, Shen F, Yang G, Zhang Y, Deng S, Zhang J, Zeng Y, Luo T, Mei Z (2018) Valorizing rice straw and its anaerobically digested residues for biochar to remove Pb(II) from aqueous solution. Int J Polym Sci 2018:1–11. https://doi.org/10.1155/2018/2684962

Li J, Yu G, Pan L, Li C, You F, Wang Y (2020) Ciprofloxacin adsorption by biochar derived from co-pyrolysis of sewage sludge and bamboo waste. Environ Sci Pollut Res 27(18):22806–22817. https://doi.org/10.1007/s11356-020-08333-y

Li P, Zhao Z, Zhang M, Su H, Zhao T, Feng W, Zhang Z (2024a) Exploring the potential of biochar derived from Chinese herbal medicine residue for efficient removal of norfloxacin. Molecules 29(9). https://doi.org/10.3390/molecules29092063

Li Z, Xiao X, Xu T, Chu S, Wang H, Jiang K (2024b) Removal of Pb(II) and Cd(II) from a monometallic contaminated solution by modified biochar-immobilized bacterial microspheres. Molecules (Basel, Switz) 29(19). https://doi.org/10.3390/molecules29194757

Liu L, Li Y, Fan S (2019) Preparation of KOH and H_3PO_4 modified biochar and its application in methylene blue removal from aqueous solution. Processes 7(12). https://doi.org/10.3390/PR7120891

Liu X, Li G, Chen C, Zhang X, Zhou K, Long X (2022) Banana stem and leaf biochar as an effective adsorbent for cadmium and lead in aqueous solution. Sci Rep 12(1). https://doi.org/10.1038/s41598-022-05652-7

Mansouri FE, Farissi HE, Zerrouk MH, Cacciola F, Bakkali C, Brigui J, Lovillo MP, Esteves da Silva JCG (2021) Dye removal from colored textile wastewater using seeds and biochar of barley (*Hordeum vulgare* L.). Appl Sci (Switz) 11(11). https://doi.org/10.3390/app11115125

Moharm AE, El El Naeem GA, Soliman HMA, Abd-Elhamid AI, El-Bardan AA, Kassem TS, Nayl AA, Bräse S (2022) Fabrication and characterization of effective biochar biosorbent derived from agricultural waste to remove cationic dyes from wastewater. Polymers 14(13):2587. https://doi.org/10.3390/polym14132587

Mohd Faizal AN, Putra NR, Abdul Aziz AH, Agi A, Ahmad Zaini MA (2024) Giant mud crab shell biochar: a promising adsorbent for methyl violet removal in wastewater treatment. J Clean Prod 447:141637. https://doi.org/10.1016/J.JCLEPRO.2024.141637

Mokrzycki J, Michalak I, Rutkowski P (2021) Tomato green waste biochars as sustainable trivalent chromium sorbents. Environ Sci Pollut Res 28(19):24245–24255. https://doi.org/10.1007/s11356-019-07373-3

Nguyen VT, Vo TDH, Tran T, Nguyen TN, Le TNC, Bui XT, Bach LG (2021) Biochar derived from the spent coffee ground for ammonium adsorption from aqueous solution. Case Stud Chem Environ Eng 4. https://doi.org/10.1016/j.cscee.2021.100141

Nguyen NT, Chen PR, Ye RH, Chuang KJ, Chang CT, Hong GB (2024) Optimization of extraction conditions from Gac fruit and utilization of peel-derived biochar for crystal violet dye removal. Molecules 29(14). https://doi.org/10.3390/molecules29143435

Niazi NK, Bibi I, Shahid M, Ok YS, Burton ED, Wang H, Shaheen SM, Rinklebe J, Lüttge A (2018) Arsenic removal by perilla leaf biochar in aqueous solutions and groundwater: an integrated spectroscopic and microscopic examination. Environ Pollut 232:31–41. https://doi.org/10.1016/j.envpol.2017.09.051

Nyamunda BC, Chivhanga T, Guyo U, Chigondo F (2019) Removal of Zn(II) and Cu(II) ions from industrial wastewaters using magnetic biochar derived from water hyacinth. J Eng (UK). https://doi.org/10.1155/2019/5656983

Ocampo-Perez R, Padilla-Ortega E, Medellin-Castillo NA, Coronado-Oyarvide P, Aguilar-Madera CG, Segovia-Sandoval SJ, Flores-Ramírez R, Parra-Marfil A (2019) Synthesis of biochar from chili seeds and its application to remove ibuprofen from water. Equilibrium and 3D modeling. Sci Total Environ 655:1397–1408. https://doi.org/10.1016/j.scitotenv.2018.11.283

Peng H, Guo J, Qiu H, Wang C, Zhang C, Hao Z, Rao Y, Gong Y (2021) Efficient removal of cr (Vi) with biochar and optimized parameters by response surface methodology. Processes 9(5). https://doi.org/10.3390/pr9050889

Piscitelli L, Rivier PA, Mondelli D, Miano T, Joner EJ (2018) Assessment of addition of biochar to filtering mixtures for potential water pollutant removal. Environ Sci Pollut Res 25(3):2167–2174. https://doi.org/10.1007/s11356-017-0650-6

Ravindiran G, Sugumar P, Elias G (2021) Continuous sorption of remazol brilliant orange 3R using caulerpa scalpelliformis biochar. Adv Mater Sci Eng. https://doi.org/10.1155/2021/6397137

Rong L, Wu L, Zhang T, Hu C, Tang H, Pan H, Zou X (2024) Significant differences in the effects of nitrogen doping on pristine biochar and graphene-like biochar for the adsorption of tetracycline. Molecules 29(1). https://doi.org/10.3390/molecules29010173

Sarkar A, Ranjan A, Paul B (2019) Synthesis, characterization and application of surface-modified biochar synthesized from rice husk, an agro-industrial waste for the removal of hexavalent chromium from drinking water at near-neutral pH. Clean Technol Environ Policy 21(2):447–462. https://doi.org/10.1007/s10098-018-1649-5

Senthilkumar R, Reddy Prasad DM, Govindarajan L, Saravanakumar K, Naveen Prasad BS (2020) Synthesis of green marine algal-based biochar for remediation of arsenic(V) from contaminated waters in batch and column mode of operation. Int J Phytorem 22(3):279–286. https://doi.org/10.1080/15226514.2019.1658710

Shafiq M, Alazba AA, Amin MT (2021) Kinetic and isotherm studies of Ni^{2+} and pb^{2+} adsorption from synthetic wastewater using eucalyptus camdulensis—derived biochar. Sustainability (Switz) 13(7). https://doi.org/10.3390/su13073785

Shen X, Zhang J, Xie H, Hu Z, Liang S, Ngo HH, Guo W, Chen X, Fan J, Zhao C (2020) Intensive removal of PAHs in constructed wetland filled with copper biochar. Ecotoxicol Environ Saf 205. https://doi.org/10.1016/j.ecoenv.2020.111028

Shirvanimoghaddam K, Czech B, Tyszczuk-Rotko K, Kończak M, Fakhrhoseini SM, Yadav R, Naebe M (2021) Sustainable synthesis of rose flower-like magnetic biochar from tea waste for environmental applications. J Adv Res 34:13–27. https://doi.org/10.1016/j.jare.2021.08.001

Shoaib AGM, Van HT, Tran DT, El Sikaily A, Hassaan MA, El Nemr A (2024) Green algae ulva lactuca-derived biochar-sulfur improves the adsorption of methylene blue from water. Sci Rep 14(1). https://doi.org/10.1038/s41598-024-61868-9

Song J, Li Y, Wang Y, Zhong L, Liu Y, Sun X, He B, Li Y, Cao S (2021) Preparing biochars from cow hair waste produced in a tannery for dye wastewater treatment. Materials 14(7). https://doi.org/10.3390/ma14071690

Suhaimi N, Kooh MRR, Lim CM, Chou Chao CT, Chou Chau YF, Mahadi AH, Chiang HP, Haji Hassan NH, Thotagamuge R (2022) The use of gigantochloa bamboo-derived biochar for the removal of methylene blue from aqueous solution. Adsorp Sci Technol. https://doi.org/10.1155/2022/8245797

Sun P, Zhang K, Gong J, Khan A, Zhang Y, Islama MS, Zhang Y (2019) Sunflower stalk–derived biochar enhanced thermal activation of persulfate for high efficient oxidation of *p*-nitrophenol. Environ Sci Pollut Res 26(26):27482–27493. https://doi.org/10.1007/s11356-019-05881-w

Sun X, Guo P, Sun Y, Cui Y (2021) Adsorption of hexavalent chromium by sodium alginate fiber biochar loaded with lanthanum. Materials 14(9). https://doi.org/10.3390/ma14092224

Tam NTM, Liu YG, Bashir H, Zhang P, Liu SB, Tan X, Dai MY, Li MF (2020) Synthesis of porous biochar containing graphitic carbon derived from lignin content of forestry biomass and its application for the removal of diclofenac sodium from aqueous solution. Front Chem 8. https://doi.org/10.3389/fchem.2020.00274

Tang S-F, Zhou H, Tan W-T, Huang J-G, Zeng P, Gu J-F, Liao B-H (2022) Adsorption characteristics and mechanisms of Fe–Mn oxide modified biochar for Pb(II) in wastewater. Int J Environ Res Public Health 19(14):8420. https://doi.org/10.3390/ijerph19148420

Tariq MA, Nadeem M, Iqbal MM, Imran M, Siddique MH, Iqbal Z, Amjad M, Rizwan M, Ali S (2020) Effective sequestration of Cr(VI) from wastewater using nanocomposite of ZnO with cotton stalks biochar: modeling, kinetics, and reusability. Environ Sci Pollut Res 27(27):33821–33834. https://doi.org/10.1007/s11356-020-09481-x

Tian X, Xie Q, Chai G, Li G (2022) Simultaneous adsorption of As(III) and Cd(II) by ferrihydrite-modified biochar in aqueous solution and their mutual effects. Sci Rep 12(1). https://doi.org/10.1038/s41598-022-09648-1

Van Limbergen T, Roegiers IH, Bonné R, Mare F, Haeldermans T, Joos B, Nouwen O, Manca JV, Vangronsveld J, Thijs S (2022) Characterisation of two wood-waste and coffee bean husk biochars for the removal of micropollutants from water. Front Environ Sci 10. https://doi.org/10.3389/fenvs.2022.814267

Xu H, Wang B, Zhao R, Wang X, Pan C, Jiang Y, Zhang X, Ge B (2022a) Adsorption behavior and performance of ammonium onto sorghum straw biochar from water. Sci Rep 12(1). https://doi.org/10.1038/s41598-022-08591-5

Xu S, Li D, Guo H, Lu H, Qiu M, Yang J, Shen F (2022b) Solvent-free synthesis of MgO-modified biochars for phosphorus removal from wastewater. Int J Environ Res Public Health 19(13):7770. https://doi.org/10.3390/ijerph19137770

Xu C, Liu R, Chen L (2023) Removal of phosphorus from domestic sewage in rural areas using oyster shell-modified agricultural waste–rice husk biochar. Processes 11(9). https://doi.org/10.3390/pr11092577

Yang C, Miao S, Li T (2021) Influence of water washing treatment on Ulva prolifera-derived biochar properties and sorption characteristics of ofloxacin. Sci Rep 11(1). https://doi.org/10.1038/s41598-021-81314-4

Yu P, Hu T, Chen HH, Wu F, Liu H (2018) Effective removal of congo red by triarrhena biochar loading with TiO_2 nanoparticles. Scanning. https://doi.org/10.1155/2018/7670929

Yu S, Wu X, Ye J, Li, M, Zhang, Q, Zhang, X, Lv, C, Xie, W, Shi, K, Liu, Y (2022) Dual effect of acetic acid efficiently enhances sludge-based biochar to recover uranium from aqueous solution. Front Chem 10. https://doi.org/10.3389/fchem.2022.835959

Zhang J, Liu M, Yang T, Yang K, Wang H (2017) Synthesis and characterization of a novel magnetic biochar from sewage sludge and its effectiveness in the removal of methyl orange from aqueous solution. Water Sci Technol 75(7):1539–1547. https://doi.org/10.2166/wst.2017.014

Zhu Y, Liang H, Yu R, Hu G, Chen F (2020) Removal of aquatic cadmium ions using thiourea modified poplar biochar. Water (Switz) 12(4). https://doi.org/10.3390/W12041117

Zoghlami RI, Hechmi S, Weghlani R, Jedidi N, Moussa M (2021) Biochar derived from domestic sewage sludge: influence of temperature pyrolysis on biochars' chemical properties and phytotoxicity. J Chem. https://doi.org/10.1155/2021/1818241

Chapter 11
Environmental Compatibility of Biochar

Noopur Srivastava, Nisha Saxena, and Priya Yadav

Abstract With the ever-growing burden on the environment, it becomes essential to ponder on the economically viable and sustainable options for the present and future generations. Biochar has emerged as a cheap and easily available source that offers multiple solutions to environmental problems. It is a versatile source in reducing GHGs, removing heavy metals and pollutants from the soil, improving the flora and fauna of the soil. In addition, there have been several findings which indicate that biochar can be an alternative source of energy. This chapter puts forward a comprehensive discussion on the compatibility of biochar with the environment.

Keywords Controlled pyrolysis · Soil conditioner · GHGs removal · Drought stress · Syngas

1 Introduction

Biochar is a substance similar to charcoal that is produced by the controlled pyrolysis of organic waste from the forestry and agricultural industries (also known as biomass). Biochar is made by a secret procedure that prevents contamination and keeps the carbon in it intact, despite its gritty exterior. Wood pellets, leaf litter, and decaying plant matter are only some of the organic resources that may be pyrolyzed. Hence, there is little to zero pollution created during combustion. Once the organic matter undergoes pyrolysis, it is transformed into Biochar, a type of carbon that is

N. Srivastava (✉) · P. Yadav
Department of Chemistry & Biochemistry, Sharda School of Engineering & Sciences, Sharda University, Plot 32, 34, Knowledge Park III, Greater Noida 201310, Uttar Pradesh, India
e-mail: noopursriv@gmail.com; noopur.srivastava@sharda.ac.in

P. Yadav
e-mail: y.rithika24@gmail.com

N. Saxena
Department of Chemistry, M. R. M. College, Lalit Narayan Mithila University, Darbhanga 846008, Bihar, India
e-mail: nisha.cdri.du@gmail.com

A. Roy et al. (eds.), *Biochar Production Engineering*,
https://doi.org/10.1007/978-3-031-92594-8_11

very stable and resistant to atmospheric decomposition. During pyrolysis, energy or heat may be gathered and converted into useful forms of renewable power. Biochar is a black, porous, lightweight, fine-grained material with a wide range of possible uses. Biochar is the solid organic byproduct of biomass pyrolysis. As a soil amendment, biochar has a major impact on soil fertility because it modifies the soil on a chemical, biological, and physical level. Hence, Biochar has significantly mitigated such problems as global warming, soil nutrient loss, and production of greenhouse gases. Biochar also has the ability to trap poisonous metals. For this reason, biochar production may enhance soil quality and provide new economic possibilities (Lee and Park 2020; Hu et al. 2020; Han et al. 2016; Jin et al. 2018a, b, c).

2 Soil Remediation

Utilizing this carbon-rich substance to improve soil health, rehabilitate damaged soils, and boost agricultural productivity is known as soil remediation. The capacity of biochar to immobilize heavy metals like lead, cadmium, and arsenic lowers their toxicity and bioavailability to microorganisms and plants, which is one of its main advantages. Its vast surface area and porous structure make it easier for these metals to bind, reducing environmental concerns and crop uptake. Furthermore, pesticides and polycyclic aromatic hydrocarbons (PAHs) are among the organic pollutants that biochar efficiently adsorbs, restoring soil quality and lowering the possibility of groundwater pollution. Biochar modifies soil pH, rendering acidic soils more neutral, thereby enhancing nutrient availability and overall soil health, which promotes improved plant growth. Additionally, it enhances nutrient retention, facilitating the availability of essential nutrients such as nitrogen, phosphorus, and potassium, which in turn decreases the reliance on chemical fertilizers (Lee and Park 2020). Biochar can help reduce soil contamination by adsorbing heavy metals and other pollutants, lowering their bioavailability and potential harm to plants and ecosystems (Table 1; Fig. 1). Finally, biochar can be made from organic waste, providing a long-term waste management solution while also helping to sequester carbon. Biochar contributes to sustainable environmental management by improving soil health, sequestering carbon, and removing contaminants. Biochar improves soil qualities including pH and cation exchange capacity (CEC), increasing nutrient availability, which is critical for long-term agricultural output. Its stable carbon concentration enables long-term carbon sequestration, which helps to mitigate climate change by lowering greenhouse gas emissions. Furthermore, biochar's capacity to absorb and immobilize pollutants makes it a valuable tool for soil and water rehabilitation, solving pollution challenges. It also provides a habitat for beneficial soil bacteria, increasing microbial activity and nutrient cycling, hence improving soil health. Furthermore, the porous structure of biochar aids in water retention, especially in arid environments, which promotes plant growth by keeping soil moisture levels stable (Shoudho et al. 2024; Rao et al. 2024).

Table 1 Effects of biochar in soil remediation

Parameters	Details
Influence of sand content	Higher sand concentration in semi-arid silt-rich soils improves the efficiency of organic additions for water retention
Comparison of organic amendments	Biochar is more successful than compost at increasing soil water retention, with biochar-amended soils containing substantially more water
Quantitative improvements	In sandy loamy soils supplemented with biochar, available water capacity increased by up to 80% as compared to unamended soils
Long-lasting effect	The application of compost and biochar can lead to long-term improvements in soil qualities, which contribute to improved water resource management
Soil texture influence	The impact of organic additions on water retention depends critically on the texture of the soil, especially the percentage of sand
Waste management	Using date palm wastes to make biochar is a sustainable way to manage agricultural waste, lowering the environmental consequences of burning or discarding these residues
Increased porosity	Biochar has a very porous structure, increasing the soil's overall porosity. This porosity helps the soil to hold more water, increasing water storage capacity and slowing drainage. Compared to unamended soils, biochar's increased surface area allows for improved water retention
Enhanced capillary action	The fine pores in biochar can enhance capillary action, allowing water to travel more efficiently across the soil profile. This helps to keep moisture available to plants, particularly during dry months

2.1 Greenhouse Gases and Their Interaction with Biochar in Soil

Climate change is not a new term and the key cause of changes in the climate is directly related to the increase in greenhouse gases (GHG) that lead to global warming. "The contribution of carbon dioxide (CO_2) emission, one of the major greenhouse gas contributes above 77% to this". In addition to this, the carbon (IV) oxide emission over soil respiration is observed about 10 times higher when compared with the burning of fossil fuel-based production of the gas.

It is required to moderate climate change that carbon dioxide contaminants from agricultural soil must be decreased. Biochar is one of the essential components needed to be added in agricultural soil. In fact, it is used to increase soil carbon sequestration and this helps to reduce both nitrous oxide (N_2O) and (CH_4) emissions. "More interestingly recent researches has shown that biochar could conceivably decrease the GHGs emissions, which are accountable for global warming". The nitrous oxide and methane, from the soil were found to have posed major effects on climate change.

A wonderful demonstration of this can be shown in, "biochar obtained from paper mill waste, biosolids, and green waste poultry litter that efficiently reduces

Fig. 1 Multifaceted role of biochar in soil remediation

N_2O emissions from an acidic Ferro sol, some researchers, however discredit the facts". "These examples thus indicate that several biochar types have an effect on GHG emissions from soils in numerous ways. In this regard, it is apparent that the type of feedstocks of biochar, water content available in the soil, and temperature for biochar pyrolysis that affects biochar potential to reduce greenhouse gas emissions" (Leng and Huang 2018).

Many researches are carried out regarding the consequence of biochar on "soil carbon dioxide emission; but the major setback about it that results are not conclusive because of the alterations in research methodology used and materials chosen". There are certain steps that encompass the reduction of GHGs emissions by employing biochar, which is multifaceted and fetching clearer methodically (Leng and Huang 2018; Purakayastha et al. 2016). It is proposed that biochar application to soil can intensify the action of microorganisms that are involved in the decomposition of nitrous oxide to nitrogen. However, "it is observed that because of alkalinity of biochar the activity of nitrous oxide reducing organisms is increased" (Major et al. 2010). Further studies on biochar provide facts that great adsorption sites are available on the surface of biochar for nitrous oxides absorption, owing to its particle nature possess large surface area, thus dropping down the release of these gases from the soil based ecological unit (Novak et al. 2010; Wild and Jones 2009). Likewise, "there is a need to also specified that when there is an increase in soil potential by adding biochar, the drawbacks of it that biochar will definitely reduce the alter soil acidity and thus affects emissions".

2.2 Biochar Application for Gas Remediation

Removal of H_2S

Biochar is very dynamic substance used for the remediation of poisonous substances from gas. Biochar is found to efficiently remove H_2S, especially from sludge, camphor, rice hull, hardwood chip, bamboo, manure from pig, and from biogas with adsorption capacity between 110 and 370 mg H_2S/g biochar, which is found to have an efficiency removal of over 96% (O'Toole et al. 2013).

Studies suggested that "adsorption of H_2S was principally assisted due to various factors such as an existing surface area, biochar moisture content (> 85% v/w), its pH (> 8.0), and possibility of chemical bonding with surface radical groups. This can be understood by considering example of OH and COOH functionality present, H_2S is found to intermingle with alkali biochar surface through ionic interaction of e-OH and e-COOH functional groups specially in the presence of oxygen and water that leads to the formation of $(K, Na)_2SO_4$, that makes it bioavailable as SO_4^{2-} to plants" (Lehmann and Joseph 2015).

Removal of Trichloroethylene

Interestingly trichloroethylene abstraction by biochar obtained from peanut shells (PBC) and soybean straw (SBC) was tested at pyrolyzed temperatures between 350 and 750 °C in one of the experiments. The studies demonstrated that for "effective trichloroethylene elimination Biochar obtained at a high pyrolyzed temperature for (PBC) and (SBC) the SBC700 and P-BC700 respectively were more efficient when compared with those obtained at a low pyrolyzed temperature i.e. SBC300 and PBC300" respectively.

The effectiveness of "PBC and SBC were also compared with the marketable activated char. The rate of application of 0.28 g/L PBC700 was found better compared to that of activated biochar that has removal efficiency of > 88% for up to trichloroethylene concentration of 9 mg/L". The results indicate that the "efficiencies for elimination of trichloroethylene by biochar (PBC700 and SBC700) were due to (i) enhancement of hydrophobicity (~ 10% O removal) (ii) surface area (12–410 m^2/g), (iii) reduction in the polarity of biochar when comparison with (PBC300 and SBC300) was performed" (Lehmann et al. 2006a).

2.3 Removal of Organic Pollutants

Sources of Pollutants

The biochar application for the "removal of organic pollutants from soil is very crucial, particularly for the removal of chemicals used in agricultural lands such as fungicide, herbicides and pesticides; including insecticides that are used extensively and have many harmful effects these are chlorpyrifos, atrazine, pyrimethanil, simazine, carbofuran etc." (Spokas 2010). In addition to this few drugs also need to be carefully removed from the environment such as antibiotics/drugs (for example acetaminophen, sulfamethoxazole, tetracycline, sulfamethazine, ibuprofen, tyrosine, etc.).

"Industrial chemicals have toxic effects on living beings and need to be addressed or removed from environment they include (i) polycyclic aromatic hydrocarbons (PAHs) (for example catechol, *p*-nitrotoluene polychlorinated biphenyl, naphthalene, pyrene, *m*-dinitrobenzene, etc.), (ii) volatile organic compounds (e.g., benzene, furan, trichloroethylene, etc.), (iii) cationic aromatic dyes (e.g., methylene blue rhodanine, methyl-violet, etc. (Herath et al. 2015).

Apart from these various organic compounds are also found available in environment from various sources such as toxic organic compounds in landfill leachate, that exist in specific waste streams (estrogen compounds in animal manure and sewage, inhibitory compounds of biomass degradation such as furfural, phenolic compounds, etc." (Xia et al. 2017).

Mechanism Works for Removal of Pollutants

The cleanliness or abstraction mechanisms are often directed by the interfaces of these contaminants with several features of biochar. The removal or elimination of organic contaminants is predominantly either by chemisorption (electrophilic interaction) or by physisorption. The types of interactions possible between biochar and organic pollutants are (i) hydrophobic (ii) electrostatic attraction/repulsion via π–π electron donor–acceptor, (iii) H bonding over carboxylic, diols, alcohols functional groups, and (iv) pore diffusion (Spokas 2010).

Other pollutants can be removed by following other mechanisms which include; (i) partitioning, (ii) chemical transformation (through a reductive reaction or electrical conductivity), and (iii) the majority of the bonded contaminants are eventually mineralized through biodegradation (using diverse microorganisms present on the surface and in the micro-pores of biochar) (Yin et al. 2014).

Factors Need Attention for Removal of Pollutants

The factors that affect elimination capabilities of biochar are type of feedstock and pollutant ratios to biochar, pH, and pyrolysis temperature. These factors affect biochar interactions with organic contaminants and play a vital role in their removal.

Role of Temperature

(a) "As higher pyrolysis temperature provides higher surface area and microporosity in the biochar, that makes more vulnerable for the removal of nonpolar organic pollutants, while biochar formed at a lower temperature does not have these desirable properties required to eliminate pollutants" (Awad 2018).
(b) "The higher the pyrolysis temperature (above 500 °C), the higher the aromaticity that leads low polarity and acidity on the surface of biochar it results in loss of O– and H-containing functional groups".
(c) "The impact of temperature can be better experimented with the removal of polar insecticide and herbicide compounds, this occurs due to polar specific interactions and in the process of removing different organic contaminants (e.g., agrochemicals) present in the soil by specific biochar and the mechanisms employed for their removal. Infact Biochar has been able to reduce the bioavailability of organic contaminants in soil and their uptake by plants and microorganisms that help grow crops safely" (Spokas 2010).

Biochar in the Management of VOCs in Soil

Biochar's huge surface area and porous nature make it a good VOC adsorbent. "Adsorption involves physical and chemical interactions between VOCs and biochar, such as π–π interactions, hydrogen bonding, and van der Waals forces. According to studies, the adsorption capacity of biochar for VOCs varies greatly, ranging from 5.58 to 91.2 mg/g depending on the type of biochar and its features. Biochar provides a habitat for microorganisms that breakdown VOCs". The large surface area of biochar not only allows for pollutant adsorption but also promotes microbial colonization, which can aid in the biodegradation of VOCs in the capping layer.

Research shows that "biochar-modified capping layers can successfully reduce VOC emissions from contaminated locations. For example, a capping layer containing biochar has been demonstrated to considerably reduce the release of VOCs in petroleum-contaminated environments". Biochar's great stability allows it to keep its adsorptive qualities throughout time, making it a long-term option for VOC elimination in contaminated soils (Wang et al. 2024).

Biochar in the Management of Carbon Cycle

Biochar serves as a stable carbon reservoir in soils, significantly contributing to the carbon cycle and aiding in climate change mitigation. "When organic biomass is pyrolyzed into biochar, a significant part of the carbon in the biomass is preserved in a stable form. This stability enables biochar to store carbon in soil for extended periods of time, potentially thousands of years, lowering the amount of carbon dioxide (CO_2) released into the environment. In the carbon cycle, biochar helps to sequester carbon by increasing soil organic carbon (SOC) levels. When applied to soils, biochar not only contributes carbon but also encourages the production of organomineral complexes, which help to maintain other types of soil carbon. This stabilization process helps to reduce soil respiration rates and CO_2 emissions, effectively locking carbon in the soil. Furthermore, biochar can change microbial communities and activity, affecting carbon dynamics in the soil. For example, certain kinds of biochar can boost microbial biomass and activity, resulting in enhanced nutrient cycling and organic matter breakdown, which can either release or retain carbon depending on the conditions. Furthermore, biochar application can affect soil parameters such as pH and nutrient availability, hence improving plant growth and biomass output. Increased plant biomass adds carbon to the soil via root exudates and decaying plant debris, thus enriching the soil carbon pool".

Biochar has several substantial benefits in the context of "the carbon cycle, most notably its ability to trap carbon and prevent climate change. Biochar, with its high carbon content and resistance to degradation, can store carbon in soil for long periods of time, effectively lowering atmospheric CO_2. This feature makes it an important instrument in combating climate change because it helps to reduce greenhouse gas emissions. When applied to soils, biochar can lower the emissions of hazardous gases like methane and nitrous oxide by immobilizing inorganic nitrogen and changing soil characteristics, which affects gas production. Biochar not only sequesters carbon, but also improves soil texture, structure, and porosity. This improves water retention and nutrient cycling, which can dramatically boost agricultural production. Furthermore, biochar stimulates soil microbial activity, which improves nutrient availability and overall soil health. It also increases the soil's cation exchange capacity, allowing it to retain key elements such as calcium, magnesium, and potassium, which are required for plant growth" (Rao et al. 2024).

3 Renewable Energy from Biochar

The world is in the midst of energy crisis. Future generations cannot depend on the existing sources of energy. Careful introspection and planning are needed to quench the increasing energy needs. One such promising method is deriving energy from biomass. Table 2 lists the different sources of biochar from which fuels have been derived.

4 Enhancement of Physiology, Mitigation of Salt-Drought Stress, Effects on Flora and Fauna Due to Biochar

Water deficiency leads to the problem of drought. This in turn affects the fertility, decreased growth of plants, and poor yield of fruits. Physiology of the plant is affected as a whole in such conditions. Application of biochar to drought-affected regions has a significant effect on the flora and fauna. Table 3 lists the effect of biochar on various flora and fauna. Some reports of the biochar serving as a compost have also been reported which enables a direct release of inhibitors of plant pathogens; increase in growth of microorganisms that act antagonistic to pathogens; enhanced disease resistance; and improved plant defense mechanisms.

5 Biochar as Soil Conditioner

The discovery of *preta de* Indio-Indian black earth (in Amazonia), which is distinguished for its high content of carbon and other useful nutrients created interest in biochar as a potential soil enhancer (Macdonald et al. 2014). "Many researches carried out previously shown that, the origin of soils took place few thousand years ago as a consequence of the burning of forests via natural fires. The soil improvement was carried out by pre-Columbian natives by the use of charcoal (Beesley et al. 2011). The application of carbonization products for soil enrichment, is found extremely beneficial from the agricultural point of view because this input improves the conditions for plant growth, that leads a better yield of crops" (Zhang et al. 2015).

The process is cost-effective and possesses rapid effects thus biochar is more and more frequently used in treatment of soil for its remediation and conservation (Woolf et al. 2010).

Efficiency and Effectiveness of Biochar

How efficient and active will be "the biochar, after it is added into the soil environment, depends chiefly on the feedstocks used for its production and the methodology adopted along with parameters selected for the pyrolysis process". These conditions

Table 2 Fuels derived from biochar

S. No.	Material/feedstock	Technology/ methodology	Type of fuel	References
1.	Chestnut wood chips (*Castanea Sativa Miller*)	Integrated Pyrolysis Regenerated Plant (IPRP)	Oil and char	Paethanom et al. (2013)
2.	"Trestina, field crops: sorghum, hemp, kenaf);—Pietrafitta field crops black locust);—Beroide field crops -poplar clones);—Montelabate field crops: Jerusalem artichoke, black locust, poplar clones);—Casalina field crops: Jerusalem artichoke, poplar clones, sunflower, sorghum, miscanthus, Arundo donax and cardoon)"	Integrated pyrolysis regenerated plant (IPRP)		Bartocci et al. (2016)
3.	"Commercial wood pellets (charcoal pellet)"	Integrated pyrolysis regenerated plant (IPRP)	Syngas	Bartocci et al. (2018)
4.	"Shell samples of walnut, sunflower, hazelnut, cotton cocoon, and almond almond, cotton cocoon, hazelnut, sunflower, and walnut"	Supercritical water extraction	Hydrogen	Demirbas (2004), Calzavara et al. (2005) and Feng et al. (2004)
5.	"*Calotropis Procera*, sugarcane bagasse and *Jatropha Curcus* seed cake beechwood, nutshells olive husks, grape residues, and straw pellets"	Fast pyrolysis	Hydrogen	Blasi et al. (1999)
6.	Cedarwood, aspen	Catalytic pyrolysis/ gasification	Hydrogen	Asadullah et al. (2001, 2003), Arauzo et al. (1994), Bohlmann et al. (1998) and Dalai et al. (2003)

(continued)

Table 2 (continued)

S. No.	Material/feedstock	Technology/ methodology	Type of fuel	References
7.	"Bagasse waste, beech wood, Japanese oak without bark, Pinus pinaster, Eucalyptus globules and holm-oak, almond shells, Cuban bagasse"	Biomass gasification	Hydrogen	Kersten et al. (2003), Adhikari et al. (2006), Andersson and Harvey (2006), Meng et al. (2006), André et al. (2005), Rioche et al. (2005), Zhang et al. (2005), Dasappa et al. (2004), Fillipis et al. (2004), Kim (2003), Hanaoka et al. (2005), Warnecke (2000), Narvaez et al. (1996), Franco et al. (2003) and Chaudhari et al. (2001)
8.	Wood biochar	Steam gasification of char	Hydrogen	Bakshi et al. (2003) and Chen and Kuo 2010)
9.	"Bamboo, willow, coconut shell and wood" (*Ficus benjamina* L.)	Pyrolysis	Hydrogen	Artiola et al. (2012)

regulate the contents of macroelements, microelements as well as harmful substances persisting in biochar produced. The harmful substances maybe heavy metals.

The chemical composition of "biochar's usually heterogeneous that enables their interactions with a large variety of organic and inorganic components/compounds present in the soil". The varied properties of "biochar materials lead its reactions with mineral and organic fractions already present in the soil and thus the accumulation of combined mineral and organic complexes therefore created by such interactions added to the soil" (Li et al. 2018).

Biochar Characterization and Capacities

The characterization of biochar depends upon "(i) high stability of biochar and (ii) its resistance towards biological decomposition. Therefore, it is considered as an extremely effective and efficient medium for the sequestration of CO_2 in soil". Further, addition of biochar to soil increases carbon contents along with other biogenic compounds, e.g., nitrogen, phosphorus, "Biochar also showed large ion exchange capacities and specific surface area to perform specified functions, they also direct and produce effects such as the reduced discharge of biogenic elements from soils and a regulate emission of nitrous oxide with decreased rate" (Yuan et al. 2011).

Table 3 Effects of Biochar on flora and fauna

S. No.	Material (on which biochar is used)	Effects seen on the physiology	References
1.	"*Lactuca sativa* (Romaine lettuce) and *Cynodondactylon* (bermuda grass)"	Increased growth	Akhtar et al. (2014)
2.	"*Solanum lycopersicum* (Tomato)"	Increased fruit yield and quality, total soluble solids, and titratable acidity	Batool et al. (2015)
3.	"*Abelmoschus esculentus* (okra)"	Increased growth	Paneque et al. (2016)
4.	"*Helianthus annus* (Sunflower)"	Increased growth	Baronti et al. (2014)
5.	"*Vitis vinifera* (Grapes)"	increased the Pn and Gs of grape leaves	Xia et al. (2015)
6.	"*Arachis hypogaea* (Peanuts)"	Improved the photosynthesis	Haider et al. (2015)
7.	"*Zeamays* (Maize)"	Increased the plant height and leaf area	Olmo et al. (2014)
8.	"*Triticum aestivum* (Wheat)"	Increased biomass	Vaccari et al. (2015)
9.	"*Solanum lycopersicum* (Tomato)"	Increased growth	Githinji (2014)
10.	"*Solanum lycopersicum* (Tomato)"	Increased growth	Hansen et al. (2016)
11.	"*Hordeum vulgare* (Barley)"	Increased the shoot and root growth	GenesioL et al. (2015)
12.	"*Vitis vinifera* (Grapes)"	Increased the vine yield especially in the years receiving lowest rainfall, no effect observed on the quality parameters such as brisk, total acidity, and anthocyanins	Afshar et al. (2016)
13.	"*Silybum marianum* (Milk thistle)"	Increased yield especially in the years receiving lowest rainfall, whereas no effect was observed on the grape quality parameters such as brisk, total acidity, and anthocyanins	Mickan et al. (2016)
14.	"Jarrah (*Eucalyptus marginata*)"	Alleviate water stress in plants when applied with microorganisms	Egamberdieva et al. (2017)

(continued)

Table 3 (continued)

S. No.	Material (on which biochar is used)	Effects seen on the physiology	References
15.	"*Lupinus angustifolius* L."	Increased the growth, biomass, nitrogen (N), and phosphorus (P) uptake and nodulation	Liu et al. (2017)
16.	"*Solanum tuberosum* (Potato)"	Decreased—leaf area, WUE (water use efficiency), N, and P; whereas biomass of roots and soil pH under limited root zone water were not affected	Pressler et al. (2017)
17.	"*Zeamays* (Maize)"	No significant effect on soil biota groups (protozoa, nematodes, bacteria, fungi, and arthropods) under limited irrigation	Kammann et al. (2011)
18.	"*Chenopodium quinoa* wild"	Increased the WUE (water use efficiency)	Mulcahy et al. (2013)
19.	"*Solanum lycopersicum* (Tomato)"	Increased seedling resistance to wilting	Singh et al. (2012)
20.	"Eucalyptus saligna wood and leaves, papermill sludge, poultry litter, cow manure"	Biochars produced through slow pyrolysis can have high C sequestration values	Lehmann et al. (2006b)
21.			Rondon et al. (2007)
22.	By common beans (*Phaseolus vulgaris* L.) through biochar additions (charcoal, biomass-derived black carbon)"	Biological N_2 fixation (BNF)	Glaser et al. (2001)
23.	"Persistence of soil organic matter in Terra Preta of Brazilian Amazon region"	Longevity of black carbon in the Terra preta soil to the carbon polyaromatic structure, making it chemically and microbial resistant, able to survive in the environment over thousands of years	McCann et al. (2001)
24.	"Organic matter and anthrosols in Amazonia"	Introduction of charged (active) surfaces and the increase in soil pH, suppressing Al activity toxic to soil biota	O'Toole et al. (2013)

(continued)

Table 3 (continued)

S. No.	Material (on which biochar is used)	Effects seen on the physiology	References
25.	"Biochar provides great adsorption sites for nitrous oxides, nitrogen due to the large surface area, thus reducing the release of these gases from the soil ecological unit"		Allen (1846a)

Effect of biochar on disease control/pesticides/ GHGs in plants		References
1.	Suppression of Potato rot disease	Allen (1846b)
2.	Suppression of rust and mildew disease	Retan (1915)
3.	Reduced damping off (caused by various pathogens)	Hoitink and Fahy (1986)
4.	Acts as a compost additive	Matsubara et al. (2002)
5.	Biochar produced by fast pyrolysis of hardwood (coconut) increases in root weights and linear reductions in the percentage of root lesions caused by *Fusarium oxysporum* f. sp. *asparagi* and *F. proliferatum*	Elmer and Pignatello (2011) and Graber et al. (2010)
6.	*Trichoderma* spp. through larger populations of microorganisms isolated from the biochar-treated soils that are known to induce resistance wood-derived biochar applied on pepper (*Capsicum annuum* L.) and tomato (*Lycopersicum esculentum*) show resistance	Nerome et al. (2005)
7.	Reduced infection of tomato and reduced bacterial wilt (*Ralstonia solanacearum*) by adding wood biochar	Elad et al. (2010)
8.	Biochar induces systemic resistance to the foliar fungal pathogens *Botrytis cinerea* (graymold) and *Leveillulataurica* (powdery mildew) on pepper and tomato and to the broad mite pest (*Polyphagotarsonemus latus* Banks) on pepper	Bais et al. (2005)
10.	Increased disease symptoms, plant mortality of *Arabidopsis* and root colonization by *Pseudomonas syringae* due to the presence of activated charcoal	Andersen (1968)
11.	Pesticides and herbicides (Simazine, atrazine, Linuron, and diuron) are rendered less effective due to absorption/adsorption of biochars	Yang et al. (2006), Zheng et al. (2010) and Yu et al. (2009)

(continued)

Table 3 (continued)

Effect of biochar on disease control/pesticides/ GHGs in plants		References
12.	Biochars prepared from the pyrolysis of *Eucalyptus* spp. wood chips cause reduction in leaching and uptake of pesticides (chlorpyrifos and carbofuran) in Spring onion (*Allium cepa*)	Yu et al. (2010)
	Biochar produced from incomplete combustion of red gum woodchip could enhance the sorption of pesticide pyrimethanil when incorporated into soil	Spokas et al. (2009)
13.	Plant metabolic activities were found to be impacted; reduction in N_2O production as a consequence of the sawdust biochar addition. This reduction in observed N_2O production could easily offset the 60% reduction in CH_4 oxidation activity with 10% w/w biochar additions in the net greenhouse gas balance. Biochar also increased the sorption of two common herbicides, reducing the likelihood of leaching and runoff losses, but also reducing bioavailability, perhaps necessitating higher application rates	Matlack (2001)
Effect of biochar seen on fauna		
1.	Increased density of soil nematodes, increased density and diversity of collembolans and oribatid mites	Uvarov (2000) and Smernik (2009)
2.	Bioavailability of pollutants (such as polychlorinated biphenyls, polyaromatic hydrocarbons) or organic agrochemicals (such as herbicides and pesticides) to soil fauna may be reduced due to the strong sorption of nonpolar and semi-polar compounds to biochars	Millward et al. (2005)
3.	Lesser availability of pollutants to—reducing polychlorinated biphenyl (PCB) bioavailability to clams (*Macomabalthica*), PCB bioaccumulation in a polychaete (*Neanthes arenaceodentata*) is reduced by 82% PCB bioaccumulation in an amphipod (*Leptocheirus plumulosus*) is reduced by 70%	Bezerra et al. (2016; Medyńska-Juraszek 2016)

Silicon Release Through Biochar

In new researches, it is hinted that crop biochar is rich in biogenic silica (phytoliths) that is capable to quickly release bioavailable silicon that is important to enhance plant biomass and crucial to promote the biological silicon cycle in soil (Novak et al. 2009).

Impact of Biochar on Nitrogen and Phosphorus Cycle

"Biochar is found cable to perform one of the important functions of modifying the nitrogen and phosphorus cycle. It is found that *Biocarbon* works as an important ingredient and works as an additive to soils that has the ability to retain NH_3 and NH_4^+ that store nitrogen in great amount required plant growth. Further, this phenomenon helps to reduce N_2O emissions and eluting NO_3^- ions. All this induces the growth of nitrogen bacteria which efficiently affects the increase in soil productivity".

Addition of Other Components in Soil Through Biochar

"Biochar is categorized by holding varied potassium contents that produce and depend on the type of batch material used during its production. In fact, all macronutrients present in salts such as carbonates, potassium are the required elements for the wellbeing of plants and their growth, this proportion of biosorptive forms of soil components is up to 95% of the total content". Interestingly Biochar based fertilizers lead to an increase soil pH (Zong et al. 2016).

PH and Soil Fertility Modulation by Biochar Addition

The associated research has revealed that numerous alkaline substances are present in carbonization products e.g., calcium carbonate, it influences soil reaction, and the highest pH increase may be attained in highly acidified soils. It is important to mention that increased pH of soil is though beneficial in acid tropical soils e.g., terra preta but has potential to reduce yield in soils of those areas such as in many temperate regions (Jien and Wang 2014; Agegnehu et al. 2016).

Physical Properties Modulation by Biochar Addition

Biochar influences many physical properties of soil. It regulates the water retention capability of soil and also impacts, capacity to form aggregates, the resistance to erosion is another advantage (Yuan et al. 2011). The improved physical characteristics of soils can be achieved by addition of biochar. The properties needed in soil can be predominantly handled while determining the technology applied in production of the biochar (Das and Sarmah 2015; Cayuela et al. 2014).

For Providing Home for Microorganisms

The highly porous structure is an essential property of soil that can be introduced through addition of biochar. Carbonization products may generate favorable conditions for microorganisms, growth, and survival, and as a consequence enhance the fertility and productivity of soils. This can be exemplified that biochar addition to the soil that contains high oil organic matter level and ultimately leads to higher microbial stimulation. This consequently increases higher N mineralization (Zhuo et al. 2023).

6 Antibiotic Removal with Biochar

"Biochar can effectively adsorb leftover antibiotics by ion exchange, π-π bond interactions, hydrogen bonding, and electrostatic interactions. The inclusion of functional groups on biochar, such as carboxyl groups, improves its ability to form complexes with antibiotics and so remove them from contaminated water. Studies have revealed that biochar can considerably lower the levels of particular antibiotics in water. For example, combining biochar with waste fungal chaff-based biochar was efficient in reducing veterinary antibiotics, resulting in decreases of 40.7% for enrofloxacin and 41.9% for oxytetracycline.

Biochar's efficiency can be increased when mixed with particular bacteria. When used in conjunction with specific bacterial strains, biochar has been shown to diminish the quantity of antibiotic resistance genes (ARGs) and their spread in the environment. Unmodified biochars may have low adsorption capabilities, particularly at high pollutant concentrations. As a result, changes, such as the addition of magnetic elements, have been proposed to improve biochar's efficacy in adsorbing antibiotics and facilitate recovery from treated solutions. The use of biochar not only aids in the removal of antibiotics but also helps minimize the environmental impact of antibiotic pollution, which is linked to the growth of multidrug-resistant bacteria that risk human health and disrupt ecological equilibrium" (Galbally et al. 2014).

7 Heavy Metal Treatment

"Biochar's unique features and methods of action have made it a potential material for heavy metal (HM) remediation in contaminated environments. It promotes heavy metal absorption through a variety of mechanisms, including ion exchange, in which biochar exchanges ions with heavy metal cations, and surface complexation, in which functional groups on the biochar surface, such as carboxyl groups (–COOH), form complexes with heavy metals, enhancing adsorption. Furthermore, cation-π interactions between biochar's aromatic compounds and positively charged metal cations facilitate their removal. Biochar has been shown in studies to have high immobilization rates for several heavy metals, hence considerably lowering their bioavailability and possible harm to human health and the environment. The mineral composition of biochar also helps to with efficiency, as minerals can interact with heavy metals to produce insoluble compounds via coprecipitation, reducing metal mobility in the environment. Compared to typical heavy metal removal technologies such as chemical precipitation and membrane filtration, biochar is a more cost-effective and environmentally sustainable option since it can be created from a variety of biomass sources". However, "economic production hurdles persist, as does the need for additional study to enhance its characteristics for specific heavy metal contamination". Overall, "biochar is a versatile and effective heavy metal treatment option that offers major environmental benefits by resolving pollution issues" (Oleszczuk et al. 2013).

8 Negative Impacts of Biochar

Despite several applications of biochar, there are some drawbacks related to its use. Some "potential risks have been identified like biochar rich in heavy metal content may lead to the problem of leaching, PAH and dioxin laden biochar affects the soil flora and fauna (Hale et al. 2012). In aquatic environments problem of eutrophication has been reported" (Tsouloufa et al. 2020). A complete life-cycle assessment is required for a better understanding of the biochar.

References

Adhikari DK, Seal D, Saxena RC, Goyal HB (2006) Biomass based fuel/energy. J Petrotech Soc III(1)

Afshar RK, Hashemi M, DaCosta M, Spargo J, Sadeghpour A (2016) Biochar application and drought stress effects on physiological characteristics of *Silybum marianum*. Commun Soil Sci Plant Anal 47:743–752

Agegnehu G, Bass A, Nelson P, Bird M (2016) Benefits of biochar, compost and biochar-compost for soil quality, maize yield and greenhouse gas emissions in a tropical agricultural soil. Sci Total Environ 543:295–306

Akhtar SS, Li G, Andersen MN, Liu F (2014) Biochar enhances yield and quality of tomato under reduced irrigation. Agri Water Manag 138:37–44

Allen AB (1846a) The American agriculturalist, vol V. Saxton and Miles, New York

Allen RL (1846b) Brief compend of american agriculture. Saxton and Miles, New York

Andersen AH (1968) The inactivation of simazine and linuron in soil by charcoal. Weed Res 8:58–60. Jordan PD, Smith, LW (1971) Adsorption and deactivation of atrazine and diuron by charcoals. Weed Sci 19:541–544

Andersson E, Harvey S (2006) Comparison of pulp-mill-integrated hydrogen production from gasified black liquor with stand-alone production from gasified biomass. Energy. Available online at www.sciencedirect.com

André RN, Pintoa FC, Diasa M, Gulyurtlua I, Matosb MAA, Cabritaa I (2005) Fluidised bed co-gasification of coal and olive oil industry wastes. Fuel 84:1635–1644

Arauzo J, Radlein D, Piskorz J, Scott DS (1994) A new catalyst for the catalytic gasification of biomass. Energy Fuels 8:1192–1196

Artiola JF, Rasmussen C, Freitas R (2012) Effects of a biochar-amended alkaline soil on the growth of romaine lettuce and bermudagrass. Soil Sci 177:561–570

Asadullah M, Kaoru F, Keiichi T (2001) Catalytic performance by Rh/CeO_2 in the gasification of cellulose to synthetic gas at low temperature. Ind Eng Chem Res 40:5894–5900

Asadullah M, Tomohisa M, Shin-ichgi I, Kimio K, Keiichi T (2003) Demonstration of real biomass gasification drastically promoted by effective catalyst. Appl Cataly A Gen 8514:1–14

Awad YM, Lee SS, Kim K-H, Ok YS, Kuzyakov Y (2018) Carbon and nitrogen mineralization and enzyme activities in soil aggregate-size classes: effects of biochar, oyster shells, and polymers. Chemosphere 198:40–48

Bais HP, Prithiviraj B, Jha AK, Ausubel FM, Vivanco JM (2005) Mediation of pathogen resistance by exudation of antimicrobials from roots. Nature 434:217–221

Bakshi NN, Dalai AK, Thring RW (2003) Biomass char and lignin: potential application. Department of Chemical Engineering, University Saskatchewan

Baronti S, Vaccari FP, Miglietta F, Calzolari C, Lugato E, Orlandini S, Pini R, Zulian C, Genesio L (2014) Impact of biochar application on plant water relations in *Vitis vinifera* (L.) Euro J Agron 53:38–44

Bartocci P, Bidini G, Saputo P, Fantozzi F (2016) Biochar pellet carbon footprint. Chem Eng Trans 50:217–222

Bartocci P, Zampilli M, Bidini G, Fantozzi F (2018) Hydrogen-rich gas production through steam gasification of charcoal pellet. Appl Therm Eng 132:817–823

Batool A, Taj S, Rashid A, Khalid A, Qadeer S, Saleem AR, Ghufran MA (2015) Potential of soil amendments (biochar and gypsum) in increasing water use efficiency of *Abelmoschus esculentus* L. Moench Front Plant Sci 6:1–13

Beesley L, Moreno-Jiménez E, Gomez-Eyles J, Harris E, Robinson B, Sizmur T (2011) A review of biochars' potential role in the remediation, revegetation and restoration of contaminated soils. Environ Pollut 159:3269–3282

Bezerra J, Turnhout E, Vasquez IM, Rittl TF, Arts B, Kuyper TW (2016) The promises of the Amazonian soil: Shifts in discourses of Terra Preta and biochar. J Environ Policy Plan 46:1–13

Blasi CD, Gabriella S, Giuseppe P (1999) Countercurrent fixed bed gasification of biomass at laboratory scale. Ind Eng Chem Res 38:2571–2581

Bohlmann JT, Lorth CM, Anja D, Rainer B (eds) (1998) Microwave high pressure thermo-chemical conversion of sewage sludge as an alternative to incineration CE&T communications

Calzavara Y, Joussot-Dubien C, Boissonnet G, Sarrade S (2005) Evaluation of biomass gasification in supercritical water process for hydrogen production. Energy Convers Manage 46:615–631

Cayuela M, Van Zwieten L, Singh B, Jeffery S, Roig A, Sanchez-Monedero MA (2014) Biochar's role in mitigating soil nitrous oxide emissions: A review and meta-analysis. Agric Ecosyst Environ 191:5–16

Chaudhari ST, Bej SK, Bakshi SK, Dalai AK (2001) Steam gasification of biomass derived char for the production of carbon monoxide-rich synthesis gas. Energy Fuels 15:736–742

Chen W, Kuo P (2010) A study on torrefaction of various biomass materials and its impact on lignocellulosic structure simulated by a thermogravimetry. Energy 35:2580–2586

Dalai AK, Sasaoka E, Hikita H, Ferdous D (2003) Catalytic gasification of saw dust derived from various biomass. Energy Fuels 17:1456–1463

Das O, Sarmah A (2015) The love-hate relationship of pyrolysis biochar and water: a perspective. Sci Total Environ 512(513):682–685

Dasappa S, Paul PJ, Mukunda HS, Rajan NKS, Giridhar G, Sridhar HV (2004) Biomass gasification technology—a route to meet energy needs. Curr Sci 87(7):908–916

Demirbas A (2004) Hydrogen—rich gas from fruit shale via supercritical water extraction. Int J Hydrogen Energy 29:1237–1243

Egamberdieva D, Reckling M, Wirth S (2017) Biochar-based *Bradyrhizobium* inoculum improves growth of lupin (*Lupinus angustifolius* L.) under drought stress. Eur J Soil Biol 78:38–42

Elad Y, Rav David D, MellerHarel Y, Borenshtein M, Ben Kalifa H, Silber A, Graber ER (2010) Induction of systemic resistance in plants by biochar, a soil applied carbon sequestering agent. Phytopathology 100:913–921

Elmer, WH, Pignatello, JJ (2011) Effect of biochar amendments on mycorrhizal associations and Fusarium crown and root rot of asparagus in replant soils. Plant Dis. Published online

Feng W, Kooi HL, Arons JS (2004) Biomass conversion in subcritical and super critical water: driving force, phase equilibria and thermodynamics analysis. Chem Eng Process 43(12):1459–1466

Fillipis PD, Carlo B, Martino P, Fausto P (2004) Gasification process of Cuban bagasse in a two-stage reactor. Biomass Bioenergy 27:247–252. Perez P, Anzar PM, Caballero MA, Gil J, Martin JA, Corella J (1997) Hot gas cleaning and upgrading with a calcined dolomite located downstream a biomass fluidized bed gasifier operating with steam—oxygen mixtures. Energy Fuels 11:1194–1203

Franco C, Pinto F, Gulyurtlu I, Cabrita I (2003) The study of reactions influencing the biomass steam gasification process. Fuel 82:835–842

Galbally P, Ryan D, Finnan J, Grant J, Fagan CC, McDonnell K (2014) Biosolids and distillery effluent amendments to Irish Miscanthus plantations: impacts on overland flow and surface water quality. Sustain Water Qual Ecol 3–4:77–89

GenesioL MF, Baronti S, Vaccari FP (2015) Biochar increases vineyard productivity without affecting grape quality: results from a four years field experiment in Tuscany. Agri Ecosys Environ 201:20–25

Githinji L (2014) Effect of biochar application rate on soil physical and hydraulic properties of a sandy loam. Arch Agron Soil Sci 60:457–470

Glaser B, Guggenberger G, Haumaier L, Zech W (2001) Persistence of soil organic matter in archaeological soils (Terra Preta) of the Brazilian Amazon region. In: Rees RM, Bell BC (2001) Campbell CD, Watson CA (eds) Sustainable management of soil organic matter. CAB Int

Graber ER, Harel YM, Kolton M, Cytryn E, Silber A, David DR, Tsechansky L, Borenshtein M, Elad Y (2010) Biochar impact on development and productivity of pepper and tomato grown in fertigated soilless media. Plant Soil 337:481–496

Haider G, Koyro HW, Azam F, Steffens D, Müller C, Kammann C (2015) Biochar but not humic acid product amendment affected maize yields via improving plant-soil moisture relations. Plant Soil 395:141–157

Hale SE, Lehmann J, Rutherford D, Zimmerman AR, Bachmann RT, Shitumbanuma V, O'Toole A, Sundqvist KL, Arp HPH, Cornelissen G (2012) Quantifying the total and bioavailable polycyclic aromatic hydrocarbons and dioxins in biochars. Environ Sci Technol 46:2830–2838

Han L, Ro KS, Sun K, Sun H, Wang Z, Libra JA, Xing B (2016) New evidence for high sorption capacity of hydrochar for hydrophobic organic pollutants. Environ Sci Technol 50(24):13274–13282

Hanaoka T, Yoshida T, Fujimoto S, Kamei K, Harada M, Suzuki Y et al (2005) Hydrogen production from woody biomass by steam gasification using CO_2 sorbent. Biomass Bioenergy 28(1):63–68

Hansen V, Hauggaard-Nielsen H, Petersen CT, Mikkelsen TN, MüllerStöver D (2016) Effects of gasification biochar on plant-available water capacity and plant growth in two contrasting soil types. Soil Tillage Res 161:1–9

Herath HMSK, Camps-Arbestain M, Hedley MJ, Kirschbaum MUF, Wang T, Van Hale R (2015) Experimental evidence for sequestering C with biochar by avoidance of CO_2 emissions from original feedstock and protection of native soil organic matter. GCB Bioenergy 7(3):512–526

Hoitink HAJ, Fahy PC (1986) Basis for the control of soil-borne plant pathogens with composts. Annu Rev Phytopathol 24:93–114

Hu B, Ai Y, Jin J, Hayat T, Alsaedi A, Zhuang L, Wang X (2020) Efficient elimination of organic and inorganic pollutants by biochar and biochar-based materials. Biochar 2(1):47–64

Jien S, Wang C (2014) Effects of biochar on soil properties and erosion potential in a highly weathered soil. CATENA 110:225–233

Jin J, Li S, Peng X, Liu W, Zhang C, Yang Y, Han L, Du Z, Sun K, Wang X (2018a) HNO_3 modified biochars for uranium (VI) removal from aqueous solution. Bioresour Technol 256:247–253

Jin J, Sun K, Liu W, Li S, Peng X, Yang Y, Han L, Du Z, Wang X (2018b) Isolation and characterization of biochar-derived organic matter fractions and their phenanthrene sorption. Environ Pollut 236:745–753

Jin J, Ke S, Yang Y, Wang Z, Han L, Wang X, Wu F, Xing B (2018c) Comparison between soil-and biochar-derived humic acids: composition, conformation, and phenanthrene sorption. Environ Sci Technol 52(4):1880–1888

Kammann CI, Linsel S, Gößling JW, Koyro HW (2011) Influence of biochar on drought tolerance of *Chenopodium quinoa* Willd and on soil–plant relations. Plant Soil 345:195–210

Kersten SRA, Prins W, Drift BV, Wam PM, Swaaji V (2003) Principles of a novel multistage circulating fluidized bed reactor for a biomass gasification. Chem Eng Sci 58:725–731

Kim HY (2003) A lost cost production from carbonaceous waste. Int J Hydrogen Energy 28:1179–1186

Lee JE, Park Y-K (2020) Applications of modified biochar-based materials for the removal of environment pollutants: a mini review. Sustainability 12(15):6112

Lehmann J, Gaunt J, Rondon M (2006a) Bio-char sequestration in terrestrial ecosystems—a review. Mitig Adapt Strat Glob Change 11:403–427

Lehmann J, Joseph S (2015) Biochar for environmental management: an introduction. In: Biochar for environmental management. Routledge, pp 1–13

Lehmann J, Gaunt J, Rondon M (2006b) Bio-char sequestration in terrestrial ecosystems—a review. Mitig Adapt Strat Glob Change 11(2):395–419

Leng L, Huang H (2018) An overview of the effect of pyrolysis process parameters on biochar stability. Biores Technol 270:627–642

Li Z, Delvaux B, Yans J, Dufour N, Houben D, Cornelis JT (2018) Phytolith-rich biochar increases cotton biomass and silicon-mineralomass in a highly weathered soil. J Plant Nutr Soil Sci 181:537–546

Liu C, Liu F, Ravnskov S, Rubæk GH, Sun Z, Andersen MN (2017) Impact of wood biochar and its interactions with mycorrhizal fungi, phosphorus fertilization and irrigation strategies on potato growth. J Agron Crop Sci 203:131–145

Macdonald L, Farrell M, Van Zwieten L, Krull E (2014) Plant growth responses to biochar addition: An Australian soils perspective. Biol Fertil Soils 50:1035–1045

Major J, Lehmann J, Rondon M, Goodale C (2010) Fate of soil-applied black carbon: downward migration, leaching and soil respiration. Glob Change Biol 16(4):1366–1379

Matlack GR (2001) Factors determining the distribution of soil nematodes in a commercial forest landscape. For Ecol Manag 146:129–143

Matsubara Y, Hasegawa N, Fukui H (2002) Incidence of Fusarium root rot in asparagus seedlings infected with arbuscular mycorrhizal fungus as affected by several soil amendments. J Jpn Soc Horticult Sci 71:370–374

McCann JM, Woods WI, Meyer DW (2001) Organic matter and anthrosols in Amazonia: interpreting the Amerindian legacy. In: Rees RM, Bell BC, Campbell CD, Watson CA (eds) Sustainable management of soil organic matter. CAB Int

Medyńska-Juraszek A (2016) Biochar as a soil amendment. Soil Sci Ann 67:151–157

Meng N, Leung DYC, Leng MKH, Sumathy K (2006) An overview of hydrogen gas production from biomass. Fuel Process Technol 87:461–472

Mickan BS, Abbott LK, Stefanova K, Solaiman ZM (2016) Interactions between biochar and mycorrhizal fungi in a water-stressed agricultural soil. Mycorrhiza 26:565–574

Millward RN, Bridges TS, Ghosh U, Zimmerman JR, Luthy RG (2005) Addition of activated carbon to sediments to reduce PCB bioaccumulation by a polychaete (*Neanthes arenaceodentata*) and an amphipod (*Leptocheirus plumulosus*). Environ Sci Technol 39:2880–2887

Mulcahy DN, Mulcahy DL, Dietz D (2013) Biochar soil amendment increases tomato seedling resistance to drought in sandy soils. J Arid Environ 88:222–225

Narvaez I, Orio A, Aznar MP, Corella J (1996) Biomass gasification with air in an atmospheric bubbling fluidized bed. Effects of six operational variables on the quality of the produced raw gas. Ind Eng Chem Res 35:2110–2120

Nerome M, Toyota K, Islam TM, Nishijima T, Matsuoka T, Sato K, Yamaguchi Y (2005) Suppression of bacterial wilt of tomato by incorporation of municipal biowaste charcoal into soil. Soil Microorgan 59:9–14

Novak JM, Busscher WJ, Laird DL, Ahmedna M, Watts DW, Niandou MAS (2009) Impact of biochar amendment on fertility of a Southeastern coastal plain soil. Soil Sci 174:105–112

Novak JM, Busscher WJ, Watts DW, Laird DA, Ahmedna MA, Niandou MAS (2010) Short-term CO_2 mineralization after additions of biochar and switchgrass to a Typic Kandiudult. Geoderma 154(3–4):281–288

O'Toole A, de Zarruk K, Steffens M, Rasse DP (2013) Characterization, stability, and plant effects of kiln-produced wheat straw biochar. J Environ Qual 42(2):429–436

Oleszczuk P, Jośko I, Kuśmierz M (2013) Biochar properties regarding to contaminants content and ecotoxicological assessment. J Haz Mater 260:375–382

Olmo M, Alburquerque JA, Barrón V, del Campillo MC, Gallardo A, Fuentes M, Villar R (2014) Wheat growth and yield responses to biochar addition under Mediterranean climate conditions. Biol Fert Soil 50:1177–1187

O'Toole A, de Zarruk KK, Steffens M, Rasse DP (2013) Characterization, stability, and plant effects of kiln-produced wheat straw biochar. J Environ Qual 42:429–436

Paethanom A, Bartocci P, D'Alessandro B, D'Amico M, Testarmata F, Moriconi N, Slopiecka K, Yoshikawa K, Fantozzi F (2013) A low-cost pyrogas cleaning system for power generation: scaling up from lab to pilot. Appl Energy 111:1080–1088

Paneque M, José M, Franco-Navarro JD, Colmenero-Flores JM, Knicker H (2016) Effect of biochar amendment on morphology, productivity and water relations of sunflower plants under non-irrigation conditions. CATENA 147:280–287

Pressler Y, Foster EJ, Moore JC, Cotrufo MF (2017) Coupled biochar amendment and limited irrigation strategies do not affect a degraded soil food web in a maize agroecosystem, compared to the native grassland. GCB Bioenergy. https://doi.org/10.1111/gcbb.12429

Purakayastha TJ, Das KC, Gaskin J, Harris K, Smith JL, Kumari S (2016) Effect of pyrolysis temperatures on stability and priming effects of C3 and C4 biochars applied to two different soils. Soil till Res 155:107–115

Rao D, Rajput P, Choudhary R, Yadav S, Yadav SK, Rajput VD, Matić S et al (2024) Multifaceted characteristics of biochar and its implementation in environmental management in a sustainable way. Environ Qual Manag 34(1):e22305

Retan GA (1915) Charcoal as a means of solving some nursery problems. For Quart 13:25–30

Rioche C, Kulkarni S, Meunier FC, Breen JP, Burch R (2005) Steam reforming of model compounds and fast pyrolysis bio-oil on supported noble metal catalysts. Appl Cataly B Environ 61:144–153

Rondon MA, Lehmann J, Ramires J, Hurtado M (2007) Biological nitrogen fixation by common beans (*Phaseolus vulgaris* L.) increases with bio-char additions. BiolFertil Soils 43:699–708

Shoudho KN, Khan TH, Ara UR, Khan MR, Shawon ZBZ, Hoque ME (2024) Biochar in global carbon cycle: towards sustainable development goals. Curr Res Green Sustain Chem 100409

Singh BP, Cowie AL, Smernik RJ (2012) Biochar carbon stability in a clayey soil as a function of feedstock and pyrolysis temperature. Environ Sci Technol 46(21):11770–11778. https://doi.org/10.1021/es302545b

Smernik R (2009) Biochar and sorption of organic compounds. In: Lehmann J, Joseph S (eds) Biochar for environmental management: science and technology. Earthscan, London, pp 289–300. McLeod PB, van den Heuvel-Greve MJ, Luoma SN, Luthy RG (2007) Biological uptake of polychlorinated biphenyls by Macomabalthica from sediment amended with activated carbon. Environ Toxicol Chem 26:980–987

Spokas KA (2010) Review of the stability of biochar in soils: predictability of O:C molar ratios. Carbon Manag 1(2):289–303

Spokas KA, Koskinen WC, Baker JM, Reicosky DC (2009) Impacts of woodchip biochar additions on greenhouse gas production and sorption/degradation of two herbicides in a Minnesota soil. Chemosphere 77:574–581

Tsouloufa A, Dailianis S, Karapanagioti HK, Manariotis ID (2020) Physicochemical and toxicological assay of leachate from malt spent rootlets biochar. Bull Environ Contam Toxicol 104:634–641

Uvarov A (2000) Effects of smoke emissions from a charcoal kiln on the functioning of forest soil systems: a microcosm study. Environ Monit Assess 60:337–357
Vaccari FP, Maienza A, Miglietta F, Baronti S, Di Lonardo S, Giagnoni L, Lagomarsino A, Pozzi A, Pusceddu E, Ranieri R, Valboa G, Genesio L (2015) Biochar stimulates plant growth but not fruit yield of processing tomato in a fertile soil. Agri Ecosys Environ 207:163–170
Wang S, Song L, He H, Zhang W (2024) Volatile organic compounds (VOCs) in soil: transport mechanisms, monitoring, and removal by biochar-modified capping layer. Coatings 14(3):270
Warnecke R (2000) Gasification of biomass: comparison of fixed bed and fluidized bed gasifier. Biomass Bioenergy 18:489–497
Wild E, Jones KC (2009) Novel method for the direct visualization of in vivo nanomaterials and chemical interactions in plants. Environ Sci Technol 43(14):5290–5294
Woolf D, Amonette JE, Street-Perrott FA, Lehmann J, Joseph S (2010) Sustainable biochar to mitigate globalclimate change. Nat Commun 1:56
Xia S, Wang X, Su G, Shi G (2015) Effects of drought on cadmium accumulation in peanuts grown in a contaminated calcareous soil. Environ Sci Pollut Res 22:18707–18717
Xia T, Qi Y, Liu J, Qi Z, Chen W, Wiesner MR (2017) Cation-inhibited transport of graphene oxide nanomaterials in saturated porous media: the Hofmeister effects. Environ Sci Technol 51(2):828–837
Yang YN, Sheng GY, Huang MS (2006) Bioavailability of diuron in soil containing wheat straw-derived char. Sci Total Environ 354:170–178
Yin Y, He X, Ren GAO, Yang Y (2014) Effects of rice straw and its biochar addition on soil labile carbon and soil organic carbon. J Integr Agric 13(3):491–498
Yu X, Ying G, Kookana RS (2009) Plant uptake of pesticides with biochar additions to soil. Chemosphere 76:665–671
Yu X, Pan L, Ying G, Kookana RS (2010) Enhanced and irreversible sorption of pesticide pyrimethanil by soil amended with biochars. J Environ Sci 22:615–620
Yuan J, Xu R, Zhang H (2011) The forms of alkalis in the biochar produced from crop residues at different temperatures. Bioresour Technol 102:3488–3497
Zhang R, Cummer K, Suby A, Brown RC (2005) Biomass-derived hydrogen from an air-blown gasifier. Fuel Process Technol 86:861–874. Pengmai L, Chang J, Xiong Z, Huang H, Wu C, Chen Y (2003) Biomass air-steam gasification in fluidized bed to produce hydrogen rich gas. Energy Fuels 17:677–682
Zhang Q, Du Z, Lou Y, He X (2015) A one-year short-term biochar application improved carbon accumulation in large macro aggregate fractions. CATENA 127:26–31
Zheng W, Guo M, Chow T, Bennett DN, Rajagopalan N (2010) Sorption properties of greenwaste biochar for two triazine pesticides. J Hazard Mater 181:121–126
Zhuo Q, Liang Y, Hu Y, Shi M, Zhao C, Zhang S (2023) Applications of biochar in medical and related environmental fields: current status and future perspectives. Carbon Res 2(1):32
Zong Y, Xiao Q, Lu S (2016) Acidity, water retention, and mechanical physical quality of a strongly acidic Ultisol amended with biochars derived from different feedstocks. J Soil Sediments 16:177–190

Chapter 12
Polycyclic Aromatic Hydrocarbons and Their Remediation by Biochar

Hemavathi Brijesh, Shilpa Borehalli Mayegowda, Medini Bheemappa, N. G. Manjula, and Brijesh Ajjappala

Abstract The distribution of polycyclic aromatic hydrocarbons (PAHs) around the world is influenced by anthropogenic, natural, and global transport activities. They are pervasive environmental contaminants that are toxic, mutagenic, carcinogenic, and have negative biological effects. PAH molecules' stability and hydrophobicity are the two main factors that influence their long-term persistence in the environment. PAHs contamination of soil, water and sediments has been widely dissipated for years, necessitating effective remediation processes through research and development to achieve impactful treatment and reuse of soil, water and sediments. The biochar amendment to PAHs contaminated soil, water and sediments has been suggested as an inventive and environmentally friendly technology. The primary method of removing PAHs involves sorption onto biochar. In this chapter, types and effects of PAHs exposure on humans the mechanisms of PAH sorption to the biochar, effects of adding biochar to soil, water and sediments that have PAH contamination are thoroughly discussed.

Keywords Polycyclic aromatic hydrocarbons · Environmental contaminants · Biochar · Sorption · Remediation

---

H. Brijesh · M. Bheemappa
Department of Biotechnology, School of Applied Sciences, REVA University, Bengaluru, Karnataka, India

S. B. Mayegowda (✉)
Department of Psychology, CHRIST University, Kengeri Campus, Kumbalgodu, Kanmanike, Bengaluru, Karnataka, India
e-mail: shilpa.borehalli@christuniverity.in

N. G. Manjula
Department of Microbiology, School of Basic and Applied Sciences (SBAS), Dayananda Sagar University, Bengaluru, Karnataka, India

B. Ajjappala
Department of Plant Biotechnology, University of Agricultural Sciences, GKVK, Bengaluru, Karnataka, India

A. Roy et al. (eds.), *Biochar Production Engineering*,
https://doi.org/10.1007/978-3-031-92594-8_12

1 Introduction

Polycyclic aromatic hydrocarbons (PAHs) refer to aromatic hydrocarbons having two or more benzene rings fused together in diverse structural configurations without heteroatoms or any other substituents. PAHs with four benzene rings are referred to as light PAHs, whereas those with more than four benzene rings are referred to as heavy PAHs and are highly stable and hazardous than that of light PAHs. The components of PAHs are carbon and hydrogen which are bound in simple to complex ring structures. The arrangement of benzene ring is responsible for various physical, chemical, and toxicological properties of PAHs. Few of the notable PAHs are naphthalene, acenaphthylene, acenaphthene, fluorine, phenanthrene, anthracene, fluoranthene, pyrene, benz(a)anthracene. In general, increasing the PAH molecule size and angularity leads to a concurrent rise in hydrophobicity and electrochemical activity. PAHs are dangerous organic pollutants which are constantly present in the environment, and high-molecular-weight (HMW) PAHs are considerably worse for the environment and human health. The hydrophobic nature and stability of PAH molecules are the two key parameters contributing to the environmental persistence of HMW PAHs. The concentration ratios of various pairings of PAHs are commonly utilized for qualitatively determining the sources of PAHs. These ratios are known as molecular diagnostic ratios of PAHs, and they are routinely used to determine the quantities of PAHs in air, soils, and sediments. Few researchers have expanded the application of these diagnostic ratios to sewage sludge, implying that calculating the ratios based on particular concentrations of PAH could be just as successful when applied to soils and/or sediments. Diagnostic ratios of PAH could be a useful technique for identifying the sources of PAH pollution.

Polycyclic aromatic hydrocarbons (PAHs) are pervasive pollutants in global environmental matrices. They are classified as low-molecular weight (LMW) or high-molecular weight (HMW) (HMW) (Fig. 1). The hazardous LMW PAHs are typically found in crude oil. HMW PAHs are byproducts of continuous burning of materials, such as bushfires, fossil fuels, subterranean rock eruptions (volcanic rocks), and ingestion of toasted food, water, and air. Human cancer may be caused by HMW PAHs. In addition, PAHs which remain reactive less could be activated to become harmful due to physicochemical interactions, leading to DNA damage, mutations, and tumour formation (Fu et al. 2012). Wilson and Jones (1993), also observed that PAHs are teratogenic, that can cause impairment in the development of fetus, resulting in abnormalities in newborn children. PAHs also have other harmful impacts on people. Knowing the PAHs sources and their distribution in environment is key to understanding their movement and persistence in soils. The entry of PAHs into the water bodies, soils, and sediments is through dry and wet deposition, industrial effluents, and runoff water from roads and dissipation from trees treated with creosote (Chen et al. 2004). HMW PAHs often partition to airborne particles in the air. Because of their low vapour pressure, HMW PAHs often get separated into aerosol particles, whereas LMW PAHs get separated into both atmospheric moisture and aerosol particles (Fig. 2).

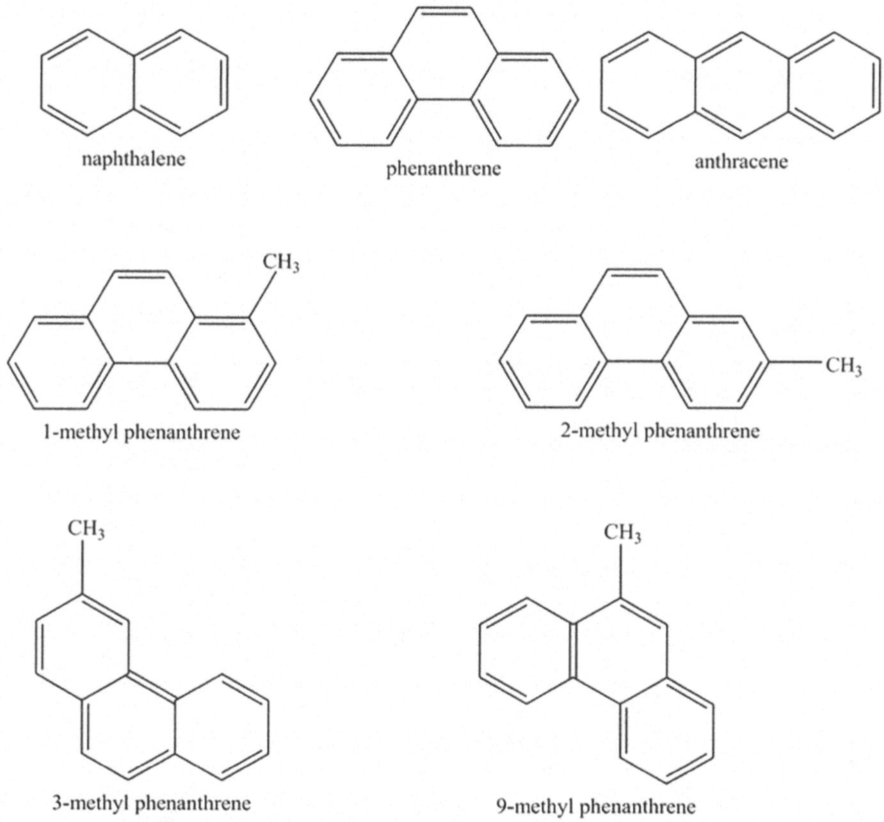

Fig. 1 Chemical structures of LMW PAHs (Anyika et al. 2015)

Different types of LMW and HMW PAHs and their molecular weights are listed in Table 1. The mixtures of pollutants at higher levels containing PAHs have been linked to causing symptoms like eye irritation, nausea, and diarrhea. Further, PAHs mixtures are known to cause inflammation and irritation of the skin. Few PAHs such as anthracene, benzo(a)pyrene, and naphthalene are known to cause skin irritation when they come in direct contact with humans. Several health risks including skin and lung cancers have been reported in employees that are exposed to PAH pollutants and other PAH compounds in the workplace. PAHs, such as Benzol(a)pyrene (BaP) and Pyrene, have been linked to malignancies in experimental animals in laboratory. Frequent exposure of skin to the naphthalene can cause reddening of skin and irritation.

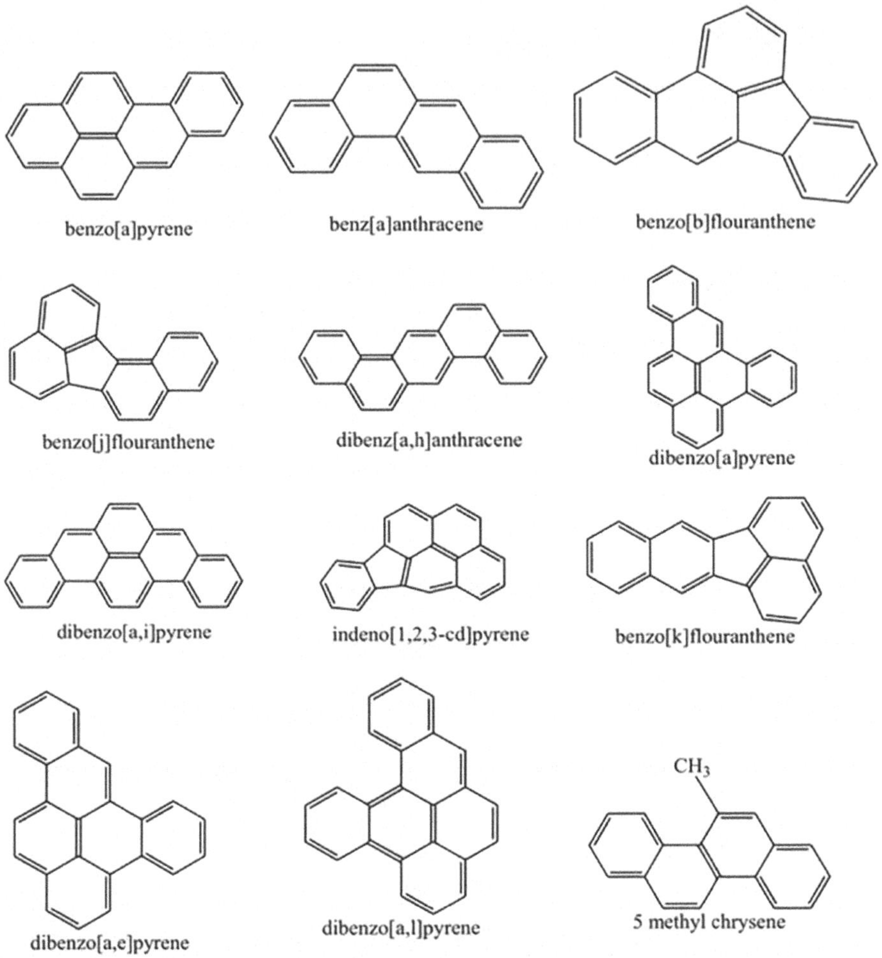

Fig. 2 Chemical structures of HMW PAHs (Anyika et al. 2015)

2 PAHs Entry into Humans

The primary source of PAH exposure among general population is through inhaling ambient indoor air, consuming foods that contain PAHs, smoking cigarettes, inhaling smoke from open fireplaces, and using fossil fuels to heat homes, cooking food, and power for the industries (Zhang et al. 2015). PAHs present in the agricultural crops and their exposure to humans and animals is another way of entry via the dietary intake (Wang et al. 2017). Exposure at workplace to PAHs can occur during the production of coke, asphaltic roofing, oil refining, and coal gasification, and among other processes. Workers involved in mining and metalworking, as well as those who operate motor vehicles or sell goods on the street, may be exposed to PAHs

Table 1 Different types of LMW and HMW PAHs

Types of LMW PAHs	Chemical formula	Molecular weight (g/mol)
Anthracene	$C_{14}H_{10}$	178.2
MeaFluorene	$C_{13}H_{10}$	166.2
Napthalene	$C_{10}H_{8}$	128.91
Penantherene	$C_{14}H_{10}$	174.2
Types of HMW PAHs		
Benzo (a) anthracene	$C_{18}H_{12}$	228.3
Benzo (b) fluoranthene	$C_{20}H_{12}$	252.3
Benzo (k) fluoranthene	$C_{20}H_{12}$	252.3
Benzo (a) pyrene	$C_{20}H_{12}$	252.3
Benzo (ghi) perylene	$C_{22}H_{12}$	276.3
Chrysene	$C_{18}H_{12}$	228.2
Dibenz (a, h) anthracene	$C_{22}H_{14}$	278.3
Fluoranthene	$C_{16}H_{10}$	202.2
Indeno (1,2,3-cd) pyrene	$C_{22}H_{12}$	276.3
Pyrene	$C_{16}H_{10}$	202.2

at work. Food consumption is the most common form of exposure, and smoking can contribute significantly. Food can be polluted by natural and largely man-made environmental factors, processing units in food industries, including some practices followed during domestic cooking. PAHs entry into the food chain could occur by air, soil and water deposition and transmission.

PAHs are often evaluated in the atmosphere to assess quality of the air, in biological tissues to monitor health impacts, in sediments and mollusks to monitor environmental quality, and in foods to ensure food safety. Emissions from automobile industries, hazardous waste sites, burning of biomass, waste burning, incinerators at municipal units, volcanic eruptions, and smoked and charcoal-broiled food consumption are few sources of BaP contamination of the environment and human exposure. Tobacco smoke, automobile traffic, e-waste and medical waste, charcoals, garbage, wild land fire, ornamental candles, and oil spills are all sources of further exposure. PAHs are ubiquitous environmental contaminants, so prevalent that no one can avoid exposure to them. Both occupational and non-occupational situations contribute to the exposure routes that include ingestion, inhalation, and skin contact. Some exposures may involve multiple pathways such as cutaneous and exposure by inhalation via polluted air at the same time, affecting the overall absorption level.

3 PAHs Sources

PAHs are generated both in natural environment and through anthropogenic activities, including burning of coal and wood, combustion of petrol and diesel oil, and industrial emissions have aided in their global dispersion. PAHs can be generated naturally, and many of them are poisonous and recalcitrant. Natural sources of PAH emissions include forest fires, waste-burning volcanoes, and hydrothermal activities. They are ubiquitous environmental pollutants with negative biological consequences with toxic and mutagenic effects. Concentration of PAH in the environment depends on the proximity of polluted sites to the source, the intensity of industrial activity, and the ways of PAH dispersion. PAHs are abundant in goods derived from fossil fuels, automobile traffic emissions and refining, wood, biomass burning, charcoals, tobacco, wood smoke, and waste. Petroleum refinery and transportation are main origin of environmental contamination by PAH. PAHs have been found in the air and sediment, as well as surface water, groundwater, and road runoff. PAHs are dispersed into the atmosphere and are found in plants, medicinal herbs, and contaminated foods. PAH concentrations in soil and sediment have been observed to range from 1 mg/kg to over 300 g/kg in contaminated and uncontaminated areas.

4 Water and Soil Contamination by PAHs

PAHs, belonging to main families of environmental contaminants arising from anthropogenic sources, were initially found and quantified in an industrialized South American estuary's coastal waters, aquatic organisms such as mussels and fish. The fact that smaller molecular weight congeners predominated in the Poyang region revealed that the primary sources of PAH intake were petroleum, fuel oil combustion byproducts, and other pyrogenic sources. Pollution pressures are always imposed on any emerging or developed country's coastal and estuarine ecosystems due to higher pollutants emitted from diverse origins such as factories, garbage, and anthropogenic activities (Mayegowda et al. 2022, 2024; Shilpa et al. 2022). The most significant sources of PAHs in coastal sediments are riverine inputs, and such soil has been identified as the main PAH reservoir in the United Kingdom (Beriro et al. 2016). When waste produced by industrial processes, including home garbage, is partially treated and disposed of in the soil–water ecosystem, it gains entry into the oceans lakes, rivers, streams, and other water bodies. Pollutants dissolved are suspended in water, or deposited on soil sediment beds (Archana et al. 2021; Ananda et al. 2022).

5 Types of PAHs

The compounds with only carbon and hydrogen atoms are specifically referred to as PAHs, whereas the term "polycyclic aromatic compounds," on the other hand, is more inclusive and includes functional derivatives such as (nitro-PAHs) and heterocyclic analogues (e.g., aza-arenes). Nearly, 100 different mixtures of PAHs are dispersed as ambient particulate matter, and over 200 have been found in tobacco smoke. PAHs, are soluble in a wide range of organic solvents and are extremely lipophilic with very low water solubility properties.

On the ring system, PAHs lack heteroatoms and substituents. Among the light and heavy PAHs, heavy PAHs are found to be more hazardous with higher stability than the light PAHs.

Patterns of ring linkage can be formed in PAHs such that the carbon atoms of tertiary structure become the center point for two or three interconnected rings. PAHs are solids that are colourless, white, or pale yellow. In general, PAHs are less soluble in water and have high melting or boiling temperatures with low vapour pressure. PAHs' aqueous solubility diminishes with each extra ring system. As a result, solubility in water reduces as the molecular weight of PAHs increases, whereas their melting and boiling temperatures rise as their molecular weight increases. Those with a linear layout are generally low in solubility than molecules with an angular or perifused configuration. Because of their lipophilic characteristics, they are soluble in organic solvents. While PAHs with four rings can exist in both the liquid and particulate phases as well as the vapour and liquid phases, PAHs with three or fewer aromatic ring structures are mostly found in the vapour phase, having boiling values in the range of 217–295 °C. Similarly, PAHs containing more than five rings having a boiling temperature higher than 375 °C can enter into the particulate matter phase. PAHs aromatic nature is determined by a number of factors, such as their cyclic structure, the conjugation of every atom in the ring, their flat planar structure, and Huckel's rule. Aromatic rings are those that contain $(4n + 2)$-electrons, according to the Huckel rule.

6 General Mechanism of PAHs Removal by Biochar

Biochar is a solid organic matter produced by converting the biomass thermochemical (pyrolysis) in an oxygen-depleted environment at temperatures from 300 to 700 °C from agricultural, industrial and forest-based residues. Due to porous structure, high surface area, and diverse functional groups, biochar is an ideal material for the remediation of sites damaged by organic pollutants (e.g., PAHs) (Xu et al. 2018). The properties and functions of biochar are listed in Table 2. Several factors influence biochar's sorption capabilities and interactions with PAHs, including pH, temperature of pyrolysis, type of biomass and pollutant-to-biochar ratios. PAH remediation using biochar involve five major mechanisms (Wang et al., 2020a, b). These are as follows:

(i) Pore-filling, where PAHs sorption efficiency is directly proportional to the surface area of micropores.
(ii) Hydrophobicity: pertaining to the increase in biochar's hydrophobicity and enhancement of sorption effectiveness caused by the H– and O– functional groups reduction occurring in pyrolysis process.
(iii) Electrostatic attraction: which is primarily controlled by π–π electron-donor–acceptor interactions.
(iv) Hydrogen bonds: the formation of intermolecular bonds between H atoms of hydrogen bonds and COOH, OH, and R–OH functional groups present on the surface of biochar.
(v) Partitioning: Occurrence of substrate polarity reduction through biochar made of carbonless organic matrix that initiates the PAHs partitioning (Fig. 3).

Biochar belonging to a carbon adsorbent group has increased molecular polarity, which may allow for more adsorption of PAHs due to intermolecular dipole–dipole interactions. Scanning electron microscope images can even show PAHs adsorbing onto biochar (Fig. 4).

There are two phases of hydrophobic interaction (entropic) mechanism, in the first phase, the PAH molecules are partitioned into the biochar's hydrophobic domain which occurs due to the partitioning coefficient, and in the second phase, PAH is loosely adsorbed onto the surface of biochar through Van der Waals forces (Tareq et al. 2019). These interactions occur mostly with biochar prepared using low-temperature pyrolysis as a result of biochar polarity (Zhu et al. 2018). Adsorption process is usually facilitated by weak intermolecular forces viz, Van der Waals forces (VDW). Derivatives of PAH contain hydroxyl groups (–OH), carboxyl groups (–COOH), and amino groups (–NH_2) are adsorbed on biochar via hydrogen bond. Hydrogen bonding between dissolved and surface-adsorbed molecules can also be used to establish multilayer PAH derivatives adsorption (Lin and Xing 2008).

Table 2 General properties and functions of biochar

Properties of biochar	Functions of biochar
Condensed charcoal-like material resistant to decomposition	Carbon storage
	Carbon sequestration
	Carbon source as energy/fuel
Material of high surface area and high porosity	Sorption of pollutants
	Water holding capacity
	Improves soil structure and drainage
Capture of macro and micronutrients	Soil benefits (nutrients)
	Improve crop yields
	May be source of pollutants (heavy metals/ organic chemicals adsorption)

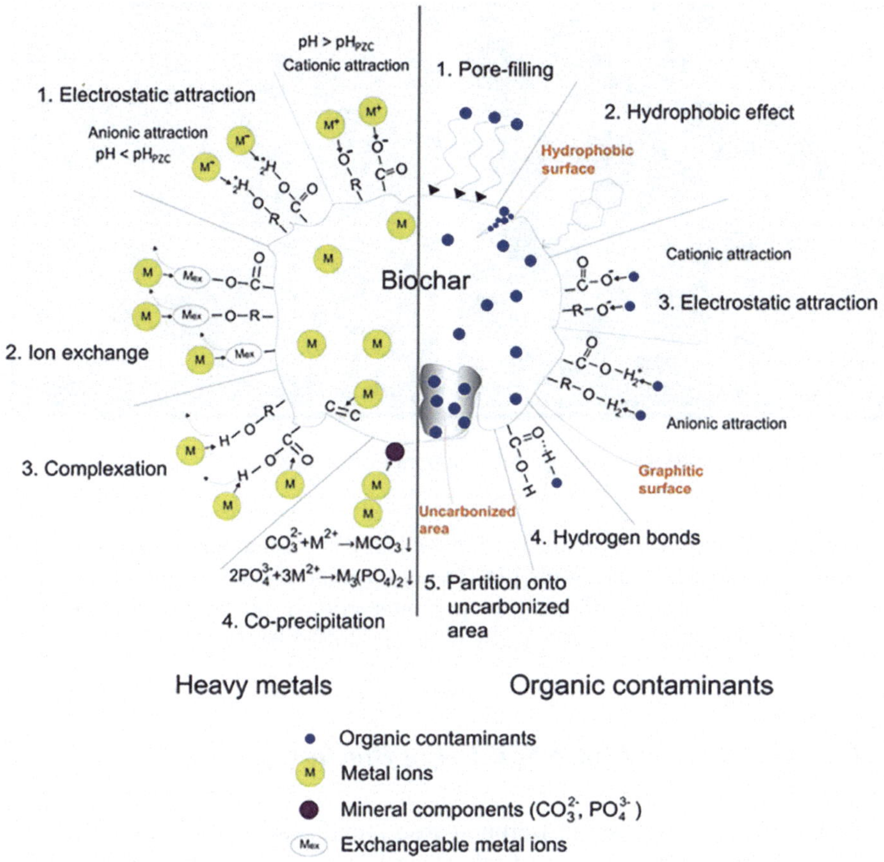

Fig. 3 Sorption mechanism during interaction between pollutants and biochar (Wang et al. 2020a, b)

PAHs aromatic rings which are stable in nature consist of multiple electrons that can be easily conjugated with aromatic rings (for example, C=C double bonds) present in biochar. A significant PAHs and biochar interaction is mediated by electron-donor acceptor (EDA) interaction (Zhang et al. 2014). Due to high electron numbers, HMW PAHs have been demonstrated to readily get adsorbed onto the biochar (Sun et al. 2013). Except for the π–π-interaction, adsorbent molecules, oxygen and nitrogen atoms can offer lone pair electron and empty orbitals, as well as establish *n*–*π* interactions with the PAH electrons (Zhang et al. 2018). Adsorption via *π* complexation between metal cation and the adsorbent is dependent on weak chemical bonds. Electron transportation across metal ions is likely accomplished via the bridging group's conjugated *π*-electron system. Metal ion-containing functional groups can receive the *π* electron that are delocalized given by PAH aromatic rings

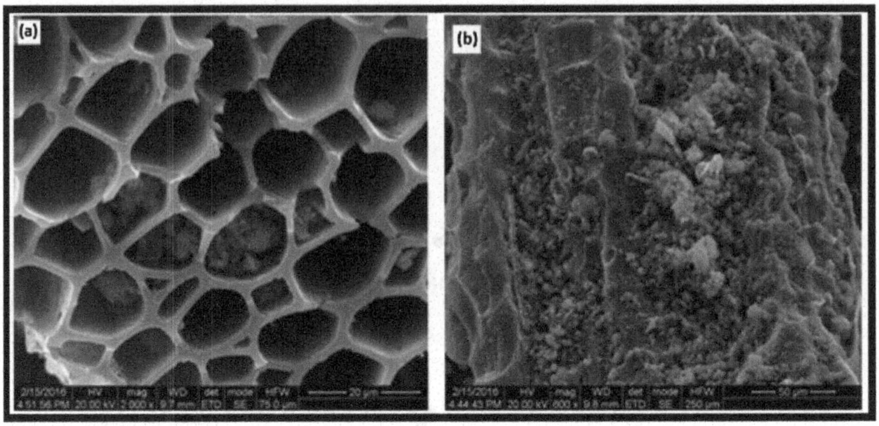

Fig. 4 Adsorption of PAHs on biochar. Adapted from Li et al. (2020)

by creating a complexation bond. The ionizable organic compound could be electrostatically adsorbed by varying the pH. The desorption of PAH from biochar is less because of high sorption affinity (e.g. pep interactions), or due to the micropores present in the biochar which physically traps the sorbed PAHs.

6.1 PAHs Removal from Soil Using Biochar

PAHs removal effectiveness or adsorption/absorption capacity is highly influenced by numerous characteristics such as adsorbent particle size, pH, temperature, solubility, salinity, and adsorbent fabrication method. Activated carbon, charcoal, and modified clay minerals have been widely employed as adsorbent mediums to extract PAHs present in the aqueous solution and to trap in soils contaminated with PAHs. Biochar is a good adsorbent because of large surface area and high porous structure that allows organic pollutants to be absorbed in PAHs removal from contaminated soil (Yargicoglu et al. 2015). It is frequently utilized as a soil amendment because it changes soil characteristics, aids in the retention of minerals and organic substances, and promotes microbial activity (Ogbonnaya and Semple 2013). Other processes such as biodegradation, volatilization and leaching could for the removal of PAHs when applied in the soil along with the biochar (Qian et al. 2023). Biochar removed $> 50\%$ of the HMW PAHs in contaminated soil and $> 40\%$ LMW PAHs, in turn reducing the toxicity effect to plants and microorganisms due to immobilization of soil PAHs (Manjula et al. 2023; Beesley et al. 2010). However, biochar along with soil properties such as particle size, pyrolysis method, and microbial community, greatly affect the removal efficiencies of PAHs (Waqas et al. 2015). For example, biomass (wood) derived biochar made through slow pyrolysis and gasification revealed that slow pyrolysis-derived biochar has better immobilization of 14^{C}-naphthalene in soil than gasification-produced biochar (Ogbonnaya et al. 2016). This efficiency could be due to the presence of more pores and organic functional groups on biochar generated through slow pyrolysis (450–500 °C) when compared to biochar generated by gasification (900–1000 °C). Soil contaminated with PAHs was applied with 5% w/w biochar for a period of six months resulted in 18% reduction of total petroleum hydrocarbon by 18 and 24–28% of bioavailable concentration Cipullo et al. (2019). In an 18-month field experiment, removing total PAHs by applying 2.5% biochar on podzolic soil for 12 months resulted in 45% reduction of PAH content (Stefaniuk et al. 2017). Field experiment conducted for more than 18-month incubation period, where the soil was supplemented with 2.5% biochar revealed 18–67% PAHs (4–6 rings) reduction (Oleszczuk et al. 2019). Three different types of biochars such as pristine biochar, steam-activated biochar, and commercial activated carbon were tested on four different soils contaminated with PAH and hetero-PAH. When compared to other types of biochar, steam-activated biochar showed better performance with 80% removal of PAH from the contaminated soil (Carlini et al. 2023). Further, fabricated biochar was loaded with humic acid and tested on PAHs contaminated soil for its removal. The results revealed that nearly 98.2, 99.3 and 90.1% of naphthalene, pyrene and benzo(a)pyrene were efficiently removed from the soil suggesting the synergistic effect of biochar loaded with humic acid (Ke et al. 2024).

6.2 PAHs Removal from Water and Sediments Using Biochar

PAHs are the most common contaminants found in wastewater arising from natural and man-made sources, such as bushfires, petroleum industries, volcanic eruptions, mining, etc. PAHs usually enter into water bodies and settle at the ground level and surface water, leading to water pollution. In water, PAHs are classified as persistent organic pollutants. PAHs adsorption from water streams has been achieved from the biochar that are made from biomass (Bedia et al. 2018). Biomass-based biochar was shown to have larger surface area with higher porous volumes than sludge-derived biochar, which is greatly attributed to the latter's high carbon content (Askeland et al. 2019). Large amount of PAHs removal using biochar generated from biomass of *Enteromorphaprolifera* at 200 °C was reported (Qiao et al. 2018). Three waste-derived raw materials, waste tyre crumb rubber (WTCR), blast furnace slag (BFS), and coconut coir fibre (CCF), biochar (BC) and iron-modified biochar (FeBC) considered as functionalized samples were used as adsorbents to extract PAHs from rainwater (Esfandiar et al. 2021). PAH partition coefficients were represented as: BC > FeBC > WTCR > CCF ≫ BFS, in the 80–390,000 L/kg range which involved the π–π hydrophobic interactions as basic elimination mechanism.

Sediments are the primary receptors of organic pollutants including PAHs, entering the aquatic ecosystem that are harmful petroleum-derived substances that are mostly released due to human activities into the environment (Chen et al. 2020; Primost et al. 2018). The main mechanism produced by biochar amendment into sediment to remove PAHs is sorption. Yang et al. (2018) observed a 34% higher sorption of PHE (Phenanthrene), than in controls without biochar (22%), prepared by pyrolyzing the nut shells of macadamia fruit (500 °C for 1 h) with a 1.25% (w/w) dose in river sediments after 100 days. Further, sorption of PHE onto biochar particles was raised by increasing the PAH partition coefficient between water and suspension, with maximizing the system's carbon content by 52% (Marchal et al. 2013).Factors such as, types of sediment, contamination level, time of remediation, dosage of biochar in sediment and size of the particle play a major role in sorption of PAH. After 28 days, 5% biochar derived from pine wood (700–1000 °C for 10 min) reduced PAH levels in marine sediments and in pore water by roughly 98% as reported by Gomez-Eyles and Ghosh (2018). Similarly, biochar produced from 1.25% macadamia nutshell (made at 500 °C), 1% mangrove plant (made at 600 °C), and 4% pine wood produced through gasification showed sorption of PYR of upto 29% after 100 days onto charcoal when applied PAH alone, but the sorption was increased to 52% after 56 days and 95% after 30 days when PYR was present in a combination with PHE and ANT PHE suggesting that sorption of PYR is less efficient when present alone as PAH (Chen et al. 2019; Jia et al. 2020; Silvani et al. 2017). Moreover, Silvani et al. (2017) reported that the high salinity of the sea sediments utilized, the higher surface area, greater dose of biochar (343 m^2 g/1 and 4%), initial PAH concentration (10 mg kg/1) that facilitate the sorption mechanism. Variations in environmental conditions (for example, pH) degradation of biochar (biological) can, however, enhance desorption of PAH from biochar to sediment. The biochar prepared from mulberry, wheat straw and pine tree sawdust was prepared to adsorb and stabilize the PAHs in the contaminated sediment. The adsorption of PAHs to the mulberry, wheat straw and pine tree sawdust-derived biochar ranged from 106.7 and 1202 μg/g, 135.1 and 1002 μg/g, and 255.6 and 909 μg/g, respectively. Further, the mulberry-derived biochar was found better PAHs stabilizing agent in the sediment (Le et al. 2023).

7 Conclusion

The remediation of soil and water bodies contaminated by PAHs has proven to be a productive and sustainable system when using biochar. The soil characteristics, biochar, and remediation conditions (origin, types of PAHs, and initial PAH concentration) have a significant impact on the sorption efficiency. Hydrophobicity, donor–acceptor, or specific interactions can cause PAHs to migrate from the contaminated soil and sediments onto the surface of biochar. Biochar derived at high pyrolysis temperatures is found to be unsuitable for bioaugmentation and phytoremediation as there will be reduction in the PAH bioavailability and pH increase which can further disturb the activity of microbes until a complete inhibition is achieved. Given that these substrates have a high-water content, hydrothermal combustion and microwave-aided pyrolysis can be effective alternatives for producing biochar in the latter case. Further, biochar application conditions should be made effective for remediation of PAH-contaminated soils. The studies in future need to be focused on executing open-field testing for longer periods of time. This will enable optimization of biochar application conditions thereby reducing the PAHs quantity in the soil. Developing nano-based biochar for efficient adsorption of PAHs in the water or sedimentation units contaminated with PAHs could be a promising way to meet the remediation requirements. Future research should focus on assessing the remediation potential of biochars made from novel feedstocks like plastics, bio-based algal biomass, animal manure, and sewage sludge.

References

Ananda A, Ramakrishnappa T, Archana S, Reddy Yadav LS, Shilpa BM, Nagaraju G, Jayanna BK (2022) Green synthesis of MgO nanoparticles using phyllanthusemblica for evans blue degradation and antibacterial activity. Mater Today Proc 49:801–810. https://doi.org/10.1016/j.matpr.2021.05.340

Anyika C, Abdul Majid Z, Ibrahim Z, Zakaria MP, Yahya A (2015) The impact of biochars on sorption and biodegradation of polycyclic aromatic hydrocarbons in soils—a review. Environ Sci Pollut Res 22(5):3314–3341

Archana S, Jayanna BK, Ananda A, Shilpa BM, Pandiarajan D, Muralidhara HB, Kumar KY (2021) Synthesis of nickel oxide grafted graphene oxide nanocomposites—a systematic research on chemisorption of heavy metal ions and its antibacterial activity. Environ Nanotechnol Monit Manag 16:100486. https://doi.org/10.1016/j.enmm.2021.100486

Askeland M, Clarke B, Paz-Ferreiro J (2019) Comparative characterization of biochars produced at three selected pyrolysis temperatures from common woody and herbaceous waste streams. PeerJ 7:e6784

Bedia J, Peñas-Garzón M, Gómez-Avilés A, Rodriguez JJ, Belver C (2018) A review on the synthesis and characterization of biomass-derived carbons for adsorption of emerging contaminants from water. C 4(4):63

Beesley L, Moreno-Jiménez E, Gomez-Eyles JL (2010) Effects of biochar and greenwaste compost amendments on mobility, bioavailability and toxicity of inorganic and organic contaminants in a multi-element polluted soil. Environ Pollut 158(6):2282–2287

Beriro DJ, Cave MR, Wragg J, Thomas R, Wills G, Evans F (2016) A review of the current state of the art of physiologically-based tests for measuring human dermal in vitro bioavailability of polycyclic aromatic hydrocarbons (PAH) in soil. J Hazard Mater 305:240–259

Carlini C, Chaudhuri S, Mann O, Tomsik D, Hüffer T, Greggio N, Marazza D, Hofmann T, Sigmund G (2023) Benchmarking biochar with activated carbon for immobilizing leachable PAH and heterocyclic PAH in contaminated soils. Environ Pollut 325:121417

Chen B, Xuan X, Zhu L, Wang J, Gao Y, Yang K, Shen X, Lou B (2004) Distributions of polycyclic aromatic hydrocarbons in surface waters, sediments and soils of Hangzhou City, China. Water Res 38(16):3558–3568

Chen Z, Chen J, Yang X, Chen C, Huang S, Luo H (2019) Biochar as an effective material on sediment remediation for polycyclic aromatic hydrocarbons contamination. IOP Conf Ser Earth Environ Sci 281(1):012016

Chen J, Fan B, Li J, Wang X, Li W, Cui L, Liu Z (2020) Development of human health ambient water quality criteria of 12 polycyclic aromatic hydrocarbons (PAH) and risk assessment in China. Chemosphere 252:126590

Cipullo S, Negrin I, Claveau L, Snapir B, Tardif S, Pulleyblank C, Prpich G, Campo P, Coulon F (2019) Linking bioavailability and toxicity changes of complex chemicals mixture to support decision making for remediation endpoint of contaminated soils. Sci Total Environ 650:2150–2163

Esfandiar N, Suri R, McKenzie ER (2021) Simultaneous removal of multiple polycyclic aromatic hydrocarbons (PAHs) from urban stormwater using low-cost agricultural/industrial byproducts as sorbents. Chemosphere 274:129812

Fu PP, Xia Q, Sun X, Yu H (2012) Phototoxicity and environmental transformation of polycyclic aromatic hydrocarbons (PAHs)—light-induced reactive oxygen species, lipid peroxidation, and DNA damage. J Environ Sci Health C 30(1):1–41

Gomez-Eyles JL, Ghosh U (2018) Enhanced biochars can match activated carbon performance in sediments with high native bioavailability and low final porewater PCB concentrations. Chemosphere 203:179–187

Jia H, Li J, Li Y, Lu H, Liu J, Yan C (2020) The remediation of PAH contaminated sediment with mangrove plant and its derived biochars. J Environ Manage 268:110410

Ke Y, Zhang X, Ren Y, Zhu X, Si S, Kou B, Zhang Z, Wang J, Shen B (2024) Remediation of polycyclic aromatic hydrocarbons polluted soil by biochar loaded humic acid activating persulfate: performance, process and mechanisms. Biores Technol 399:130633

Le QTN, Lee HH, Hwang I (2023) Evaluation of the use of biochar to stabilize polycyclic aromatic hydrocarbons and phthalates in sediment. Environ Pollut 317:120644

Li F, Chen J, Hu X, He F, Bean E, Tsang DC, Ok YS, Gao B (2020) Applications of carbonaceous adsorbents in the remediation of polycyclic aromatic hydrocarbon-contaminated sediments: a review. J Clean Prod 255:120263

Lin D, Xing B (2008) Adsorption of phenolic compounds by carbon nanotubes: role of aromaticity and substitution of hydroxyl groups. Environ Sci Technol 42(19):7254–7259

Manjula NG, Tajunnisa, Mamani V, Meghana CA, Mayegowda SB (2023) Fungal-based synthesis to generate nanoparticles for nanobioremediation. In: Policarpo Tonelli FM, Roy A, Ananda Murthy HC (eds) Green Nanoremediation: Sustainable Management of Environmental Pollution. Springer International Publishing, pp 83–108. https://doi.org/10.1007/978-3-031-30558-0_4

Marchal G, Smith KE, Rein A, Winding A, de Jonge LW, Trapp S, Karlson UG (2013) Impact of activated carbon, biochar and compost on the desorption and mineralization of phenanthrene in soil. Environ Pollut 181:200–210

Mayegowda SB, Sureshkumar K, Yashaswini R, Ramakrishnappa T (2022) Phytonanotechnology for the removal of pollutants from the contaminated soil environment. In: Shah MP, Roy A (eds) Phytonanotechnology. Springer Nature, pp 319–336. https://doi.org/10.1007/978-981-19-4811-4_15

Mayegowda SB, Chikkud V, Barua S, Manjula NG (2024) Heavy metal detection by nanotechnology-based sensors. In: Nanotechnology-based sensors for detection of environmental pollution. Elsevier, pp 237–263. https://doi.org/10.1016/B978-0-443-14118-8.00013-9

Ogbonnaya U, Semple KT (2013) Impact of biochar on organic contaminants in soil: a tool for mitigating risk? Agronomy 3(2):349–375

Ogbonnaya UO, Thomas J, Fasina SA, Semple KT (2016) Impact of two contrasting biochars on the bioaccessibility of 14C-naphthalene in soil. Environ Technol Innov 6:80–93

Oleszczuk P, Rakowska M, Bucheli TD, Godlewska P, Reible DD (2019) Combined effects of plant cultivation and sorbing carbon amendments on freely dissolved PAHs in contaminated soil. Environ Sci Technol 53(9):4860–4868

Primost MA, Commendatore M, Torres PJ, Bigatti G (2018) PAHs contamination in edible gastropods from north Patagonian harbor areas. Mar Pollut Bull 135:828–831

Qiao K, Tian W, Bai J, Dong J, Zhao J, Gong X, Liu S (2018) Preparation of biochar from enteromorphaprolifera and its use for the removal of polycyclic aromatic hydrocarbons (PAHs) from aqueous solution. Ecotoxicol Environ Saf 149:80–87

Qian S, Zhou X, Fu Y, Song B, Yan H, Chen Z, Sun Q, Ye H, Qin L, Lai C (2023) Biochar-compost as a new option for soil improvement: Application in various problem soils. Sci Total Environ 870:162024

Shilpa BM, Rashmi R, Manjula NG, Sreekantha A (2022) Bioremediation of heavy metal contaminated sites using phytogenic nanoparticles. In: Shah MP, Roy A (eds) Phytonanotechnology. Springer Nature, pp 227–253. https://doi.org/10.1007/978-981-19-4811-4_11

Silvani L, Di Palma PR, Riccardi C, Eek E, Hale SE, Viotti P, Papini MP (2017) Use of biochar as alternative sorbent for the active capping of oil contaminated sediments. J Environ Chem Eng 5(5):5241–5249

Stefaniuk M, Oleszczuk P, Różyło K (2017) Co-application of sewage sludge with biochar increases disappearance of polycyclic aromatic hydrocarbons from fertilized soil in long term field experiment. Sci Total Environ 599:854–862

Sun Y, Yang S, Zhao G, Wang Q, Wang X (2013) Adsorption of polycyclic aromatic hydrocarbons on graphene oxides and reduced graphene oxides. Chem Asian J 8(11):2755–2761

Tareq R, Akter N, Azam MS (2019) Biochars and biochar composites: low-cost adsorbents for environmental remediation. In: Biochar from biomass and waste. Elsevier, pp 169–209

Wang L, Ok YS, Tsang DC, Alessi DS, Rinklebe J, Wang H, Mašek O, Hou R, O'Connor D, Hou D (2020a) New trends in biochar pyrolysis and modification strategies: feedstock, pyrolysis conditions, sustainability concerns and implications for soil amendment. Soil Use Manag 36(3):358–386

Wang X, Guo Z, Hu Z, Zhang J (2020b) Recent advances in biochar application for water and wastewater treatment: a review. PeerJ 8:e9164

Wang C, Wang Y, Herath HMSK (2017) Polycyclic aromatic hydrocarbons (PAHs) in biochar–Their formation, occurrence and analysis: A review. Org Geochem 114:1–11

Waqas M, Li G, Khan S, Shamshad I, Reid BJ, Qamar Z, Chao C (2015) Application of sewage sludge and sewage sludge biochar to reduce polycyclic aromatic hydrocarbons (PAH) and potentially toxic elements (PTE) accumulation in tomato. Environ Sci Pollut Res 22(16):12114–12123

Wilson SC, Jones KC (1993) Bioremediation of soil contaminated with polynuclear aromatic hydrocarbons (PAHs): a review. Environ Pollut 81(3):229–249

Xu S, Zhang L, Huang S, Zeeman G, Rijnaarts H, Liu Y (2018) Improving the energy efficiency of a pilot-scale UASB-digester for low temperature domestic wastewater treatment. Biochem Eng J 135:71–78

Yang X, Chen Z, Wu Q, Xu M (2018) Enhanced phenanthrene degradation in river sediments using a combination of biochar and nitrate. Sci Total Environ 619:600–605

Yargicoglu EN, Sadasivam BY, Reddy KR, Spokas K (2015) Physical and chemical characterization of waste wood derived biochars. Waste Manage 36:256–268

Zhang M, Ahmad M, Lee SS, Xu LH, Ok YS (2014) Sorption of polycyclic aromatic hydrocarbons (PAHs) to lignin: effects of hydrophobicity and temperature. Bull Environ Contam Toxicol 93(1):84–88

Zhang Y, Cui B, Zhang Q, Liu X (2015) Polycyclic aromatic hydrocarbons in the food web of coastal wetlands: distribution, sources and potential toxicity. Clean-Soil Air Water 43(6):881–891

Zhang K, Chen B, Mao J, Zhu L, Xing B (2018) Water clusters contributed to molecular interactions of ionizable organic pollutants with aromatized biochar via π-PAHB: sorption experiments and DFT calculations. Environ Pollut 240:342–352

Zhu X, Wang Y, Zhang Y, Chen B (2018) Reduced bioavailability and plant uptake of polycyclic aromatic hydrocarbons from soil slurry amended with biochars pyrolyzed under various temperatures. Environ Sci Pollut Res 25(17):16991–17001

Chapter 13
Biochar a Futuristic Approach for Heavy Metal Remediation from Wastewater

Shilpa Borehalli Mayegowda, Adarsh Patil Kulkarni, Hemavathi Brijesh, and N. G. Manjula

Abstract Heavy-metal-induced contamination has been one of the major pollutants, and currently being one of the most serious concerns for its ill effects on the environment, and toxicity towards humans and plants. Being persistent in the milieu for decades, these heavy metals can also seep into underground waters polluting the soil and aquifers. Contaminated water causes the physiochemical and biochemical changes affecting the plant growth and yield. Additionally, it also has a major concern with the human health effects. Hence, it is very important to find ways to remove these heavy metals efficiently from water. In recent years, different ways of remediating heavy metals from waste water have been undertaken. Biochar, a carbon-rich source is derived from the decomposition of organic wastes that has been competent adsorbent for the heavy metal from water thus, helping as an effective treatment option. An adsorbent for heavy metals, biochar has porous structure, pH value, surface function and high cation exchange. Several methods have been used to synthesise biochar with a significant physiochemical property and have been used for various chemical applications. Being an exceptional heavy metal adsorbent, biochar is also ecofriendly, least expensive and effective treatment that has a greater potential in the near future as a best option for wastewater treatment.

Keywords Heavy metals · Biochar · Adsorbent · Eco-friendly

S. B. Mayegowda
Department of Psychology, CHRIST University, Kengeri Campus Mysore Road Kumbalgodu, Kanmanike, Bengaluru, Karnataka 560074, India

A. P. Kulkarni · N. G. Manjula (✉)
Department of Microbiology, School of Basic and Applied Sciences (SBAS), Dayananda Sagar University, Bengaluru, Karnataka 560111, India
e-mail: manjulang-sbas@dsu.edu.in

H. Brijesh
Department of Biotechnology, School of Applied Sciences, REVA University, Bangalore, Karnataka, India

A. Roy et al. (eds.), *Biochar Production Engineering*,
https://doi.org/10.1007/978-3-031-92594-8_13

1 Introduction

An increased industrialisation and urbanisation have led to a massive scale of water pollution all over the world, seriously challenging the ill effects on both human beings and environment. With the addition of numerous dangerous pollutants, notably the polycyclic aromatic hydrocarbon, dyes etc. the heavy metal pollutants are prevalent and have been persistent in nature (Adarsha et al. 2022; Alhothali et al. 2021). Heavy metals are referred to density greater than 5 g/cm^3 like silver (Ag), mercury (Hg), cadmium (Cd), chromium (Cr), nickel (Ni), zinc (Zn) etc. and other metalloids. They are water-soluble and have strong biological activity even at very low concentrations. However, the nature of these heavy metals, being non-biodegradable leads to concern as it gets accumulated in the environment over a long period and are transferred into living cells causing various disorders in the human system. It has been observed that these heavy metals like lead (250–500 ppm), cadmium (3–6 ppm) and chromium ($\geq$150 ppm) are the foremost hazardous substances to the atmosphere and humans via the food chain (Shilpa et al. 2022; Mayegowda et al. 2024; Inyang et al. 2012). Following with chromium levels as low as 0.05 mg/L can be toxic to humans and also affect the productivity of agricultural crops (Brunetti et al. 2011).

The maximum allowable contaminant concentration for heavy metals has been described by the World's current trends due to the growing necessity, in order to enhance the ecological norms despite being used in least resources feasible for treating wastewater. However, cost-effective and specialized application remains two major critical considerations while choosing efficient methods, such as the use of adsorption designed for the elimination of harmful heavy metals for wastewater treatment. With the advanced technology and recent developments with biochar applications, it has been extensively accepted for dealing with wastewater treatment too. The principle of adsorption being very common due to its many benefits, like use of green adsorbents that are cost-effective and are accessible for the production of a variety of proven industrial and agricultural waste as it is simple to rejuvenate with low energy requirements, that has made it as a potential remediator (Brijesh et al. 2024; Hassanet al. 2020).

There is a need for appropriate treatment of waste waters from the industries, urban sewage and from agricultural production being most common source of inclusion for waste heavy metals that are disposed into the natural water resources causing serious global environmental pollutions (Fig. 1). There are various techniques that has been used like evaporation, chemical precipitation, ion exchange, electrolysis, membrane separation (Archana et al. 2021). Comparatively, adsorption is more economical and efficient treatment using wide adsorbing resources as a treatment option for heavy metals (Fig. 1) (Elbasiouny et al. 2021).

Biochar with carbon matrix having extreme degree of porosity and wide surface area that permits the sorbent activities, thus playing a key role in remediating the pollutants from the environment (Beesley et al. 2011; Chen et al. 2011). With abundant carbon sources, it is characteristically made by thermal treating of the biomass

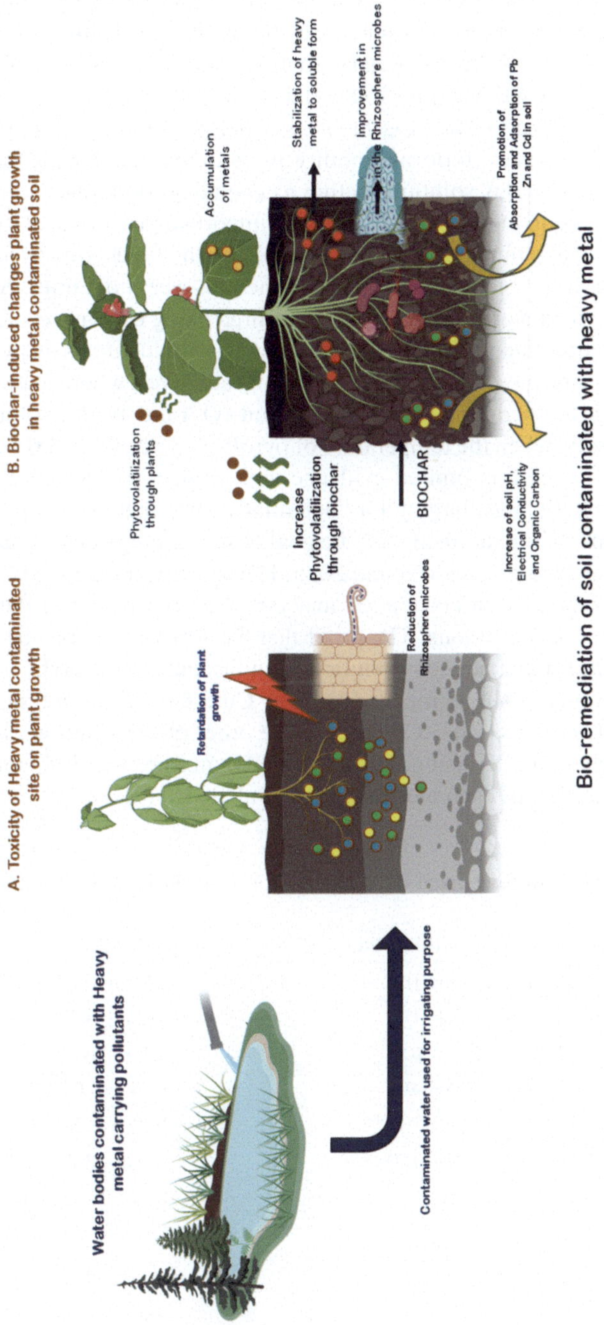

Fig. 1 Effect of heavy metals on the environment and their bioremediation using biochar

from various waste substrates like wood, leaves and manure under anaerobic conditions. Owing to its distinct physiochemical properties and its varied applications, it has gained immense popularity in numerous fields like agriculture, remediation of organic wastes including heavy metals, energy production and climate change mitigation (Mayegowda and Manjula 2025; Halaswamy et al. 2025).

More precisely, adsorption process involves a porous solid that has both large surface area to volume/mass ratio and ability to associate itself with the surface functional groups in aqueous solutions acting as chemisorption, physisorption and various amalgamations to metals. One of the most important properties of the metal's purpose is its capacity to bind. However, it depends on the types, functional groups at a specific region of the biochar surface, and other adsorbent qualities. So far, the drawbacks include the materials like reagents required being expensive and difficult in manufacturing requirements. The best way of biochar production is by pyrolysis in additional to bio-oil and gases, without the use of any additives, where in the biomass is heated to a high temperature. Recently, oxygen (O_2) has been explored as an adsorbent particularly, when the temperature of pyrolysis is between 300 and 900 °C (Papari et al. 2015). There is limited evidence with respect to "designed biochar" to help accomplish pollutants, largely for "potentially toxic metals," despite the fact that biochar remains to be examined with the goal of scavenging contaminants in the aquatic bodies (Mayegowda and Manjula 2025; Halaswamy et al. 2025).

There are fewer studies on correlation analysis that are present in the aqueous solutions at similar concentrations. The proof that the sorption of specific metals is carried in accordance in the aqueous solution as in instance, the metals' adsorption capabilities from aqueous single metals like Pb > Cd > Cr > Zn > and multi metal solutions of Pb > Cu > Cr > Zn > Cd are in same order of the solution obtained in sesame straw biochar. While, using multiple metals, mixture was frequently made with equal concentration of each element (Park et al. 2016) (Table 1).

Table 1 pH, electrical conductivity, and the proportion of heavy metal in the used effluent (Ra et al. 2016)

Parameters	Critical limit	Effluent
Cu (mgL^{-1})	<0.01	2.5 ± 0.0
Zn (mgL^{-1})	<0.05	1383 ± 4.2
Cr (μgL^{-1})	<0.10	510 ± 36.8
pH	5.5–9.5	1.25 ± 0.02
COD (mgO_2L^{-1})	<75	10.2 ± 0.24
EC (mSm^{-1})	70–150	212 ± 13.8

2 Effects of Heavy Metals on Human Health and Environment

In disparity to the organic pollutants, heavy metals are non-degradable, gets accumulated and transferred through the food web leading to numerous health issues or disorders over the period of time. Few heavy metals like cobalt, iron, copper, magnesium, nickel, zinc, selenium, chromium, molybdenum are important trace elements required for the biochemical and physiological functioning of the cells and are essential nutrients. While, others such as barium, silver, gold, aluminium, mercury, lead, tin, titanium, indium, antinomy, vanadium, uranium have no familiar role in biological systems and are considered as nonmetals (Table 2). These heavy metals are considered as trace elements that are usually present in concentration of 100 ppb to <10 ppm in environmental sources (Tchounwou et al. 2012).

Heavy metals induced toxicity and carcinogenicity needs further investigation for their roles in physiological behaviour and their unique physiochemical properties with its active ecological mechanism (Tchounwou et al. 2012). The higher concentrations of heavy metal exposure in humans are extremely risky leading to diseases such as cancers, Parkinson's, Alzheimer's and anaemia while, longer exposure can also affect the liver, gastrointestinal, kidneys, dermal and nervous systems (Table 2)

Table 2 Heavy metals with their toxicity on humans and environment

Heavy metals	Maximum contaminant level by IS (mg/L)	Toxicity	
		Acute	Chronic
Mercury (Hg)	0.001	Elementary fever Nausea Diarrhoea Gastroenteritis Acute lung damage	Metallic taste Neuropathy Hypersensitivity
Cadmium (Cd)	0.003	Pneumonitis Bone disease Coughing Skin irritation	Proteinuria Lung cancer Kidney diseases
Lead (Pb)	0.01	Diarrhoea Stomach discomfort Emesis	Anaemia Copper deficiency Neurologic disorders Osteoporosis
Nickel (Ni)	0.02	Myocarditis Dermatitis	Tumours in nasopharynx and pulmonary fibrosis
Chromium (Cr)	0.05	Haemolysis Gastrointestinal bleeding Acute renal failure	Lung cancer Pulmonary fibrosis
Arsenic (As)	0.05	Neuropathy, Emesis Diarrhoea	Diabetes Hypopigmentation Cancer of skin, lungs

(Jorgensen 2012; Thomas et al. 2021). Following the disposition of wastes derived from industries, municipal and agrochemical in open water systems like rivers etc., it has created problems where in these heavy metals reach to an alarming situation with the deterioration of natural water quality and aquatic resources (Ali et al. 2016). With the above aforementioned causes there has been a growing demand for clean water globally with the remediation of water contaminants that has gained immense attention due to its it serious issues (Yap et al. 2019). Thus, heavy metals are prerequisite to be removed from water in order to reduce their contamination and address the water pollution, followed by their toxicity reduction and accumulation in plants and animals (Elbasiouny et al. 2021).

2.1 Biochar

Biochar, a porous carbonaceous solid formed with the help of thermochemical conversion of organic substrates in anaerobic conditions. Biochar with its physiochemical charatcerstics is safe carbon source that can be potentially used for varied applications. The electroplating business of ES and LC manufacturing company, at Pietermaritzburg, South Africa, generated the untreated industrial wastewater while, being focused on creating and electroplating brackets for satellites. However, the subsequent wastewater was tested using standard rules in accordance to electric conductivity, pH and chemical oxygen demand (COD) that had a bad impact on the environment. The use of biochar for their carbon matrix structures with a very high porosity and surface played an important role in controlling these pollutants especially the heavy metals in the milieu (Goswami et al. 2022).

2.2 Mineral Elements and Ash in Biochar

With a carbon content of more than 60%, biochar is mostly made of aromatic hydrocarbon, carbon singly or carbon with graphite, other elements such as hydrogen (H), oxygen (O), nitrogen (N) and sulphur (S) that are present in varied substrates. With increasing temperature in pyrolysis treatment, there is a substantial massive outcome on the biochar qualities along with volatile components, through gradual decrease in hydrogen and oxygen content with increased concentration of carbon and ash. The presence of mineral elements like Ca, K, Na and Mg in large quantities and nutrient elements (nitrogen and phosphorous) that are rich in wastes derived from the animals and sludge/effluents that do not decrease with temperature but rather rise during the pyrolysis process result in high ash contents (Cao and Harris 2010).

2.3 Biochar based substrates

Researchers have opted numerous approaches to increase the efficiency towards heavy metal remediation and also other organic pollutants using biochar both in the contaminated soils and waste waters (Mayegowda et al. 2022). To enhance the process using biochar for removal of heavy metals chemical activation, composting etc. has been relied on.

2.4 Pristine Biochar

It is usually the biomass collected at low cost for the production of biochar, that includes various organic substrates like agricultural waste, food wastes, sludge and manure that are readily accessible in huge quantities, that are sustainable in nature (Fig. 2). The final characteristics and yield of biochar varies with respect to the biomass, parameters and the technique used as in pyrolysis or hydrothermal carbonisation, with pyrolysis being most commonly employed (Mayegowda and Manjula 2025; Halaswamy et al. 2025). The process involving fast and slow pyrolysis treatment use varied temperatures ranging from 850–1250 to 450–500 °C respectively (Amusat et al. 2021). These following process yields high biochar that can be used variedly for different applications. However, slow pyrolysis is environmentally friendly as it produces fewer toxins that is released in the atmosphere and highly sustainable hence, considered as the best method. This biochar is useful for remediating the heavy metal contaminants for waste water treatment by adsorption method (Shilpa et al. 2022; Zhang et al. 2017).

2.5 Acid and Alkali Based Activated Biochar Substrates

Biochar with its exceptional surface adsorbent characteristics can be changed or modified by implementing with the use of an acid or alkali that helps to expand specific area and pore structures that can effectively adsorb the contaminants physically (Cheng et al. 2021). The process involves the C–OH and C–H functional groups as in acid, base activated in chemical process in order to change the removal capacity of biochar. Acidic modification of biochar during pyrolysis is by acidic solution such as, phosphoric acid, hydrochloric acid, oxalic acid, nitric acid and citric acid that alters the physiochemical properties to help aid pollutant adsorption (Hemavathy et al. 2020; Yang et al. 2021a, b; Lonappan et al. 2020; Li and Li 2019; Liu et al. 2021a). The phosphoric acid activated biochar with eucalyptus has been found to have superior removal efficiency of Cr(VI) as compared to individually (Zeng et al. 2021). While, the alkaline process involves the use of basic solutions that modify the biochar structures bypre or post pyrolysis treatment using bases that include sodium

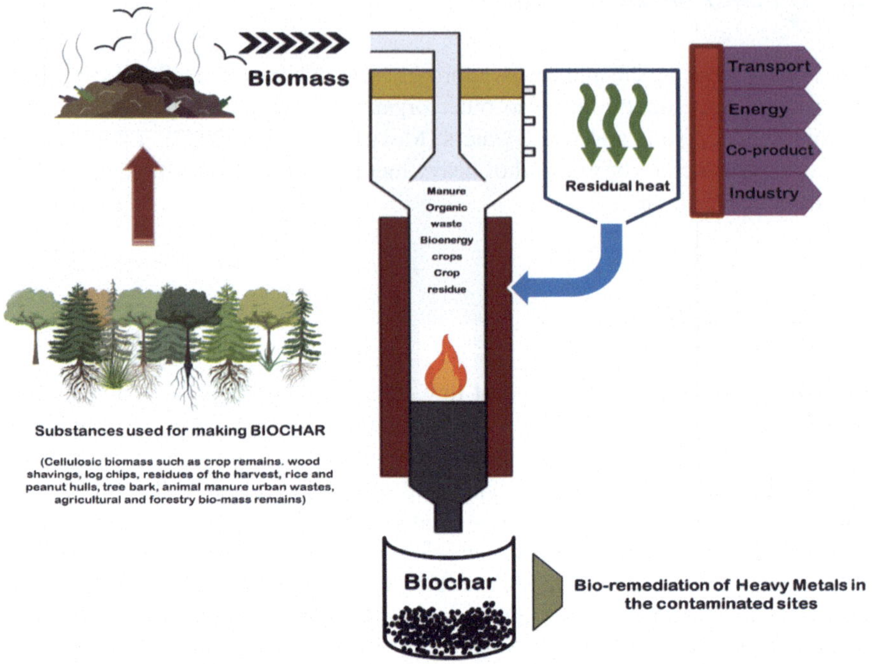

Fig. 2 Production of Biochar using various substrates for the bioremediation of heavy metals from contaminated waste water

hydroxide, potassium hydroxide to increase the pH, ash content, hydrophily etc. (Amusat et al. 2021; Xu et al. 2020).

2.6 *Organic Based Biochar*

The functionality can be enhanced by combining biochar with organic matter along with the increased the adsorption sites (Qiuet al. 2021a, b). Chitosan derived from crustacean shells has been most effective against the textile waste water treatment which is abundant in –OH, –O, and $-NH_2$ groups that elevate the surface adsorption of Cd (Zubair et al. 2021).

2.7 *Biochar Derived Composites*

Biochar engineered nanoparticles is an emergent area for nanocomposites to be used for the bioremediation benefits with wide-ranging combined effects of both the materials (Ravishankar et al. 2025; Amusat et al. 2021). They have properties that help benefit the nanomaterials with functional groups in abundance specially, in pyrolyzed biochar like carboxyl, hydroxy and amino groups that have the ability to remediate heavy metals and pollutants from atmosphere. Clays, a hydrous aluminosilicate is an easily available, nontoxic, low cost employed to remove toxic pollutants from waste water systems (Arif et al. 2021). Further, this high surface energy comprising nano-metal oxides with specific surface area when combined with biochar compounds have distinct advantages as novel adsorbent particles (Yadav et al. 2021). Following metals, such as iron, magnesium, aluminium and manganese are commonly used in metal oxide-biochar composites (Zhang et al. 2021). The compatibility exists between the functional groups of carbonaceous adsorbents creating strong bonds with biochar due to their physiochemical properties that help to remediate heavy metals in waste waters.

3 Methods for Biochar Production

3.1 Physical Methods

A physical alteration could improve the general volume, surface area, pore structure of the biochar. Additionally, chemical properties that include the hydrophobicity along with surface functional groups were increased through physical activation necessitated by using physical mediators such as, carbon dioxide, nitrogen and steam grouping, polarity. Interestingly, chemicals have not been used for this low-cost and any straightforward alteration procedures (Kołtowski et al. 2017). The pore size and surface area would both increase when treated with these gases. Further, biochar treatment with oxygen gas has a higher carbon content compared to nitrogen gas. Additionally, biochar's porous structure, large surface area, and surface characteristics could be physically activated using steam too (Feng et al. 2023).

3.2 Chemical Method

Chemical technique for activation is also referred as wet oxidation. Resulting chemical modification is mainly concerned with altered metal salts, acids, oxidizing agents and alkali. In order to get the described properties for designing biochar, acids as (H_2SO_4), alkalis (NaOH, KOH) and oxidizing reagents (H_2PO_4, HNO_3 and HCl) has been used in chemical manufacturing technique (Zhou et al. 2021). With the use of following acids, bases or inorganic salts it helps to activate essential physiochemical properties of the biochar altering the functional groups on the surface to increase the binding and adsorbing capacity (Feng et al. 2023). While, surface area and porosity were observed to have variable impacts, treatment with oxidizing agents and acids actually helped to aid in increased oxygen hydrophilicity and oxygen comprising functional groups (Yang et al. 2019). The chemical modification process has been employed more frequently than physical due to their distinguishing characteristics as activated biochar has larger surface area, high production, short disposal time, low activation temperatures with well-developed and evenly distributed microporous structures (Panwar and Pawar 2022).

4 Effects of Parameters on Biochar Production

4.1 Effect of pH

The efficacy involving the adsorption capacity of the substrate is by altered surface charges using varying pH in an aqueous solution, a crucial factor in the adsorption process (Yang et al. 2021a, b). A number of surface functional groups are present in biochar with most important being the oxygen-containing groups, that includes hydroxyl (–OH) and carboxylate (–COOH). The positively charged functional group are protonated at a low pH at the biochar surface below pH pzc (point of zero change) which facilitates the adsorption of anions. The negatively charged surface of the biochar in an acidic environment, on the other hand (when pH > pHpzc) can effectively collect cations at higher pH values (Tan et al. 2015).

4.2 Effect of Contact Time

A crucial parameter in adsorption process is the contact time, essential for eliminating organic contaminants. The relationship between biochar dosage and reaction time is considered in order to check the impact of contaminated time with pH. Following biochar dosage at 0.25 g L1, 48.3% elimination of polycyclic aromatic hydrocarbons (PAHs) was achieved in 30 min of fixed reaction time, with subsequent extension of contact period from 30 to 180 min, further reduction of PAHs was observed. A maximal 73.8% clearance of PAH was successful within two hours. However, further increasing the reaction time to 180 min the elimination of PAH dropped to 65.2%. For COD elimination, similar results were observed, with initial 30 min of interaction79.6% of COD reduction was denoted. With increased reaction time from 30 to 180 min, it grew steadily and peaked up at 92.7%, with finally decreasing after 180 min of contact time. The maximum availability of binding sites during the 30 min of initial reaction time might be the cause of this tendency, which improvised the interaction between the adsorbent and pollutants, resulting in dramatically increased adsorption at 30 min. Furthermore, some of the adsorbed pollutants may be released into the solution when the contact duration increased after equilibrium. At about

180 min of contact time a greatest amount of pollution was eliminated efficaciously (Dong et al. 2018).

5 Performance and Mechanism of Heavy Metal Adsorption on Biochar

One of the main causes of ecological deterioration is the heavy metal pollutants, posing a serious risk to the ecosystem, human health, and other living things (Shilpa et al. 2022). With heavy metals remediation taken a priority, biochar has attracted a major attention in the present scenario due to its inexpensive, safe, environmentally friendly with extremely adsorbent, porous and surface-area abundance in helping to bioremediate them.

More critically, carbon-rich materials like biochar are able to effectively remove heavy metals. In certainty, a number of parameters, including the biochar amount with the parameters like pH, types of heavy metals to bioremediate, water temperature, initial conditions affected the heavy metal adsorption with respect to the additional cation functional groups of the biochar. The adsorption process reveals that chemical adsorption was a dominating process, and the isotherm and a second-order kinetic model could both explain the same (Yaashikaa et al. 2019).

5.1 Mechanism in Heavy Metal Remediation

The heavy metal adsorbing mechanism by using biochar is intricate with the use of other techniques involving ion exchange, precipitation, electrostatic adsorption, complexation, or physical (Wang et al. 2020). The following entitled, biochar adsorbed heavy metals via three different ways has an effective remediation process. Historically, electrostatic forces linked negatively charged charcoal to positively charged metals while, the electron donor present in biochar lowered the metals. Finally, reduced metal still in solution were linked with functional groups from biochar to form the complexes. The porosity and surface area of the biochar also contribute to the heavy metal adsorption, further enhancing the ability of biochar specially made from various feedstocks in order to immobilize the heavy metals (Lu et al. 2018).

The non-mediated mechanisms being well documented includes the complexation, ion exchange, precipitation, and electrostatic attraction. The positively and negatively charged heavy metal ions in solution were physically transferred during an ion exchange process due to their coulombic forces that occurs amongst the ions and the charged functional groups present on the biochar surface. The non-specific adsorption technique with adsorption capacity has the way of metal ions fixed by functional groups via an exchange process comprising of carboxyl and oxygen-containing groups (–COOH). The high cation exchange capacity may contribute to the increased capacity of charcoal in order to absorb the heavy metals (Tang et al. 2021).

Currently, major research is being done on heavy metal cleanup, specifically involving As, Pb, Cd, Zn, Cr, Cu, and Hg. As opposed to various organic contaminants, these heavy metals are harder to degrade, making remediation more challenging (Shilpa et al. 2022). Biochar, a novel substance, having extensive sources, a porous structure, with a greater convinced amount of heavy metal pollution that can be efficiently repaired by functional groups and surface area (Park et al. 2016).

5.2 Ion Exchange and Cationic Adsorption Are Some of Its Functions

The main key features in deterioration of heavy metal activities are the ion exchange adsorption on the biochar surface that helps to increase the retention of heavy metals with the amount of cation exchangers. The process involved in ion exchange is an electrical interaction amongst ions, where the contaminants positive charges and the biochar surface negative charges gets complexed. This type of reaction falls under the nonspecific adsorption and has a lower adsorption energy which is clearly reversible. The aromatization of biochar is necessary for the cationic action. The ability of function groups to lose electrons grows as the amount of conjugate aromatic structure increases, and the adsorption impact becomes more important resulting in increased negative charge in orbital alterations (Li & Li 2019).

5.3 Co-Precipitation

The process of adsorbing and dissolution- precipitating of constituting mineral components of the biochar have been effectively reducing the activities of heavy metals. Both the interface binding of the heavy metal ions with –OH, $PO_4$and the additional biochar rises the pH of the effluents (Rees et al. 2017).

5.4 Complexation

Heavy metal ions with highly linear structures can remain fixed due to the complexation process. Numerous studies also have demonstrated that the interactions between oxygenic functional groups like hydroxyl (–OH), carboxyl (–COOH), and amino (–NH_2) groups and heavy metal ions result in significant contributions on the biochar surface in order to adsorb the heavy metal ions (Chao et al. 2018).

5.5 Electrostatic Adsorption

Larger surface areas with greater surface energies allow biochar to effectively adsorb and eliminate the heavy metal contaminants. With different biochar's, the adsorption process for same heavy metal ion differs however, with numerous variables include biochar's starting materials, pyrolysis temperature, soil pH, physical and chemical characteristics of the heavy metal ions, and the amount of biochar added, that affects the heavy metal ion adsorption by them. According to studies, animal feces biochar has a stronger adsorption impact on heavy metal ions than sludge biochar and plant biochar under the same conditions, although with a larger surface area. This is due to the abundance of phosphorous in the biochar made from animal manure. The H^+ content in the effluents vary depending on the water used to evaluate the acidic and alkaline conditions. Since the role of pH affects the surface charges on biochar that helps heavy metals to gets adsorbed or distributed (Li et al. 2020).

6 Remediation of Waste Water Contaminated by Heavy Metals Using Biochar

The limited potential of pristine biochar has been variously modified in order to enhance the selective eradication of heavy metals from waste waters. There has been continual research to adapt the biochar surface adsorption, pore structure, functional groups and surface area to increase the capability to remove pollutants. Engineered biochar's has the role of high efficacy in the removal of waste water heavy metals by several mechanisms involving ion exchange, hydrogen binding, precipitations and electrostatic reactions. The biochar adsorbents have greatest probable to help remove the heavy metals like Pd, Zn, Cd, Cu, U etc., with influence of various features like water, environmental and biochar characteristics such as pyrolysis temperature, feedstock, pH, temperature etc., with the reaction mechanism comprising of ion exchange, electrostatic interaction and reduction. The lethal heavy metal ion cadmium (Cd) has a huge apprehension due to the difficulties in degradation and it gets easily accumulated in human beings (Zhang et al. 2021). With the adsorption efficiency in blue algae derived biochar was highest at 137.5%, as compared with biochar derivatives of rice husk and corn straw with 66.9 and 85.9% respectively (Liu et al. 2021b). The eliminating mechanism of fresh and aged ramie biochar for Cd(II) resulted in physisorption and chemisorption while, in addition cation exchange, coprecipitation and cation interactions were stronger in fresh biochar as compared to older ones

(Su et al. 2021). Biochar-based waste water treatment for lead (Pb) by adsorption is another promising aspect with the most important factor being synthetic approach, type of feedstock and temperature for its removal by means of microwave-assisted pyrolysis method using canola straw biochar using three temperatures at highest adsorption rate of 165 mg/g (Nzediegwu et al. 2021) (Table 3).

An aqueous Cr(VI) immobilised or transformed to fewer toxic components like Cr(III) has reduced effects on the environment. A biochar consisting of porous corn straw with definite area by activated pyrolysis of 2 step process using KOH exhibited a high-efficiency absorbance of 116.97 mg/g (Li et al. 2021a; Qu et al. 2021). Iron oxide composites with unique biochar was effective for eliminating Cr(VI) (Zhou et al. 2021; Tang et al. 2021). Uranium (U), a radioactive material being extremely toxic has an everlasting impact on the environment. The remediation can be processed by the sorption-reduction-solidification method that immobilises and extracts from waste water system (Wang et al. 2022b; Cheng et al. 2021; Li et al. 2021d). Several biochar has been studied and tested for their efficacy in adsorbing U(VI) for waste water treatment with various biochar substrates like modified moso bamboo biochar; orange peel composite biochar helped to eliminate uranium (Wang et al. 2020; Ying et al. 2020). Most of the research targets a single pollutant while, variety needs to be targeted in one-step remediation process in aqueous solutions during field applications. A greater significance needs to investigate and help stimulate the multiple pollutants by adsorption using biochar-based materials for waste water treatment and environmental management for heavy metal reduction.

7 Bioremediation of Heavy Metals by Microbial Sources

It is possible to eliminate or render harmless variety of toxins using the ecosystem's inherent biological activity through the use of bioremediation (Siddiqui et al. 2015). To breakdown or detoxify potentially dangerous chemicals for humans, various microbes such as bacteria, fungus, or plants have been used. Accordingly, bioremediation is the biological degradation or conversion of organic or inorganic waste into typically harmless compounds. A nutrient, electron acceptor, or other elements may be added to the process to improve it or make it work more naturally. As a result, it employs low-tech, low-cost methods that are typically well-liked by the general public and may be used on the spot. Microorganisms must enzymatically attack the contaminants and transform them into harmless compounds. Bioremediation has been successfully supported by the microbial growth and activity in environments, its application frequently entails changing environmental parameters to promote the remediation process (Su et al. 2021). These microbes can be applied to the polluted site after being separated from another location or an indigenously contaminated area. Microbes' breakdowns the contaminated materials by biochemical reactions catalyzed by enzymes from their metabolic activities that can be applied in varied ways, such as composting, biofilters, bioventing, biosorption, land farming, biostimulation, bioaugmentation, and even using a bioreactor (Siddiqui et al. 2015).

Table 3 Adsorption effect of different biochar on different heavy metals

Heavy metals	Types of biochar	Absorption mechanism	Remark	References
Cu^{2+}	Biochar	$Cu_3(CO_3)_2(OH)_2$, CuO is formed through complexation with functional group	pH is 8–9 pH is 6–7	Ippolito et al. (2012)
Zn^{2+}	Water hyacinth hardwood	Electrostatic adsorption, ion exchange		Chen et al. (2011)
Cd^{2+}	Corn straw	Precipitation and electrostatic adsorption		Zhang et al. (2012)
	Sugarcane leaves, tapioca stem, rice straw, silkworm excrement	Electric charge absorption Precipitate with CO_3^{2-}, OH^-	Under the same circumstance: silkworm excrement > rice straw > tapioca stem > sugarcane leaves	Abbas et al. (2023)
As^{5+}	Pine needle, maize straw, dairy manure	Electric charge absorption	Under the same circumstance: dairy manure > pine needle > maize straw	Guan et al. (2015)
Hg^{2+}	Sugarcane, walnut wood chips	Electrostatic adsorption and a complicated process involving hydroxyl (-OH) and carboxyl (-COOH)	Under the same circumstance: sugarcane > walnut wood chips	Xu et al. (2015)
	Beanpoles	Formation of Hg $(OH)_2$, $HgCl_2$, and precipitation		Kong et al. (2011)
	Crispus, rice husk, rice straws	Ion exchange and a complicated process involving hydroxyl (–OH) and carboxyl (–COOH)	Crispus is better when pH = 5	Zhao et al. (2015)
	Brazilian pepper	Complex reaction involving hydroxyl (–OH) and carboxyl (–COOH)	Pyrolytic temperature: low	Dong et al. (2018)
		To create Hg-π, react with an aromatic structure	Pyrolytic temperature: high	
Pb^{2+}, Cu^{2+}, Zn^{2+}, Cd^{2+}	Rice husk	Compound on the surface with phenolic hydroxyl groups	Four heavy metals can be most effectively absorbed by dairy manure, and Pb^{2+} is the most readily absorbed of these	Xu et al. (2020)
	Dairy manure	Oxygenic functional groups and form a PO_4^{3-}, CO_3^{2-}		

(continued)

Table 3 (continued)

Heavy metals	Types of biochar	Absorption mechanism	Remark	References
Pb^{2+}	Walnut green husk	Oxygen functional groups react with aromatic structure and exchange ions		Xie et al. (2016)
	Diary manure, rice straw	Electrostatic adsorption, ion exchange	Under the same circumstance: dairy manure > rice straw	Liang et al. (2014)
	Diary manure	Adsorption, ion exchange and precipitation with PO_4^{3}, CO_3^{2-}		Cao and Harris (2010)
	Peanut shell, Chinese medicine residue	The PB^{2+}-carbonate bounded state changes into a Pb^{2+}-insoluble phosphate and silicate state through ion exchange, electrostatic absorption, complexation, and an increase in pH		Liu et al. (2024)
	Sludge biochar	Perception, Complexation	Absorption rate: 57.7–61.8%	Lu et al. (2012)
		Hydroxyl (–OH), carboxyl (–COOH) and complex reaction (–COOH)	Absorption rate: 38.2–42.3%	

There are various aspects or parameters that must be taken into account when controlling and optimizing the bioremediation processes. These include environmental factors like soil, pH, temperature, aerobic or anaerobic conditions, electron acceptors, nutrients, and the presence of a microbial population capable of degrading pollutants. The elements also include the availability of contaminants to the microbial population as energy or nutrient source. Some prospective bacteria have shown to be exceptional to tolerate these heavy metals, either by eliminating them from the environment or breaking down into absolutely harmless forms and utilize for their metabolic processes and growth. These microbes including bacteria, fungus, and algae are potential bioremediators with microbial resistance and tolerance against contaminants, particularly heavy metals that are very crucial for their survival and clean up technology for these pollutants (Borehalli Mayegowda et al. 2023).

8 Future Approaches

Biochar plays a significant role in heavy metal remediation, however the need for further investigations has to be accounted by the researchers. The use of biochar has been limited currently on the laboratory use, in real environment with large contaminated sites needs to be scaled up to check the efficiency of biochar for remediation process. The study of different production using feed stocks, parameters, biochar production, properties, cost, heavy metal remediation, efficiency and its acceptability needs to be further studied. The biochar use and its long-term effects on soil, water, microbes, toxicity and minerology are probable concerns. Understanding these with the remediation of heavy metals with biochar is relatively important as some biochar might also contain relatively high concentration from the feed stocks.

9 Conclusion

Biochar has an excellent property of adsorbing heavy metals and other pollutants effectively. Hence, it is a potential environmental remediating application for heavy metals specially for waste water treatment. However, the efficiency prospects lie in the characteristics of the biochar with the substrates, parameters and the methods involved in the productions. Being an excellent adsorbent of toxic metals from waste waters they are also an excellent environmentally friendly, cost-effective and nontoxic in nature. However, along with playing important role in bioremediation of waste water at this moment, the understanding of biochar adsorption to treat heavy metal mechanism is important so that it will help to utilise the available biomass feed stocks for its production. Agricultural waste biochar can be a promising factor for bioremediation of water contaminated with toxic hazardous metals in the near future. Another flourishing way is the biochar and biological membranes for heavy metal adsorption by the use of microorganisms that can be explored.

References

Abbas R, Zaib M, Haider K, Ali M, Abbasi R, Hassan M (2023) Remediation of heavy metal contaminated sites by application of biochar: a review with future prospects. Int J Sci Res Eng Dev 5:59–73

Adarsha JR, Ravishankar TN, Ananda A, Manjunatha CR, Shilpa BM, Ramakrishnappa T (2022) Hydrothermal synthesis of novel heterostructured $Ag/TiO_2/CuFe_2O_4$ nanocomposite: characterization, enhanced photocatalytic degradation of methylene blue dye, and efficient antibacterial studies. Water Environ Res 94(6):744. https://doi.org/10.1002/wer.10744

Alhothali A, Haneef T, Mustafa MRU, Moria KM, Rashid U, Rasool K, Bamasag OO (2021) Optimization of micro-pollutants' removal from wastewater using agricultural waste-derived sustainable adsorbent. IJERPH 18(21):11506. https://doi.org/10.3390/ijerph182111506

Ali MM, Ali ML, Islam MS, Rahman MZ (2016) Preliminary assessment of heavy metals in water and sediment of Karnaphuli River, Bangladesh. Environ Nanotechnol Monit Manag 5:27–35

Amusat SO, Kebede TG, Dube S, Nindi MM (2021) Ball-milling synthesis of biochar and biochar–based nanocomposites and prospects for removal of emerging contaminants: a review. J Water Process Eng 41:101993. https://doi.org/10.1016/j.jwpe.2021.101993

Archana S, Jayanna BK, Ananda A, Pandiarajan D, Muralidhara HB, Kumar KY (2021) Synthesis of nickel oxide grafted graphene oxide nanocomposites—a systematic research on chemisorption of heavy metal ions and its antibacterial activity. Environ Nanotechnol Monit Manag 16:100486. https://doi.org/10.1016/j.enmm.2021.100486

Arif M, Liu G, Yousaf B, Ahmed R, Irshad S, Ashraf A, Zia-ur-Rehman M, Rashid MS (2021) Synthesis, characteristics and mechanistic insight into the clays and clay minerals-biochar surface interactions for contaminants removal: a review. J Clean Product 310:127548. https://doi.org/10.1016/j.jclepro.2021.127548

Beesley L, Moreno-Jiménez E, Gomez-Eyles JL, Harris E, Robinson B, Sizmur T (2011) A review of biochars' potential role in the remediation, revegetation and restoration of contaminated soils. Environ Pollut 159(12):3269–3282. https://doi.org/10.1016/j.envpol.2011.07.023

Borehalli-Mayegowda S, Roy A, Pandit S, Alghamdi S, Almehmadi M, Allahyani M, Awwad NS, Sharma R (2023) Eco-friendly synthesized nanoparticles as antimicrobial agents: an updated review. Front Cell Infect Microbiol 13:1224778. https://doi.org/10.3389/fcimb.2023.1224778

Brijesh H, Ajjappala B, Chatterjee A, Manjunath R, Siddappa HB, Mayegowda SB (2024) Plant-derived nanocellulosic material: a promising technology application in environmental bioremediation. Sustainable Nanoremediation 391–409

Brunetti P, Zanella L, Proia A, De Paolis A, Falasca G, Altamura MM, Sanità di Toppi L, Costantino P, Cardarelli M (2011) Cadmium tolerance and phytochelatin content of arabidopsis seedlings over-expressing the phytochelatin synthase gene AtPCS1. J Exper Bot 62(15):5509–5519. https://doi.org/10.1093/jxb/err228

Cao X, Harris W (2010) Properties of dairy-manure-derived biochar pertinent to its potential use in remediation. Bioresou Technol 101(14):5222–5228. https://doi.org/10.1016/j.biortech.2010.02.052

Chao X, Qian X, Han-Hua Z, Shuai W, Qi-Hong Z, Dao-You H, Yang-Zhu Z (2018) Effect of biochar from peanut shell on speciation and availability of lead and zinc in an acidic paddy soil. Ecotoxicol Environ Saf 164:554–561

Chen X, Chen G, Chen L, Chen Y, Lehmann J, McBride MB, Hay AG (2011) Adsorption of copper and zinc by biochars produced from pyrolysis of hardwood and corn straw in aqueous solution. Bioresour Technol 102(19):8877–8884. https://doi.org/10.1016/j.biortech.2011.06.078

Cheng G, Zhang A, Zhao Z, Chai Z, Hu B, Han B, Ai Y, Wang X (2021) Extremely stable amidoxime functionalized covalent organic frameworks for uranium extraction from seawater with high efficiency and selectivity. Sci Bull 66(19):1994–2001. https://doi.org/10.1016/j.scib.2021.05.012

Dong Q, Li H, Niu M, Luo C, Zhang J, Qi B, Li X, Zhong W (2018) Microwave pyrolysis of moso bamboo for syngas production and bio-oil upgrading over bamboo-based biochar catalyst. Bioresour Technol 266:284–290

Elbasiouny H, Darwesh M, Elbeltagy H, Abo-alhamd FG, Amer AA, Elsegaiy MA, Khattab IA, Elsharawy EA, Ebehiry F, El-Ramady H, Brevik EC (2021) Ecofriendly remediation technologies for wastewater contaminated with heavy metals with special focus on using water hyacinth and black tea wastes: a review. Environ Monit Assess 193(7):449. https://doi.org/10.1007/s10661-021-09236-2

Elbehiry F, Elbasiouny H, El-Ramady H, Brevik EC (2019) Mobility, distribution, and potential risk assessment of selected trace elements in soils of the Nile Delta. Egypt. Environ Monit Assess 191(12):713. https://doi.org/10.1007/s10661-019-7892-3

Feng L, Gao Z, Hu T, He S, Liu Y, Jiang J, Zhao Q, Wei L (2023) Performance and mechanisms of biochar-based materials additive in constructed wetlands for enhancing wastewater treatment efficiency: a review. Chem Eng J 471:144772

Feng Z, Zhu L (2018) Sorption of phenanthrene to biochar modified by base. Front Environ Sci Eng 12(2):1. https://doi.org/10.1007/s11783-017-0978-7

Goswami L, Kushwaha A, Kafle SR, Kim B-S (2022) Surface modification of biochar for dye removal from wastewater. Catalysts 12(8):817. https://doi.org/10.3390/catal12080817

Guan X, Zhou J, Ma N, Chen X, Gao J, Zhang R (2015) Studies on modified conditions of biochar and the mechanism for fluoride removal. Desalination and Water Treatment 55(2):440–447

Halaswamy D, Mayegowda SB, Brijesh H, Bheemappa M (2025) Economic aspects of marine biopolymers. In: Marine biopolymers. Elsevier, pp 989–1007. https://doi.org/10.1016/B978-0-443-15606-9.00033-4

Hassan M, Naidu R, Du J, Liu Y, Qi F (2020) Critical review of magnetic biosorbents: their preparation, application, and regeneration for wastewater treatment. Sci Total Environ 702:134893. https://doi.org/10.1016/j.scitotenv.2019.134893

Hemavathy RV, Kumar PS, Kanmani K, Jahnavi N (2020) Adsorptive separation of Cu(II) ions from aqueous medium using thermally/chemically treated *Cassia fistula* based biochar. J Clean Product 249:119390. https://doi.org/10.1016/j.jclepro.2019.119390

Home|CODEXALIMENTARIUS FAO-WHO. https://www.fao.org/fao-who-codexalimentarius/en/. Accessed 14 Nov 2022

Inyang M, Gao B, Yao Y, Xue Y, Zimmerman AR, Pullammanappallil P, Cao X (2012) Removal of heavy metals from aqueous solution by biochars derived from anaerobically digested biomass. Bioresour Technol 110:50–56. https://doi.org/10.1016/j.biortech.2012.01.072

Ippolito JA, Strawn DG, Scheckel KG, Novak JM, Ahmedna M, Niandou MAS (2012) Macroscopic and molecular investigations of copper sorption by a steam-activated biochar. J Environ Qual 41(4):1150–1156

Jorgensen SE (ed) (2012) Encyclopedia of environmental management, four volume set (print version), 1st ed. CRC Press, Boca Raton, FL

Kołtowski M, Hilber I, Bucheli TD, Charmas B, Skubiszewska-Zięba J, Oleszczuk P (2017) Activated biochars reduce the exposure of polycyclic aromatic hydrocarbons in industrially contaminated soils. J Chem Eng 310:33–40

Kong H, He J, Gao Y, Wu H, Zhu X (2011) Cosorption of phenanthrene and mercury (II) from aqueous solution by soybean stalk-based biochar. J Agric Food Chem 59(22):12116–12123

Li B, Li K (2019) Effect of nitric acid pre-oxidation concentration on pore structure and nitrogen/oxygen active decoration sites of ethylenediamine -modified biochar for mercury(II) adsorption and the possible mechanism. Chemosphere 220:28–39. https://doi.org/10.1016/j.chemosphere.2018.12.099

Li M, Guo X, Wei Y, Liu A, Lu J, Niu X, Ma Y, Li S, Shang Z, Liu X (2020) Adsorption mechanism and structure-performance relationship of chromium ions by biochar. Water Air Soil Pollut 231:1–12

Li S, Hu Y, Shen Z, Cai Y, Ji Z, Tan X, Liu Z, Zhao G, Hu S, Wang X (2021a) Rapid and selective uranium extraction from aqueous solution under visible light in the absence of solid photocatalyst. Sci China Chem 64(8):1323–1331. https://doi.org/10.1007/s11426-021-9987-1

Li Y, Xiong W, Wei X, Qin J, Lin C (2021b) Transformation and immobilization of hexavalent chromium in the co-presence of biochar and organic acids: effects of biochar dose and reaction time. Biochar 3(4):535–543. https://doi.org/10.1007/s42773-021-00117-y

Liang Y, Cao X, Zhao L, Arellano E (2014) Biochar-and phosphate-induced immobilization of heavy metals in contaminated soil and water: implication on simultaneous remediation of contaminated soil and groundwater. Environ Sci Pollut Res 21:4665–4674

Liu C, Tian J, Chen L, He Q, Liu X, Bian R, Zheng J et al (2024) Biochar boosted high oleic peanut production with enhanced root development and biological N fixation by diazotrophs in a sand-loamy Primisol. Sci Total Environ 932:173061

Liu L, Fang W, Yuan M, Li X, Wang X, Dai Y (2021a) Metolachlor-adsorption on the walnut shell biochar modified by the fulvic acid and citric acid in water. J Environ Chem Eng 9(5):106238. https://doi.org/10.1016/j.jece.2021.106238

Liu P, Rao D, Zou L, Teng Y, Yu H (2021b) Capacity and potential mechanisms of Cd(II) adsorption from aqueous solution by blue algae-derived biochars. Sci Total Environ 767:145447. https://doi.org/10.1016/j.scitotenv.2021.145447

Lonappan L, Liu Y, Rouissi T, Brar SK, Surampalli RY (2020) Development of biochar-based green functional materials using organic acids for environmental applications. J Clean Product 244:118841

Lu H, Zhang W, Yang Y, Huang X, Wang S, Qiu R (2012) Relative distribution of Pb2+ sorption mechanisms by sludge-derived biochar. Water Res 46(3):854–862

Lu HP, Li ZA, Gasco G, Mendez A, Shen Y, Paz-Ferreiro J (2018) Use of magnetic biochars for the immobilization of heavy metals in a multi-contaminated soil. Sci Total Environ 622:892–899

Manjula NG, Sarma G, Shilpa BM, Suresh Kumar K (2022) Environmental applications of green engineered copper nanoparticles. In: Shah MP, Roy A (eds) Phytonanotechnology. Springer, Singapore, pp 255–276

Mayegowda SB, Sureshkumar K, Yashaswini R, Ramakrishnappa T (2022) Phytonanotechnology for the removal of pollutants from the contaminated soil environment. In: Shah MP, Roy A (eds) Phytonanotechnology. Springer, Singapore, pp 319–336

Mayegowda SB, Manjula NG (2025b) Marine microbial biopolymers and their applications. In: Marine biopolymers. Elsevier, Amsterdam, pp 933–954. https://doi.org/10.1016/B978-0-443-15606-9.00031-0

Mayegowda SB, Chikkud V, Barua S, Manjula NG (2024) Heavy metal detection by nanotechnology-based sensors. In: Nanotechnology-based sensors for detection of environmental pollution. Elsevier, Amsterdam, pp 237–263. https://doi.org/10.1016/B978-0-443-14118-8.00013-9

Nzediegwu C, Naeth MA, Chang SX (2021) Lead(II) adsorption on microwave-pyrolyzed biochars and hydrochars depends on feedstock type and production temperature. J Hazard Mater 412:125255. https://doi.org/10.1016/j.jhazmat.2021.125255

Panwar NL, Pawar A (2022) Influence of activation conditions on the physicochemical properties of activated biochar: a review. Biomass Conv Bioref 12(3):925–947. https://doi.org/10.1007/s13399-020-00870-3

Papari S, Hawboldt K, Helleur R (2015) Pyrolysis: a theoretical and experimental study on the conversion of softwood sawmill residues to biooil. Ind Eng Chem Res 54(2):605–611. https://doi.org/10.1021/ie5039456

Park J-H, Cho J-S, Ok YS, Kim S-H, Heo J-S, Delaune RD, Seo D-C (2016) Comparison of single and competitive metal adsorption by pepper stem biochar. Arch Agron Soil Sci 62(5):617–632. https://doi.org/10.1080/03650340.2015.1074186

Qiu B, Tao X, Wang H, Li W, Ding X, Chu H (2021a) Biochar as a low-cost adsorbent for aqueous heavy metal removal: a review. J Anal Appl Pyrolysis 155:105081. https://doi.org/10.1016/j.jaap.2021.105081

Qiu L, Wu J, Qian Y, Nafees M, Zhang J, Du W, Yin Y, Guo H (2021b) Impact of biochar-induced vertical mobilization of dissolved organic matter, sulfamethazine and antibiotic resistance genes variation in a soil-plant system. J Hazard Mater 417:126022. https://doi.org/10.1016/j.jhazmat.2021.126022

Qu J, Wang Y, Tian X, Jiang Z, Deng F, Tao Y, Jiang Q, Wang L, Zhang Y (2021) KOH-activated porous biochar with high specific surface area for adsorptive removal of chromium (VI) and naphthalene from water: Affecting factors, mechanisms and reusability exploration. J Hazard Mater 401:123292. https://doi.org/10.1016/j.jhazmat.2020.123292

Ra J-S, Jeong T-Y, Lee S-H, Kim SD (2016) Application of toxicity identification evaluation procedure to toxic industrial effluent in South Korea. Chemosphere 143:71–77. https://doi.org/10.1016/j.chemosphere.2015.05.004

Ravishankar TN, Ananda A, Shilpa BM, Adarsh JR (2025) Eco-friendly synthesis of NiO and Ag/NiO nanoparticles: applications in photocatalytic and antibacterial activities. R Soc Open Sci 12(2):241733

Rees F, Watteau F, Mathieu S, Turpault MP, Le Brech Y, Qiu R, Morel JL (2017) Metal immobilization on wood-derived biochars: distribution and reactivity of carbonate phases. J Environ Qual 46(4):845–854

Shilpa BM, Rashmi R, Manjula NG, Sreekantha A (2022) Bioremediation of heavy metal contaminated sites using phytogenic nanoparticles. In: Shah MP, Roy A (eds) Phytonanotechnology. Springer, New York, pp 227–253. https://doi.org/10.1007/978-981-19-4811-4_11

Siddiqui AR, Nazeer S, Piracha MA, Saleem MM, Siddiqi I, Shahzad SM, Sarwar G (2016) The production of biochar and its possible effects on soil properties and phosphate solubilizing bacteria. J Appl Agric Biotechnol 1(1):27–40

Su Y, Wen Y, Yang W, Zhang X, Xia M, Zhou N, Xiong Y, Zhou Z (2021) The mechanism transformation of ramie biochar's cadmium adsorption by aging. Bioresour Technol 330:124947. https://doi.org/10.1016/j.biortech.2021.124947

Tan X, Liu Y, Zeng G, Wang X, Hu X, Gu Y, Yang Z (2015) Application of biochar for the removal of pollutants from aqueous solutions. Chemosphere 125:70–85. https://doi.org/10.1016/j.chemosphere.2014.12.058

Tang J, Zhao B, Lyu H, Li D (2021) Development of a novel pyrite/biochar composite (BM-FeS2@BC) by ball milling for aqueous Cr(VI) removal and its mechanisms. J Hazard Mater 413:125415. https://doi.org/10.1016/j.jhazmat.2021.125415

Tchounwou PB, Yedjou CG, Patlolla AK, Sutton DJ (2012) Heavy metal toxicity and the environment. In: Luch A (ed) Molecular, clinical and environmental toxicology. Springer, New York, pp 133–164

Thomas C, Mayegowda-Shilpa B, Babu-Mythri R (2021) Disease modifying potential of functional foods for neurodegenerative disorders: status update on regulatory compliance. In: Sajid-Arshad M, Haseeb-Ahmad M (eds) Functional foods—phytochemicals and health promoting potential. IntechOpen. https://doi.org/10.5772/intechopen.97546

Wang X, Feng J, Cai Y, Fang M, Kong M, Alsaedi A, Hayat T, Tan X (2020) Porous biochar modified with polyethyleneimine (PEI) for effective enrichment of U(VI) in aqueous solution. Sci Total Environ 708:134575. https://doi.org/10.1016/j.scitotenv.2019.134575

Wang C, Wang X, Li N, Tao J, Yan B, Cui X, Chen G (2022a) Adsorption of lead from aqueous solution by biochar: a review. Clean Technol 4(3):629–652. https://doi.org/10.3390/cleantechnol4030039

Wang S, Shi L, Yu S, Pang H, Qiu M, Song G, Fu D, Hu B, Wang X (2022b) Effect of *Shewanella oneidensis* MR-1 on U(VI) sequestration by montmorillonite. J Environ Radioact 242:106798. https://doi.org/10.1016/j.jenvrad.2021.106798

Xie T, Sadasivam BY, Reddy KR, Wang C, Spokas K (2016) Review of the effects of biochar amendment on soil properties and carbon sequestration. J Hazard Toxic Radioact Waste 20(1):04015013

Xu CY, Hosseini-Bai S, Hao Y, Rachaputi RC, Wang H, Xu Z, Wallace H (2015) Effect of biochar amendment on yield and photosynthesis of peanut on two types of soils. Environ Sci Pollut Res 22:6112–6125

Xu Y, Bai T, Yan Y, Ma K (2020) Influence of sodium hydroxide addition on characteristics and environmental risk of heavy metals in biochars derived from swine manure. Waste Manag 105:511–519. https://doi.org/10.1016/j.wasman.2020.02.035

Yaashikaa PR, Kumar PS, Varjani SJ, Saravanan A (2019) Advances in production and application of biochar from lignocellulosic feedstocks for remediation of environmental pollutants. Bioresour Technol 292:122030

Yadav LSR, Shilpa BM, Suma BP, Venkatesh R, Nagaraju G (2021) Synergistic effect of photocatalytic, antibacterial and electrochemical activities on biosynthesized zirconium oxide nanoparticles. Eur Phys J plus 136(7):764. https://doi.org/10.1140/epjp/s13360-021-01606-6

Yang X, Ng W, Wong BSE, Baeg GH, Wang C-H, Ok YS (2019) Characterization and ecotoxicological investigation of biochar produced via slow pyrolysis: effect of feedstock composition and pyrolysis conditions. J Hazard Mater 365:178–185. https://doi.org/10.1016/j.jhazmat.2018.10.047

Yang CX, Zhu Q, Dong WP, Fan YQ, Wang WL (2021a) Preparation and characterization of phosphoric acid-modified biochar nanomaterials with highly efficient adsorption and photodegradation ability. Langmuir 37(30):9253–9263. https://doi.org/10.1021/acs.langmuir.1c01468

Yang Y, Ye S, Zhang C, Zeng G, Tan X, Song B, Zhang P, Yang H, Li M, Chen Q (2021b) Application of biochar for the remediation of polluted sediments. J Hazard Mater 404:124052. https://doi.org/10.1016/j.jhazmat.2020.124052

Yap HC, Pang YL, Lim S, Abdullah AZ, Ong HC, Wu CH (2019) A comprehensive review on state-of-the-art photo-, sono-, and sonophotocatalytic treatments to degrade emerging contaminants. Int J Environ Sci Technol 16:601–628

Ying D, Hong P, Jiali F, Qinqin T, Yuhui L, Youqun W, Zhibin Z, Xiaohong C, Yunhai L (2020) Removal of uranium using MnO_2/orange peel biochar composite prepared by activation and in-situ deposit in a single step. Elsevier Ltd

Zeng H, Zeng H, Zhang H, Shahab A, Zhang K, Lu Y, Nabi I, Naseem F, Ullah H (2021) Efficient adsorption of Cr (VI) from aqueous environments by phosphoric acid activated eucalyptus biochar. J Clean Product 286:124964. https://doi.org/10.1016/j.jclepro.2020.124964

Zhang H, Chen C, Gray EM, Boyd SE (2017) Effect of feedstock and pyrolysis temperature on properties of biochar governing end use efficacy. Biomass Bioenergy 105:136–146. https://doi.org/10.1016/j.biombioe.2017.06.024

Zhang G, Liu X, Sun K, He F, Zhao Y, Lin C (2012) Competitive sorption of metsulfuron-methyl and tetracycline on corn straw biochars. J Environ Qual 41(6):1906–1915

Zhang D, Zhang K, Hu X, He Q, Yan J, Xue Y (2021) Cadmium removal by MgCl2 modified biochar derived from crayfish shell waste: batch adsorption, response surface analysis and fixed bed filtration. J Hazard Mater 408:124860. https://doi.org/10.1016/j.jhazmat.2020.124860

Zhao L, Cao X, Zheng W, Wang Q, Yang F (2015) Endogenous minerals have influences on surface electrochemistry and ion exchange properties of biochar. Chemosphere 136:133–139

Zhao C, Hu L, Zhang C, Wang S, Wang X, Huo Z (2021a) Preparation of biochar-interpenetrated iron-alginate hydrogel as a pH-independent sorbent for removal of Cr(VI) and Pb(II). Environ Pollut 287:117303. https://doi.org/10.1016/j.envpol.2021.117303

Zhao C, Wang B, Theng BKG, Wu P, Liu F, Wang S, Lee X, Chen M, Li L, Zhang X (2021b) Formation and mechanisms of nano-metal oxide-biochar composites for pollutants removal: a review. Sci Total Environ 767:145305. https://doi.org/10.1016/j.scitotenv.2021.145305

Zhou Y, Qin S, Verma S, Sar T, Sarsaiya S, Ravindran B, Liu T, Sindhu R, Patel AK, Binod P, Varjani S, Rani Singhnia R, Zhang Z, Awasthi MK (2021) Production and beneficial impact of biochar for environmental application: a comprehensive review. Bioresour Technol 337:125451. https://doi.org/10.1016/j.biortech.2021.125451

Zubair M, Adnan Ramzani PM, Rasool B, Khan MA, ur-Rahman M, Akhtar I, Turan V, Tauqeer HM, Farhad M, Khan SA, Iqbal J, Iqbal M (2021) Efficacy of chitosan-coated textile waste biochar applied to Cd-polluted soil for reducing Cd mobility in soil and its distribution in moringa (*Moringa oleifera* L.). J Environ Manag 284:112047. https://doi.org/10.1016/j.jenvman.2021.112047

Chapter 14
Trends and Challenges in the Production of Biochar

Prakash Baburao Rathod, Parshuram Maske, Mahendra Pratap Singh, and Akshay Chandrakant Dhayagude

Abstract Biochar, a by product of pyrolysis, is widely recognized for its potential to enhance soil health, boost fertility, and aid carbon sequestration. However, the environmental impact of its production, particularly the carbon dioxide emissions and thermal energy generated during pyrolysis, remains uncertain. Research into the effectiveness of biochar in mitigating climate change and improving soil fertility, especially in alkaline soils, is needed. Additionally, studies must address the interaction between biochar's properties and its role in nutrient delivery, microbial activity, and adsorption of pollutants. Future research should focus on optimizing pyrolysis conditions, understanding the relationship between biochar's physicochemical properties and its environmental applications, and exploring innovative modification techniques.

Keywords Hydrothermal carbonization · Flash carbonization · Biofuels · Alkaline soils · Environmental stability

---

P. B. Rathod
School of Chemical Sciences, Kavayitri Bahinabai Chaudhari North Maharashtra University, Jalgaon, Maharashtra 425001, India

P. B. Rathod · M. P. Singh
Department of Chemistry, Shri Sadgurur Saibaba Science and Commerce College, Ashti, Maharashtra 442707, India

P. Maske (✉)
Department of Chemistry, S.G.V. and S.S.P'S, Art's Commerce and Science College, Onde, Maharashtra 401605, India
e-mail: pdmaske@gmail.com

A. C. Dhayagude
Department of Chemistry, MVP's S. S. S. M. ASC College, Saikheda, Niphad, Nashik, Maharashtra 422210, India

A. Roy et al. (eds.), *Biochar Production Engineering*,
https://doi.org/10.1007/978-3-031-92594-8_14

1 Introduction

Carbon-containing biomass (plants and manure) when heated in absence of air the product obtained is biochar. Biochar has a spongy assembly with an alcoholic, phenolic, carboxyl, and aldehyde/ketone groups, etc. functionality. There are different plant precursors are used for the production of biochar such as pine needles, bamboo, Brazilian pepperwood, sugarcane biogases, wheat straw, hardwood, softwood, and grass. Invasive plants are one of the important precursors for biochar (Haider et al. 2022).

The type of biomass and the carbonized condition have many effects on biomass preparation and properties. Production of biochar by different processes like pyrolysis, gasification, hydrothermal carbonization, flashes carbonization, etc. It has many applications in the arena of organization of waste, production of energy, improvement of soil, environmental climate change, and so on.

To increase soil quality, plant development, and soil productivity, soil amendment treatments frequently employ biochar, a carbon-rich sustainable solid substance. Moreover, it is used as a key animal feed supplement to promote digestion, toxin adsorption, nutrition intake effectiveness, and blood values in cattle. As the world's population grows, so does the demand for crops, which has prompted farmers to become more stead fast in their use of biochar to improve crop yields. During the projected span, this can be one of the key aspects of increasing the biochar market. Also, mixing the biochar with the animal's food has had promising results. This feed supplement is used in a variety of ruminants, including cattle and goats, to enhance blood profiles, growth performance, and disease resistance. Additionally, it holds substantial quantities of calcium, potassium, and magnesium, which inferior the likelihood of broken eggs in chickens. These elements have enlarged the use of biochar in animal nutrition, which has increased the demand for it (Rawat et al. 2019).

This chapter emphasizes the difficulties of producing biochar. The chapter also provides a summary of the key details found in the literature reports and summaries all of the studies presented. The subsections that follow are organized according to various preparation methods and challenges encountered throughout those methods. We also look at the potential use of this chapter in the field of research.

2 Production of Biochar from Different Techniques

However, producing biochar is an expensive process since it needs a variety of high-end, pricey pyrolysis equipment. Additionally, it costs money to store and ship the pyrolysis gas produced during the manufacturing of biochar since it needs an oxygen-free environment maintained at a certain temperature. In contrast, expanding R&D efforts have advanced the potential uses of biochar in several end-use industries. For instance, biochar is now being used as a fabric addition in the textile sector.

Moreover, it is fairly good at adsorbing color from the wastewater used during textile cleaning. Additionally, while plastering bricks and concrete components, it is frequently utilized as an ingredient in the mortar that also holds clay, lime, and cement. These elements are likely to create new chances in the market for biochar on a worldwide scale.

Analysis of the worldwide biochar market is conducted based on production technique, application, and geographic location. The market is segmented into pyrolysis, gasification, and other types of manufacturing technologies. It is divided into several categories according to its intended use, including industrial, animal feed, soil amendment and others.

Production of biochar takes place in anaerobic (low oxygen) conditions. The manufacture of biochar underwent several alterations and changes. Eventually, several strategies for producing biochar have been discovered. There are numerous ways to produce biochar, but there is no accurate classification of them in the literature. Thus, many methods for producing biochar are highlighted in the next section. Based on their developments and modernization, the biochar production techniques are primarily divided into (Gabhane et al. 2020).

2.1 Traditional Approaches

Archaeological data indicate that biochar production and use dates back more than 2500 years in human history. The first piece of proof was found in the Terra Preta region of South America's Amazon Basin, which has three times as much soil organic matter as other regions (Glaser 2007; Glaser et al. 2001). In slow-burning dirt trenches with little to no air, the ancient inhabitants would pile covered wood and burn it (Thines et al. 2017; Barrow 2012). A further technique comprised burning the biomass outside and immediately burying the partially burned biomass in the dirt (Reilly 1925). In ancient times, biochar was not the only soil amendment method used. Among other things, the liquid result of burning wood was used for shipbuilding, wall painting, filling to seal wooden drums, and fixing barbs to spear shafts. Additionally, it was used to preserve meat and corpses. Because of the growth of mankind and the advancement of science, many antiquated techniques for making biochar have been replaced by more modern ones. Various homemade apparatuses, like firebrick pits, sand burners, brick kilns, and iron retorts, were used to make biochar (Reilly 1925; Emrich 1985). Each technique utilized a pit surrounded by bricks, metal, and clay. The volatile compounds generated during pyrolysis can be recovered and used thanks to these improved biochar manufacturing methods (Reilly 1925). These methods were widely used up until the development of labor- and time-saving steel ovens at the end of the nineteenth century (Emrich 1985). While raising the bar for biochar quality, using a steel oven to make biochar does speed up output. Additionally, it aids in the retrieval of volatile substances and bio-oil (Gabhane et al. 2020).

2.1.1 Pyrolysis

Under anaerobic conditions or with little oxygen present, biomass is heated during pyrolysis, a thermal breakdown process that yields a variety of gases, liquid chemicals, and residues of biochar. Silica, lignin, cellulose, and hemicelluloses make up agricultural biomass. Typically, the cellulose is pyrolyzed at 350 °C, whereas the melting point of lignin is higher (Czernik and Bridgwater 2004). Consequently, it was found that pyrolysis can happen between 300 and 700 °C (Suárez-Abelenda et al. 2017). Pyrolysis is the process of heating the substance in absence of air, but it also requires heat transfer media and other heating techniques (Li et al. 2016; Xie et al. 2015). Pyrolysis procedures have been developed over time. It is further separated into distinct modes, such as slow and rapid pyrolysis, based on the factors and variables such as temperature, heating rate, and residence duration. In contrast to conventional procedures, pyrolysis can happen swiftly or slowly. Microwave, vacuum, and flash pyrolysis are just a few of the more recent techniques that have changed as a result of modern technology. As a result, slow pyrolysis and quick pyrolysis have been separated into separate sections under the heading "traditional procedures." In contrast, the "modern techniques" section has been expanded to cover microwave, vacuum, and flash pyrolysis.

2.1.2 Slow Pyrolysis

As the name implies, biochar is the primary byproduct of slow pyrolysis, and it is created over many hours. Slow pyrolysis, also known as conventional pyrolysis, involves 5–7 °C per minute heating of biomass between 300 and 600 °C. Along with bio-oil (25–35%), syngas (20–30%), and biochar (35–45%) are the main byproducts of slow pyrolysis (Lai et al. 2013; Méndez et al. 2015; Verma et al. 2012). In slow pyrolysis auger/screw pyrolyzer reactor is used (Verma et al. 2012). Lai et al. heated the woodcuts to prepare biochar at a rate of 3 °C per minute while keeping a mean temperature between 290 and 700 °C (Lai et al. 2013). Mendez et al. (2015) used deinking sludge like this to make biochar in closed steel cups over two hours using an electric furnace with a heating rate of 10 °C min^{-1}. Using slow pyrolysis, a variety of biomasses have been employed to create biochar, including Conocarpus wood wastes (Yuan et al. 2016), cotton stalks (Shen et al. 2015), coconut shells, palmyra nutshells, and rice husks (Jegajeevagan et al. 2016). According to a different study (McBeath et al. 2014), the kind of pyrolysis and the make-up of the feedstock have a big impact on the chemical composition of biochar. This study employed slow pyrolysis using sawdust, bull dung, pinewood, oak wood, dairy waste with rice hulls, hazelnut shells, maize, food waste, and white paper mill sludge in a Daisy Reactor at Best Energies Inc. USA. Roberts et al. 21 conducted a similar study with 4 distinct feedstocks and 60 min of slow pyrolysis in an inert environment inside a muffle furnace with a sealed stainless-steel retort. Roberts et al. (2015) discovered that the quantity of biochar produced depends on the temperature of pyrolysis and the pretreatment of the biomass. Ulva produced the most effective biochar when heated flakes at 300 °C; higher temperatures produced less biochar. According to a study (Moralı and Şensöz 2015), a moderate heating rate is superior to a fast heating rate while producing biochar.

2.1.3 Fast Pyrolysis

Fast pyrolysis is a very effective thermochemical method for creating biofuels sourced from biomass (Dai et al. 2017). Fast pyrolysis has the advantages of a short retention time and rapid product recovery. However, when exposed to the advancement method for the creation of fuel extracts, the principal outputs are bio-oil and syngas, not charcoal (Huang et al. 2016). The fast pyrolysis is performed in absence of air, with a temperature greater than 500 °C at a rate of 300 °C min^{-1} of heating. Fast pyrolysis is a quick method that produces biochar in very few minutes. According to reports, in fast pyrolysis 60% bio-oil, 20% biochar, and 20% syngas are produced (Dai et al. 2017; Huang et al. 2016). It is typically used for industrial-scale biochar synthesis. By quickly pyrolyzing biomass in a fixed-bed quartz processor outfitted with a temperature-controlling unit. Liu et al. created biochar, with a rate of heating of 15 °C min^{-1}, the furnace was heated to the desired temperature (200–330 °C) and maintained there for <20 min. In comparison to bio-oil and syngas, the higher charcoal yield was achieved because of the low temperature (Liu et al. 2016). A group (Karunanayake et al. 2017) observed that the biochar from fast pyrolysis is extra beneficial for adsorptive effects when using it to remove salicylic acid, 4-nitroaniline, phthalic acid, and benzoic acid from water. Another paper (Essandoh et al. 2015; Al-Rahbi and Williams 2017) states that pinewood pyrolysis biochar was examined for the adsorption of ibuprofen and salicylic acid from water. Biochar was produced by rapid pyrolysis at a temperature of 425 °C and a residence time of 20–30 min. In addition, a team (Dong et al. 2015) created a low-cost catalyst for pre-esterification in the manufacture of biodiesel employing fast pyrolysis in an Auger pyrolysis reactor at 600 °C for 1 min. Another group (Laird et al. 2017) produced biochar from hardwood by quick pyrolysis, and the results were encouraging in terms of soil quality and crop biomass yield. Efficiencies of slow and fast pyrolysis process at different temperatures and residue time are depicted in Fig. 1.

2.2 *Modern Approaches*

Eventually, several methods for producing biochar were developed, including torre faction, gasification, electro-modified procedures, hydrothermal carbonization, vacuum, flash, and microwave pyrolysis, among others. The biochar preparation techniques and their conditions reported by various scientists are presented in Table 1.

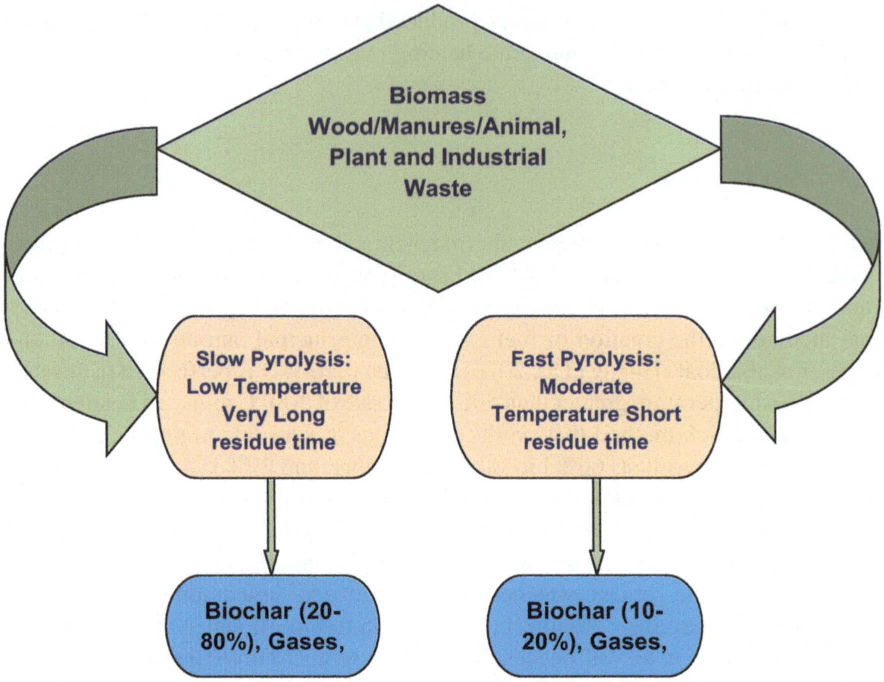

Fig. 1 Represents the efficiencies of pyrolysis approaches to produce biochar (Dafalla et al. 2024)

2.2.1 Gasification

Syngas is typically produced via gasification from a variety of solid fuel sources. Gasification produces more syngas volume and less levelized emissions than other typical processes like combustion, pyrolysis, and fermentation. The main byproduct of gasification is hydrogen. However, the gasification process can also produce a sizable biochar amount. Additionally, biochar produced during gasification is regarded as a waste and has several crucial applications, including the sequestration of carbon, adsorption of pollutants, and use as a soil amendment agent (Al-Rahbi and Williams 2017; González et al. 2008). Biochar is a result of the efficient thermochemical conversion of biomass into energy fuel known as gasification (Guan et al. 2016).

2.2.2 Torre Faction

Torre faction is a thermochemical process that is used to treat biomass under pressure and in absence of oxygen. The Torre faction method typically uses a fixed-bed reactor (Abdulyekeen et al. 2024). Torre faction and densification techniques, as shown by many researchers, can produce excellent-quality torrefied wood pellets when

Table 1 Representing the biochar preparation techniques and their conditions

Technique	Temperature (°C)	Residence time	Biochar yield (%)	Bio-oil yield (%)	Syngas yield (%)	References
Pyrolysis	300–700 (slow)	<2 s (slow)	35 (slow)	30 (slow)	35 (slow)	
	500–1000 (fast)	Hour-day (fast)	12 (fast)	75 (fast)	13 (fast)	Cantrell et al. (2012), Shao et al. (2024)
Hydrothermal carbonization	180–300	1–16 h	50–80	5–20	2–5	Funke and Ziegler (2010), Hang et al. (2024)
Gasification	750–900	10–20 s	10	5	85	Klinghoffer et al. (2015); Akgül et al. (2024)
Torre faction	290	10–60 min	80	0	20	Bergman et al. (2005), Kaewtrakulchai et al. (2024)
Flash carbonization	300–600	<30 min	37	0	0	Nunoura et al. (2006), Muthu Kumar and Varunkumar (2024)

associated with raw controlled pellets. Studies on the Torre faction of agricultural and forestry wastes abound (Prins et al. 2006; Chen et al. 2012).

2.2.3 Flash Pyrolysis

Fast pyrolysis has been refined and improved into flash pyrolysis. In flash pyrolysis, biomass decomposes quickly, typically in less than a minute, at more than 1000 °C temperature. Sometimes, the heating rate exceeds 1000 °C/s. Operating temperatures for flash pyrolysis are between 900 and 1200 °C, which can be reached in less than a second (Demirbas and Arin 2002; Li et al. 2013; Kaur et al. 2024).

2.2.4 Hydrothermal Carbonization (HTC)

Using the hydrothermal carbonization (HTC) process, high humidity content feedstocks like animal waste, sewage sludge, and compost are transformed into biochar (Titirici et al. 2007). In the HTC method, the wet biomass combination is heated between 220 and 240 °C temperatures under high pressure (2–10 MPa) in a reactor for some hours. This eliminates the need to dry the biomass before treating it. To operate the HTC process, rotary drums, stoves, and kilns are used. Most organic materials used in HTC procedures are either kept in dissolved form or converted into brown coal (Libra et al. 2011).

2.2.5 Electro-Modified Biochar

The ability of biochar to absorb contaminants helps to clean up the air, water, and soil. Additionally, it helps with nutrient adsorption. Biochar must be treated so that it adsorbs a specific set of chemicals while quickly eliminating undesirable moieties from a specific environment. Chemical processing is used to modify biochar in this way, and the resulting biochar is known as modified biochar (Jung et al. 2015; Fang et al. 2014). The chemical process involves mixing biochar with Fe, Mg, or Al for 2–12 h while an electric current is present. This could change the functionality of the pore's surface and improve specific adsorption (Zhang et al. 2013; Wan et al. 2020; Hu et al. 2015).

2.2.6 Magnetic Biochar

Numerous researchers from all around the world have been drawn to the improved adsorption capacity with a total recovery of adsorbent material, which can accelerate research into novel adsorbents. Magnetic Nanoparticles (MNP's) show the most impact in the adsorption and catalysis (Rathod et al. 2016, 2018, 2019). MNP's with combination of biochar shows most greater potential for sorption. Magnetic biochar has exceptional adsorption capability, full separation from water or pollutant sites, and recovery. The magnetic property of magnetic biochar, which is produced using a variety of biomasses, is good, and it has a considerable shape and large surface area. By adding Fe-ions to the biochar's surface and using a chestnut shell-derived binding agent at 450 °C while being heated in a microwave, magnetic biochar has been created. Gelatin and iron were added to magnetic biochar by a group (Zhou et al. 2017).

2.2.7 Microwave Pyrolysis

Microwave heating is a well-known sophisticated technology for producing biochar. The advantage of microwave heating over conventional heating is that it produces thermal energy by dielectric heating and introduces the energy into the reactor remotely, avoiding direct contact between the energy source and the reaction mixture (Intanakul et al. 2003; Gabhane et al. 2011; Wu et al. 2024).

3 Challenges in Biochar Preparation and Applications

Although engineering obstacles persist throughout the production process, understanding biochar is developing quickly. Woody material from forests makes up a large portion of the Pacific Northwest's (PNW) potential biomass for biochar production. As a result, the first task is to enhance the collection and handling of this material so that producers of biochar may acquire feedstock more effectively while advancing other land-management goals. This involves abandoning the habit of gathering biomass burning in slash piles outside. This also entails effectively gaining access to the sizable amount of "stranded biomass" that is now inaccessible due to access problems or the high cost of harvesting and transporting it from the terrain.

Another key issue is coming up with new biochar-based systems that are more efficient in terms of the environment, emit less soot and CH_4, and perform better economically than current ones. Generally speaking, medium- to large-scale equipment is being developed. The basis for the development is U.S. patent 2018/0010043 A1. 30,000 tonnes /year of feedstock facilities are more cost-effective to run (Han et al. 2018) and frequently can change production modes depending on market conditions from complete bioenergy through a mixture of biochar and bioenergy. The large-scale technique has been created, but because of the high capital costs, it is most likely to be applied in areas where there is a consistent supply of affordable biomass. The main task is creating small-scale (<20,000 TPY feedstock) biochar manufacture schemes that perform as well as large-scale systems in terms of technology and cost. According to studies of small-scale biochar producers, there is a considerable demand for enhanced small-scale systems.

The integration of the biomass-harvested system with biochar manufacturing systems, especially those situated in the field at forest landings, is a perfect illustration of how design may have a direct impact on economics. Approximately 50% of the harvested forest biomass is presently left at the landing due to transportation costs and market circumstances (Dooley et al. 2018). Therefore, increasing the amount of biomass turned to biochar overall will be greatly facilitated by the development of efficient low-level production systems that can be lucrative at forest landings (Amonette et al. 2002).

Before creating policy incentives, it is essential to fully comprehend the climate benefits associated with the production and use of a given biochar. These benefits come from both the GHGs released (or avoided) during production as well as the climate impacts once the biochar has been applied to the soil. Therefore, accurate measurements of GHG emissions are required for the manufacture, transportation, and application of biochar as well as finally used soil system. These emissions must then be compared to those linked with the other potential outcomes of the biomass to determine the overall climate gain for a particular used scenario and production.

For biochar systems, achieving economic viability continues to be difficult (Sohi et al. 2010). Costs associated with the acquisition of feedstock, capital investments, operations, and transportation of feedstocks and products, as well as revenue streams related to energy and biochar products, climate offsets, and renewable energy subsidies, are significant factors determining economic viability, according to (Whitman et al. 2010). Right now, biomass-produced bioenergy is more profitable. Even quick pyrolysis systems, which produce more bioenergy and less charcoal than slow pyrolysis systems, exhibit this phenomenon (Brown et al. 2011). When contrasting the biochar and bioenergy schemes created on their ability to reduce global warming, the situation is reversed (Woolf et al. 2010). Biochar systems will continue to be economically inferior to bioenergy systems until the successful implementation of government policies that sufficiently value and monetize the generally higher climatic welfare of biochar accordingly (Shackley et al. 2015).

Before the benefits of biochar can be asserted on a scientific basis, there are large information gaps that must be filled.

- For instance, crucial elements consist of; the type of biomass feedstock utilized (Gholizadeh et al. 2024), or a combination of feedstocks; the pace, temperature, and type of pyrolysis, and the sorts of additives utilized, such as manure, flue-scrub leftovers, compost, and synthetic fertilizers.
- The type of soil to which it is mixed, including the presence of soil microorganisms; cultivation patterns and previously used soil; consequences for soil nutrients and nutrient leaking; various used crops, including trees, grains, pasture, and fodder; the environment in which it will be used—including rainfall, temperature, and native vegetation; and the biochar particles size found and used and their effect on microbes, among other factors.
- How much should be added in each situation; how it should be added—scattered on the surface or turned in; how long it will take for biochar to truly trap carbon in the soil; and the socioeconomic circumstances of biochar production and use.

The method of pyrolysis utilized to create biochar raises a lot of questions. Speedy (5–10 s for 400–550 °C) provides low biochar (15%) whereas slow (<400 °C for half an hour to many more hours) found a high percentage (35%) of biochar. Speedy pyrolysis uses less energy than slow pyrolysis, which requires more energy. The properties of biochar, such as its structure, chemistry, composition, and mode of action, might alter depending on the pyrolysis temperature and the feedstock used. The texture of the biochar produced by different pyrolysis stages, as well as its properties and dependence on various biomass sources, provide a variety of challenges on the soil and soil biota in terms of their effects (Paul 2009).

Critics of biochar and its by products claim that the development of biomass feedstock (Rajput et al. 2024) could lead to the loss of habitat and biodiversity as well as the removal of crop wastes from soils. The advantages of using biochar may be countered by significant greenhouse gas emissions from land use changes, feedstock collection, and transportation. Additionally, poorly designed pyrolysis plants may release powerful greenhouse gases that will worsen air quality and contribute to global warming and climate change (Laird 2008). Environmentalists and skeptics who oppose biochar and biofuels have utilized the arguments extensively in their advocacy.

There are three main areas where biochar technology faces scientific obstacles. The first is how soil–plant interactions are impacted by biochar amendments. To assess the potential financial advantages of producers utilizing biochar in agriculture and forestry, it is crucial to comprehend this component. The second category is concerned with the overall implications of biochar technology on the global climate system, from biomass harvesting to biochar production and finally to biochar application. Because this seemingly favorable feature sets biochar apart from other uses of biomass, it is crucial to encourage the development of carbon (C) markets that can provide the financial support necessary for widespread adoption. The utility of biochar in composting methods is the last category. Here, significant variation in emissions and plant responses is observed, necessitating further research to identify the situations in which biochar can have a positive impact.

Regardless of size and use, locating a thermochemical conversion facility is difficult. Numerous factors must be coordinated to produce a certain form of biochar that can be used to create a particular product from a specific feedstock. Is the feedstock offered all year long? Is it cumbersome to transport or heavy? Where is the output's market demand? Will that continue over the long run? These are only a handful of the many inquiries that planners and entrepreneurs must make to assess every possibility that can affect a site's success.

4 Future Scope of the Chapter

For many years, biochar has been hailed as a valuable by product of pyrolysis that can enhance/conserve soil health and fertility while enhancing the advantages of carbon sequestration from the atmosphere. However, the kilns needed for its manufacturing must be kept at 400–700 C for several hours, producing carbon dioxide and a significant amount of thermal energy. Are biochar production processes able to capture more carbon than they release into the atmosphere? Can you offer proof that producing biochar helps to mitigate climate change on a global scale? The answer to these questions has a promising scope of study in the biochar's future.

Now, there is a noticeable increase in the usage of biochar in soil treatments (Alburquerque et al. 2014). Single and multiple nutrients can be delivered to plant roots more effectively with the help of biochar and other low-activated fertilizers (Ding et al. 2016). It is currently unknown, though, how adding biochar to alkaline soils might impact soil fertility and crop productivity.

To create biochar for certain uses, such as the adsorption of pollutants and the mineralization desorption of nutrients, modification, research, and standardization regarding pyrolysis conditions and feedstock type are required (Agegnehu et al. 2017). The chemical interactions between biochar and soil surface also require in-depth study, particularly about micronutrients and ion exchange processes. Future studies on biochar must assess a larger range of distinctive traits and entail collaborative efforts to close knowledge gaps since studies yet have concentrated on a small number of criteria. Knowledge must be based on thorough data that is trustworthy and repeated in the field, like the qualities of biochar that relate to the target application. To get the best benefits, it's important to pick the right amount of biochar and application techniques. The most efficient rates, application windows, and methodology should be the focus of future studies, considering the cropping system and crop growth stage.

It is unclear how microorganisms that change nutrients in soil treated with charcoal help plants develop. There hasn't been much research done on the connections between bacteria and the pH, nutritional content, microporosity, and particle size of biochar. The use of waste-derived biochar to address the major problem of soil nutrient insufficiency opens up new opportunities for the application of micronutrient-containing biochar to starved soils of nutrients. The best use of biochar in the soil will depend on future research into the relationships between its traits and qualities and its manufacturing process. Along with agricultural uses to increase nutrient uptake in crops and minimize environmental impact, biochar may be additionally used in the industrial sector to control waste. Furthermore, a detailed understanding of how biochar might mitigate negative effects on the atmosphere and soil is required. Models can be used to assess how alternative bioenergy systems affect carbon footprints at regional, national, and global to guide decisions on incentives and restrictions. Biochar has unquestionably become a significant potential tool for environmental applications. Various challenges should be looked upon to permit additional biochar applications.

Extrapolating from wastewater treatment outcomes to soil remediation/fertility enhancement is necessary. The secret to this expansion will be the same goals shared by the repair and modification systems.

Further in situ or field tests should be conducted over a longer length of time to simulate the real environment and determine the genuine impact of biochar before large-scale applications.

More research is needed to determine the dangers connected with using biochar made from diverse sludge or other contaminated biomass types, as well as the effectiveness of different mitigation strategies (e.g., co-pyrolysis effects).

New statistical exploration, such as machine learning, can be used to "predict" the properties of biochar. More organized studies should be done to examine the connection between feedstock/pyrolysis situations and biochar physiochemical properties (e.g., the relationship between biochar structure and stability; the role of cellulose, hemicellulose, and lignin content).

Further efforts should be made to further expand the modification techniques and environmental applications of engineered biochar to streamline the modification stages, reduce the cost, and maximize the applicability of a specific type of engineered biochar to assure sustainability.

Future research should concentrate on the processes that modify biochar use in environmental applications. For instance, even though numerous studies have shown that biochar activated by acid or alkaline can stimulate plant development, the reactions underlying governing those processes are still poorly known.

Large research should be done to determine whether it is feasible to combine cutting-edge pyrolysis techniques with emerging modification techniques to create highly effective biochar (e.g., co-pyrolysis and magnetic changes to found sludge-containing magnetic biochars with minimum atmospheric risk; microwave-related pyrolysis and activation to develop biochars with outer high specific surface layer constricting that of activated carbon).

5 Conclusion

In the current chapter, we assess the data about biochar and its preparation techniques. Additionally, the chapter provides information on the challenges in the production of biochar. A thorough analysis of difficulties in the preparation of useful and specific biochar was also highlighted. The content in this chapter will be evaluated as a single source of knowledge for students and researchers who are interested in learning more about biochar.

References

Abdulyekeen KA, Daud WMAW, Patah MFA (2024) Torrefaction of wood and garden wastes from municipal solid waste to enhanced solid fuel using helical screw rotation-induced fluidised bed reactor: effect of particle size, helical screw speed and temperature. Energy 293:130759. https://doi.org/10.1016/j.energy.2024.130759

Agegnehu G, Srivastava AK, Bird MI (2017) The role of biochar and biochar-compost in improving soil quality and crop performance: a review. Appl Soil Ecol 119:156–170

Akgül A, Akın SŞ, Güleç F, Kazanç F (2024) Biochar gasification: Insights from pyrolysis atmospheres and gasification heating rates. Fuel 360:130469. https://doi.org/10.1016/j.fuel.2023.130469

Alburquerque JA, Calero JM, Barrón V, Torrent J, del Campillo MC, Gallardo A, Villar R (2014) Effects of biochars produced from different feedstocks on soil properties and sunflower growth. J Plant Nutr Soil Sci 177:16–25

Al-Rahbi AS, Williams PT (2017) Hydrogen-rich syngas production and tar removal from biomass gasification using sacrificial tyre pyrolysis char. Appl Energy 190:501–509

Amonette JE, Archuleta JG, Fuchs MR (2002) Biomass to biochar|MaximizingCarbon Value 25

Barrow CJ (2012) Biochar: potential for countering land degradation and for improving agriculture. Appl Geogr 34:21–28

Bergman PCA, Boersma AR, Zwart RWR, Kiel HA (2005) Energy research centre of the Netherlands (ECN), 71

Brown TR, Wright MM, Brown RC (2011) Estimating profitability of two biochar production scenarios: slow pyrolysis versus fast pyrolysis. Biofuels Bioprod Biorefining 5:54–68

Cantrell KB, Hunt PG, Uchimiya M, Novak JM, Ro KS (2012) Impact of pyrolysis temperature and manure source on physicochemical characteristics of biochar. Bioresour Technol 107:419–428

Chen S, Liu C, Peng C, Liu H, Hu M, Zhong G (2012) Biodegradation of chlorpyrifos and its hydrolysis product 3,5,6-trichloro-2-pyridinol by a new fungal strain Cladosporiumcladosporioides Hu-01. PLoS ONE 7:e47205

Czernik S, Bridgwater AV (2004) Overview of applications of biomass fast pyrolysis oil. Energy Fuels 18:590–598

Dafalla M, Inayat A, Jamil F, Ghenai C (2024) Prospective of biochar material production and process optimization using co-pyrolysis approach: a mini-review. J Phys Conf Ser 2751:012024. https://doi.org/10.1088/1742-6596/2751/1/012024

Dai L, Fan L, Liu Y, Ruan R, Wang Y, Zhou Y, Zhao Y, Yu Z (2017) Production of bio-oil and biochar from soap stock via microwave-assisted co-catalytic fast pyrolysis. Bioresour Technol 225:1–8

Demirbas A, Arin G (2002) An overview of biomass pyrolysis. Energy Sour 24:471–482

Ding Y, Liu Y, Liu S (2016) Biochar to improve soil fertility: a review. Agron Sustain Dev 36:36

Dong T, Gao D, Miao C, Yu X, Degan C, Garcia-Pérez M, Rasco B, Sablani SS, Chen S (2015) Two-step microalgal biodiesel production using acidic catalyst generated from pyrolysis-derived bio-char. Energy Convers Manag 105:1389–1396

Dooley JH, Wamsley MJ, Perry JM (2018) Moisture content of baled forest and urban woody biomass during long-term open storage. Appl Eng Agric 34:225–230

Emrich W (1985) Handbook of charcoal making. Springer, Dordrecht, pp 12–15

Essandoh M, Kunwar B, Pittman CU, Mohan D, Mlsna T (2015) Sorptive removal of salicylic acid and ibuprofen from aqueous solution using pine wood fast pyrolysis biochar. Chem Eng J 265:219–227

Fang C, Zhang T, Li P, Jiang R, Wang Y (2014) Application of magnesium modified corn biochar for phosphorus removal and recovery from swine wastewater. Int J Environ Res Public Health 11:9217–9237

Funke A, Ziegler F (2010) Hydrothermal carbonization of biomass: a summary and discussion of chemical mechanisms for process engineering. Biorefin 4:160–177

Gabhane J, Prince William SPM, Vaidya AN, Mahapatra K, Chakrabarti T (2011) Influence of heating source on the efficacy of lignocellulosic pretreatment: a cellulosic ethanol perspective. Biomass Bioenergy 35:96–102

Gabhane JW, Bhange VP, Patil PD, Bankar ST, Kumar S (2020) Recent trends in biochar production methods and its application as a soil health conditioner: a review. SN Appl Sci 2:1307

Gholizadeh M, Meca S, Zhang S et al (2024) Understanding the dependence of biochar properties on different types of biomass. Waste Manag 182:142–163. https://doi.org/10.1016/j.wasman.2024.04.011

Glaser B (2007) Prehistorically modified soils of central Amazonia: a model for sustainable agriculture in the twenty-first century. Philos Trans R Soc B Biol Sci 362:187–196

Glaser B, Haumaier L, Guggenberger G, Zech W (2001) The 'Terra Preta' phenomenon: a model for sustainable agriculture in the humid tropics. Naturwissenschaften 88:37–41

González JF, Román S, Bragado D, Calderón M (2008) Investigation on the reactions influencing biomass air and air/steam gasification for hydrogen production. Fuel Process Technol 89:764–772

Guan G, Kaewpanha M, Hao X, Abudula A (2016) Catalytic steam reforming of biomass tar: prospects and challenges. Renew Sustain Energy Rev 58:450–461

Haider FU, Coulter JA, Cai L, Hussain S, Cheema SA, Wu J, Zhang R (2022) An overview on biochar production, its implications, and mechanisms of biochar-induced amelioration of soil and plant characteristics. Pedosphere 32:107–130

Han HS, Jacobson A, Bilek EM, Sessions J (2018) Waste to wisdom: utilizing forest residues for the production of bioenergy and biobased products. Appl Eng Agric 34:5–10

Hang J, Guo Z, Zhong C et al (2024) A super magnetic porous biochar manufactured by potassium ferrate-accelerated hydrothermal carbonization for removal of tetracycline. J Clean Prod 435:140470. https://doi.org/10.1016/j.jclepro.2023.140470

Hu X, Ding Z, Zimmerman AR, Wang S, Gao B (2015) Batch and column sorption of arsenic onto iron-impregnated biochar synthesized through hydrolysis. Water Res 68:206–216

Huang YF, Chiueh PT, Kuan WH, Lo SL (2016) Microwave pyrolysis of lignocellulosic biomass: heating performance and reaction kinetics. Energy 100:137–144

Intanakul P, Krairiksh M, Kitchaiya P (2003) Enhancement of enzymatic hydrolysis of lignocellulosic wastes by microwave pretreatment under atmospheric pressure. J Wood Chem Technol 23:217–225

Jegajeevagan K, Mabilde L, Gebremikael MT, Ameloot N, De Neve S, Leinweber P, Sleutel S (2016) Artisanal and controlled pyrolysis-based biochars differ in biochemical composition, thermal recalcitrance, and biodegradability in soil. Biomass Bioenergy 84:1–11

Jung KW, Hwang MJ, Jeong TU, Ahn KH (2015) A novel approach for preparation of modified-biochar derived from marine macroalgae: dual purpose electro-modification for improvement of surface area and metal impregnation. Bioresour Technol 191:342–345

Kaewtrakulchai N, Wisetsai A, Phongaksorn M et al (2024) Parametric study on mechanical-press torrefaction of palm oil empty fruit bunch for production of biochar. Carbon Resour Convers 12:100285. https://doi.org/10.1016/j.crcon.2024.100285

Karunanayake AG, Toddo A, Crowley ML, Ricchetti LB, Pittman CU, Anderson R, Mlsna TE (2017) Rapid removal of salicylic acid, 4-nitroaniline, benzoic acid and phthalic acid from wastewater using magnetized fast pyrolysis biochar from waste Douglas fir. Chem Eng J 319:75–88

Kaur R, Kumar VT, Krishna BB, Bhaskar T (2024) Sweet sorghum bagasse pyrolysis: unravelling thermal degradation via slow and flash pyrolysis investigations. J Chem Sci 136:70. https://doi.org/10.1007/s12039-024-02293-3

Klinghoffer NB, Castaldi MJ, Nzihou A (2015) Influence of char composition and inorganics on catalytic activity of char from biomass gasification. Fuel 157:37–47

Lai WY, Lai CM, Ke GR, Chung RS, Chen CT, Cheng CH, Pai CW, Chen SY, Chen CC (2013) The effects of woodchip biochar application on crop yield, carbon sequestration and greenhouse gas emissions from soils planted with rice or leaf beet. J Taiwan Inst Chem Eng 44:1039–1044

Laird DA (2008) The charcoal vision: a Win–Win–Win scenario for simultaneously producing bioenergy, permanently sequestering carbon, while improving soil and water quality. Agron J 100:178–181

Laird DA, Novak JM, Collins HP, Ippolito JA, Karlen DL, Lentz RD, Sistani KR, Spokas K, Van Pelt RS (2017) Multi-year and multi-location soil quality and crop biomass yield responses to hardwood fast pyrolysis biochar. Geoderma 289:46–53

Li L, Rowbotham JS, Christopher HG, Dyer PW (2013) New and future developments in catalysis. Elsevier, Amsterdam, pp 173–208

Li J, Dai J, Liu G, Zhang H, Gao Z, Fu J, He Y, Huang Y (2016) Biochar from microwave pyrolysis of biomass: a review. Biomass Bioenergy 94:228–244

Libra JA, Ro KS, Kammann C, Funke A, Berge ND, Neubauer Y, Titirici MM, Fühner C, Bens O, Kern J, Emmerich KH (2011) Hydrothermal carbonization of biomass residuals: a comparative review of the chemistry, processes and applications of wet and dry pyrolysis. Biofuels 2:71–106

Liu S, Xie Q, Zhang B, Cheng Y, Liu Y, Chen P, Ruan R (2016) Fast microwave-assisted catalytic co-pyrolysis of corn stover and scum for bio-oil production with CaO and HZSM-5 as the catalyst. Bioresour Technol 204:164–170

McBeath AV, Smernik RJ, Krull ES, Lehmann J (2014) The influence of feedstock and production temperature on biochar carbon chemistry: a solid-state ^{13}C NMR study. Biomass Bioenergy 60:121–129

Méndez A, PazFerreiro J, Gil E, Gascó G (2015) The effect of paper sludge and biochar addition on brown peat and coir based growing media properties. Sci Hortic 193:225–230

Moralı U, Şensöz S (2015) Pyrolysis of hornbeam shell (*Carpinusbetulus* L.) in a fixed bed reactor: characterization of bio-oil and bio-char. Fuel 150:672–678

Muthu Kumar K, Varunkumar S (2024) Ultra-rich carbonization through flash devolatilization for synthesis of biochar from biomass. Biomass Convers Biorefinery 14:12001–12012. https://doi.org/10.1007/s13399-023-03817-6

Nunoura T, Wade SR, Bourke JP (2006) Studies of the flash carbonization process. 1. Propagation of the flaming pyrolysis reaction and performance of a catalytic after burner. Ind Eng Chem Res 45:585–599

Paul H (2009) Biochar for environmental management: science and technology B. Biochar Environ Manag Sci Technol

Prins MJ, Ptasinski KJ, Janssen FJJG (2006) Torre faction of wood: part 2 analysis of products. J Anal Appl Pyrol 77:35–40

Rajput NA, Laghari M, Mangio H-R, Soothar RK (2024) Enhanced phosphorus and potassium recovery through co-pyrolysis of nutrient-rich biomass feedstocks for engineered biochar production. Bioresour Technol Rep 26:101818. https://doi.org/10.1016/j.biteb.2024.101818

Rathod PB, Pandey AK, Meena SS, Athawale AA (2016) Quaternary ammonium bearing hyper-crosslinked polymer encapsulation on Fe_3O_4 nanoparticles. RSC Adv 6:21317–21325. https://doi.org/10.1039/c6ra01543c

Rathod PB, Chappa S, Ajish Kumar KS et al (2018) Poly(ethylenimine) functionalized magnetic nanoparticles for sorption of Pb, Cu, and Ni: potential application in catalysis. Sep Sci Technol 00:1–11. https://doi.org/10.1080/01496395.2018.1520731

Rathod PB, Chappa S, Kumar KSA et al (2019) Cadmium(II)-loaded Fe_3O_4 @MPTS nanoparticles: preparation and application as catalyst for C-N coupling reactions. Chem Select 4:11796–11800. https://doi.org/10.1002/slct.201902432

Rawat J, Saxena J, Sanwal P (2019) Biochar: an imperative amendment for soil and the

Reilly J (1925) The technology of wood distillation: with special reference to the methods of obtaining the intermediate and finished products from the primary distillate. Nature 116:779–780

Roberts DA, Cole AJ, Paul NA, de Nys R (2015) Algal biochar enhances the re-vegetation of stockpiled mine soils with native grass. J Environ Manag 161:173–180

Shackley HPS, Simon S, Clare A, Joseph S, McCarl B (2015) Economic evaluation of biochar systems: current evidence and challenges. BiocharEnviron Manag 1:813–852

Shao M, Zhang C, Chen Q et al (2024) Hydrothermal-enhanced pyrolysis for efficient NOX reduction and biochar valorization from food waste digestate. Waste Manag 183:112–122. https://doi.org/10.1016/j.wasman.2024.05.007

Shen B, Chen J, Yue S, Li G (2015) A comparative study of modified cotton biochar and activated carbon based catalysts in low temperature SCR. Fuel 156:47–53

Sohi SP, Krull E, Lopez CL (2010) A review of biochar and its use and function in soil. R Bol 105:47–82

Suárez-Abelenda M, Kaal J, McBeath AV (2017) Translating analytical pyrolysis fingerprints to thermal stability indices (TSI) to improve biochar characterization by pyrolysis-GC-MS. Biomass Bioenergy 98:306–320

Thines KR, Abdullah EC, Mubarak NM, Ruthiraan M (2017) Synthesis of magnetic biocharfrom agricultural waste biomass to enhancing route for waste water and polymer application: a review. Renew Sustain Energy Rev 67:257–276

Titirici MM, Thomas A, Antonietti M (2007) Back in the black: hydrothermal carbonization of plant material as an efficient chemical process to treat the CO_2 problem? New J Chem 31:787–789

Verma M, Godbout S, Brar SK, Solomatnikova O, Lemay SP, Larouche JP (2012) Biofuels production from biomass by thermochemical conversion technologies. Int J Chem Eng 12:542426

Wan Z, Sun Y, Tsang DCW, Hou D, Cao X, Zhang S, Gao B, Ok YS (2020) Sustainable remediation with an electroactivebiochar system: mechanisms and perspectives. Green Chem 22:2688–2711

Whitman T, Scholz SM, Lehmann J (2010) Biochar projects for mitigating climate change: an investigation of critical methodology issues for carbon accounting. Carbon Manag 1:89–107

Woolf D, Amonette JE, Street-Perrott FA, Lehmann J, Joseph S (2010) Sustainable biochar to mitigate global climate change. Nat Commun 1:56

Wu Q, Deng L, Lan T et al (2024) Outstanding enhancement of caproate production with microwave pyrolyzed highly reductive biochar addition. Bioresour Technol 413:131457. https://doi.org/10.1016/j.biortech.2024.131457

Xie T, Reddy KR, Wang C, Yargicoglu E, Spokas K (2015) Characteristics and applications of biochar for environmental remediation: a review crit. Rev Environ Sci Technol 45:939–969

Yuan H, Lu T, Wang Y, Chen Y, Lei T (2016) Sewage sludge biochar: nutrient composition and its effect on the leaching of soil nutrients. Geoderma 267:17–23

Zhang M, Gao B, Varnoosfaderani S, Hebard A, Yao Y, Inyang M (2013) Preparation and characterization of a novel magnetic biochar for arsenic removal. Bioresour Technol 130:457–462

Zhou Z, Liu Y, Liu S, Liu H, Zeng G, Tan X, Yang C, Ding Y, Yan Z, Cai X (2017) Sorption performance and mechanisms of arsenic(V) removal by magnetic gelatin-modified biochar. Chem Eng J 314:223–231

Chapter 15
The Economic and Environmental Significance of Biochar Application

Tuybia Bilal, Tawseef Ahmad Mir, Muatasim Jan, Irtizah Mushtaq, Benatul Behar, Bilkees Ayoob, and Muskan Un Nisa

Abstract Biochar, a carbon-rich material produced through the pyrolysis of organic waste, has gathered attention for its potential in soil decontamination and bioeconomy. The properties of biochar depend on pyrolysis conditions and feedstock types. This review explores various biochar production methods, including pyrolysis, gasification, and hydrothermal carbonisation, with a focus on their application in soil remediation. The utilisation of biochar has expanded significantly, contributing to environmental benefits such as reducing greenhouse gas emissions, mitigating global warming, minimising nutrient leaching, sequestering carbon, and enhancing agricultural productivity. Additionally, biochar's ability to decrease the bioavailability of contaminants positions it as a valuable bioproduct in the bioeconomy, which encompasses the use of biotechnology for sustainable production of bioproducts. Biochar offers economic opportunities across the energy, industrial, and agricultural sectors, improving soil quality and creating potential revenue streams. This review discusses the production methods, agronomic benefits, and financial advantages of biochar.

Keywords Biochar · Bioeconomy · Environmentally sound · Soil quality · Carbon rich

T. Bilal · I. Mushtaq · B. Behar · B. Ayoob · M. U. Nisa
Faculty of Forestry, Sher-E-KashmirUniversity of Agricultural Sciences and Technology of Kashmir, Benhama, Ganderbal, Jammu and Kashmir, India

T. A. Mir (✉) · M. Jan
Centre of Research for Ethnobotany, Government Model Science College, Jiwaji University, Gwalior 474009, India
e-mail: tawseefbot@gmail.com

T. A. Mir
Assistant Professor, Department of Botany, BFIT College, Dehradun 248015, India

Commission Member, IUCN-CEESP, Gland, Switzerland

NEET Educator, Kashmir Institute of Excellence, Kupwara, Jammu and Kashmir 193222, India

M. Jan
Assistant Professor, Department of Botany, BFIT College, Dehradun 248015, India

A. Roy et al. (eds.), *Biochar Production Engineering*,
https://doi.org/10.1007/978-3-031-92594-8_15

1 Introduction

Biochar is a solid organic substance produced as a by-product of pyrolysis. It serves as a soil amendment, enhancing the soil's chemical, physical, and microbial properties, which significantly improves soil fertility, as noted by Awad et al. (2018). By acting as a soil enhancer, biochar improves soil structure and promotes plant growth, thus boosting agricultural yields. Its effectiveness, however, depends on various factors such as the type of crop, the biochar source, production method, soil composition, and environmental conditions. Typically, biochar is made from feedstocks such as agricultural waste, animal manure, and paper products, turning these materials into valuable resources (Brewer et al. 2014). Biochar can be produced on both small and large scales using pyrolysis equipment or kilns. Pyrolysis is a thermochemical process that occurs at temperatures between 350 and 700 °C, converting biomass into biochar, bio-oil, and syngas in the absence of oxygen (Varma et al. 2018). Pyrolysis and gasification are the key thermochemical processes that result in biochar production (Lehmann et al. 2011). There are two types of pyrolysis: slow pyrolysis and fast pyrolysis, both of which are influenced by the residence time and heating rate. Time-consuming transformation produces more sorbent, whereas rapid pyrolysis yields more oils and liquids. The production of biochar is also increased by time-consuming pyrolysis (36%) compared to high-speed pyrolysis (~17%) or conversion of gas (12%) (Uchimiya et al. 2011). Time-consuming transformation, sometimes referred to as traditional cementation, produces biochar by heating biomass at a relatively low temperature for an extended period of time (this could take days) (Cao et al. 2009). However, biochar is synthesised at a privileged warmth with a very petite-abodetime for rapid pyrolysis. The fundamental difference between fast and slow pyrolysis methods is the yields they produce. Here, the former produces more biochar while the latter produces more bio-oil. The initial stage in creating biochar is drying the biomass, beyond which the particle is heated further to allow volatile compounds to release from the solid (Rhodes et al. 2008). The unstable substances produced may either be organic condensable substances like acetic acids and methanol, or they may be (CO_2, CH_4, and H). In the gas phase, cracking and polymerisation events change the total product range (Cetin et al. 2005). Biochar's enhanced porosity and surface area are the results of gasification. It primarily transforms biomass into syngas, a gaseous form of the fuel that produces a little amount of oxidising agent at a high temperature (about 700 °C) (Brewer et al. 2014). The oxidising agents used in the gasification process are oxygen, water, or their combinations. The carbonaceous solid result is one of the main interests in the manufacturing of biochar. The increased fixed carbon contents of the material are a result of the water evaporation and volatile components produced (Novak et al. 2009), and various pore diameters (Lu et al. 2014). Megaspore plays a significant role in biochar's mass soil texture, hydropower, circulation, and root movement, while microparticles are in charge of the material's tremendous absorptive capacity and surface area. Processes involving liquid–solid adsorption require to use mesopores as well (Freddo et al. 2012). The outline and range of the particles in the biochar are compared to the mixture (raw material materials and pressure) used throughout its manufacture. Different feedstocks for biochar are examined using scanning electron microscopy (SEM) to determine their shape and pore size distribution (Quilliam et al. 2013). Additionally, various functional groups and a number of aromatic compounds from lignin-based biomasses can be found in the permeable formation of biochar (Zimmerman et al. 2011).

1.1 Biochar as a Soil Enhancement for Cleaning up Contaminated Soils

The use of biochar accelerates the bioremediation of organic substances, which attempts to increase the microbial activity in biochar-amended soils that breaks down petroleum hydrocarbons. Even though substances may not be entirely eliminated through soil, they can change forms, particularly when their concentrations go from large to small (Wu et al. 2016). While mitigating metals, some considerations should be made, including extracting metals in woody plants and biofuels plants on polluted farms, removing metalloids by capturing the metalloids stored biomass, and converting metalloids into less hazardous products (Gurwick et al. 2013). Biochar is the best substance for cleaning polluted soils because it has opposite charges, a huge interior surface area, and is resistant to deterioration (Budzianowski 2017). Generally, when untreated substance is pyrolyzed at elevated temperatures, the biochar surface area increases. A source is where the pollutant is present, whereas a receptor is where damage is done. The route, which serves as the process by which toxins move from the origin to the receiver, is involved. When a pollutant travels from origin to receiver along the route in an amount sufficient to cause harm, it is referred to as a pollutant, and the soil is measured as having been contaminated (Sun et al. 2014). The most practical, time-consuming, and cost-effective methods for getting rid of pollutant soil are to remove the origin or the receiver (Laird et al. 2010). When significant soil contamination occurs or when the polluting site is being actively used, these recommendations can be unreasonable and expensive. Pollutants may travel from sources to receptors, but many entail ending up in the soil solution via a variety of paths (Luo et al. 2011). By adsorbing contaminants on its surface and lowering their concentration in the soil solution, biochar disrupts source pathway-receptor links (El-Naggar et al. 2018). If biochar eternally eliminate contaminants that reach the soil result, totally blocking the passage to receiver, remediation is possible. Pollutants can no longer pose a threat to health or the environment after sorption on the surface of biochar. The appropriate biochar surface area is crucial since it increases the material's ability to bind metal ions and organic molecules (Luo et al. 2011).The conversion of biomass into a carbon compound is increasing daily to reform the degraded soil.

1.2 Biochar Raw Material

The type of raw material is crucial in determining the properties of biochar. Cellulose and hemicellulose are responsible for bio-oil production, while lignin contributes to a higher yield of biochar (Joshi et al. 2022). Gymnosperms contain 47–50% carbon content, while angiosperms have a relatively higher cellulose-to-lignin ratio, making both types suitable as raw materials for biochar production. Additionally, residues from mills and forests have been identified as promising raw materials for biochar. Agricultural lignocellulosic biomass residues, such as sorghum, sugarcane bagasse, bamboo, and rice husk, are also used in biochar production (Ramola et al. 2021a, b).

The properties of biochar are heavily influenced by the pyrolysis conditions and feedstock used, with factors such as temperature, residence time (RT), and heat transfer rate playing a crucial role (Sun et al. 2014). Mašek et al. (2018) found that different feedstocks produce biochars with varying characteristics, such as surface area, porosity, and functional groups, all of which affect the sorption properties of biochar. Common feedstocks for pyrolysis include wood bark, sugar beet, rice husks, poultry manure, and dairy manure. The quality and quantity of the feedstock, as well as the production methods, can also influence the presence of heavy metals and other pollutants, including phenol, dioxins, and furans (Al-Wabel et al. 2013). Biochar produced from animal manure tends to have a smaller surface area compared to biochar made from plant materials, primarily due to its higher inorganic content and ash levels (Ok et al. 2015). The composition of biochar is also determined by factors such as the type of substrate, pyrolysis time, pressure, and the levels of elements like carbon, nitrogen, potassium, and calcium. For example, biochar made from wood, which generally has a higher carbon content, will have lower potassium levels than biochar derived from feedstocks with higher potassium, such as animal litter (Gurwick et al. 2013). Since pyrolysis conditions can significantly alter the nutrient content of biochar, each batch should be analysed for its specific properties. The purity of the feedstock can also influence the presence of perfluoro chemicals and other contaminants, and the presence of heavy metals and pollutants may affect the composition of polycyclic aromatic hydrocarbons in the final biochar (Hale et al. 2015).

2 Biochar Relevance for Gas Reclamation

Biochar is particularly effective in removing hazardous gas constituents. H_2S is efficiently removed from biogas by biochar made from camphor, rice hulls, bamboo, mud, hardwood chips, and boar excrement. This biochar has an adsorption capability between 110 and 370 mg H_2S/g, and its effective elimination is over 96% (Lehmann and Joseph 2015). Biochar's moisture level (>85% v/w), pH (>8.0), outside region, and compound interaction by plane major groups like OH and COOH all helped by the absorption of H_2S. H_2S reacts with the surface of alkali biochar by ionising the eoH and eCOOH efficient group in the existence of O_2 and H_2O, forming (K, Na)$_2$SO$_4$ that may be accessible as SO_4 2 to plants (Lehmann and Joseph 2015). Trichloroethylene abstraction was investigated in biochar made from peanut shells (PBC) and soybean straw (SBC) at pyrolyzed degree between 350 and 750 °C. When compared to commercially available activated char, biochar of (PBC) and (SBC) that is obtain at a soaring pyrolyzed warmth (SBC700 and PBC700) was additional effective than that formed at a small down pyrolyzed heat (SBC300 and PBC300). When compared to activated biochar, the rate of use of 0.28 g/L PBC700 had a clearance efficiency >88% up to 9 mg/L of trichloroethylene. The hydrophobicity (10% O removal) and surface area (12–410 m^2/g) of biochar were improved by the efficiency of removal of biochar (PBC700 and SBC700), and the split of biochar was decreased in contrast through (PBC300 and SBC300) (Lehmann et al. 2006).

2.1 *Biochar and Greenhouse Gases*

Climate change is primarily caused by enlarge in greenhouse gases (GHG) and global warming, but (CO_2) emissions account for more than 77% of these changes. Comparing the emission of carbon (IV) oxide from soil respiration to that from burning fossil fuels, the difference is around ten times greater (Nguyen et al. 2014). In order to moderate climate change, it is also crucial to reduce carbon dioxide pollutants in agricultural soil. In essence, biochar is old to promote soil carbon confiscation and decrease emissions of nitrous oxide (N_2O) and carbon monoxide (CO) (Leng and Huang 2018). Recent studies have indicated that biochar may be able to reduce the emission of N_2O and CH_4 from the soil, two GHGs that contain a significant impact on climate change and are responsible for global warming. According to Novak et al. (2010), adding biochar in the direction of soil increases the activity of bacteria concerned in converting nitrous oxide to nitrogen, offering the fundamental mechanisms. On the other hand, the alkalinity of biochar increase the activity of organisms that reduce nitrous oxide. Similarly, Wild and Jones (2009) stated that the downsides of biochar will unquestionably minimise the ameliorate soil acidity and emissions as there is a rise in soil impending H. According to research, biochar has a huge surface area that makes it an excellent adsorption site for nitrous oxide and nitrogen, which reduces the discharge of these gases from the soil biological part (O'Toole et al. 2013).

3 Removal of Organic Pollutants

The application of biochar is crucial for eliminating organic pollutants from soil, especially fungicides, herbicides, and pesticides such insecticides like chlorpyrifos, carbofuran, and simazine (Spokas 2010), industrial chemicals such as polycyclic aromatic hydrocarbons (PAHs) volatile organic compounds and cationic fragrant dyes ((Herath et al. 2015). However, a number of crude chemicals, such as furfural, phenolic compounds, estrogenic compounds found in animal dung and manure, poisonous raw compound found in landfill leachate, etc., hinder the decomposition of biomass and are present in particular waste streams (Xia et al. 2017).

The relation between these pollutants and diverse properties of biochar frequently influence the abstraction mechanisms. Chemisorption and physisorption, such as electrostatic attraction/repulsion via—electron donor–acceptor, pore diffusion, and H bonding, over carboxylic, diol, and alcohol functional groups, are the main methods used to remove organic pollutants (Spokas 2010). The bulk of the bound pollutants is sooner or later mineralised by biodegradation (by various bacteria present on the surface and in the micropores of biochar), partitioning, chemical transformation (through a reductive reaction or electrical conductivity), and other mechanisms (Yin et al. 2014). The interactions of biochar with crude pollutants are influenced by pH, pyrolysis temperature, feedstock type, and pollutant ratios. For the elimination of nonpolar organic contaminants, biochar created at a higher pyrolysis temperature is preferable because it has a greater surface area and microporosity.

4 Biochar as a Soil Modification for Remediation

4.1 Effect of Biochar on Soil Properties

The depth, structure, porosity, texture, and consistency of the topsoil are all significantly impacted by the presence of biochar through altering the distribution of surface area, packings, pore sizes, bulk density, and particle sizes. However, it modifies the soil's characteristics, which directly impacts the well-being of plant grows (Saxena et al. 2014). A soil's permeability, reactivity to water, swelling, and shrinking, as well as its coagulation and land preparation ability reaction to variations in air temperature, are all influenced by the presence of biochar. It modifies the soil's physical features, increasing the soil's total specific surface area, which enhances the soil's aeration and structure (Oleszczuk et al. 2013). The following are some ways by which biochar stimulates the activity of mycorrhizal fungi: (i) altering physico-chemical composition of soil; (ii) tediously altering the mycorrhizae, which impacts the soil microbes in the ecosystem; (iii) preventing the signalling between plants and fungus and allelochemical cleansing; (iv) as long as protection against fungus crofter (Ameloot et al. 2013). The soil's quality and mycorrhiza fungi's ability to live there are both enhanced by biochar's porosity (Kim et al. 2015). It improves soil properties, elevates pH and total P and N levels while lowers any potential aluminium levels by increasing the soil's capacity to interchange cations and anions. But because biochar increases soil moisture, it not only avoids drought but also land degradation and nutrient depletion (Ma et al. 2014). In heavy metal-polluted environments, the biochar surface has a significant potential for the adsorption of dangerous substances like (Pb), (Cd),

(As), (Ni), and (Cu). This is because the biochar surface has a variety of chemically vigorous group, such as ketones, diols, and carboxylic. (Ma et al. 2014). Some biochar ingredients cause the soil's porosity to dramatically increase, which promotes breeze during the dust heap wrap and, in turn, enhances oxygen dispersal within the dust heap wrap, which may cause severe microbial degradation. The application of biochar to soil provides several benefits. Biochar acts as a soil conditioner by enhancing the biophysical characteristics of the soils, such as their capacity to store water and retain nutrients and promoting plant development (Harvey et al. 2012). The usage of biochar has various benefits, including (a) lowering soil flexible strength, (b) raising pH and soil structure, (c) improving fertiliser use efficiency, (d) Reduced aluminium virulent to plant roots and microbiota, (e) providing better soil quality for earthworm residents (Sun et al. 2014).

4.2 Effect of Biochar on Soil Physico-Chemical Characteristics

Despite the fact that biochar can boost soil's ability to retain water, the hydrophobic nature of biochar can have a detrimental impact on this ability. High pH biochar would significantly improve the properties of basic to neutral soil but only slightly improve those of acidic soil (Ameloot et al. 2013). According to the collision of biochar on the soil's interchangeable cationic ability value, variations in the amount of Ca^{2+} in the soil frequently coincide with a rise in pH. While pH increases are beneficial for neutral soils, such those in temperate regions, acidic soils, like peat, prevent the intensification of pH-sensitive bacteria (Ameloot et al. 2013).

4.3 Effect of Biochar on Soil Biological Properties

Influence on Microbial Population the deployment of biochar enhances not only the physical and chemical properties of the soil but also its microbiological characteristics. Due to its extensive surface area and reactivity, biochar attracts ions and organic molecules, creating numerous possible sites for relationships between microbes and substrates. Consequently, biochar can modify the composition of the soil microbial community, impact microbial activity, and influence soil enzyme activity, thereby impacting various biogeochemical processes (Zhu et al. 2017). Biochar is an excellent organic amendment for improving the carbon content of soil, water-holding, and microbial habitats. Many researchers has observed that the pores in biochar support the growth of microbes from adverse environmental conditions, providing them with essential nutrients for growth and multiplication (Niraj and Mahat 2024).

5 Characteristics and Application of Biochar

5.1 Characteristics of Biochar

The ability of biochar to absorb metals is prejudiced by its porosity and surface area. Because of the water loss during the dehydration phase during the pyrolysis of waste material, small pores are created in this material (biochar) (Yin et al. 2014). There are several pore sizes in biochar, from micro- (2.00 nm) to macro- (>50.00 nm) to nano- (0.900 nm), in that order. But in addition to these volatile molecules, biochar also contains, ash components, moisture, labile carbon and fixed carbon. When biochar is heated, the carbon's chemical make-up changes to an aromatic form that is mostly resistant to micro-organism deterioration.

5.2 Biochar's Stability

Although biochar is a crucial long-term soil addition (for purposes including recycling nutrients, carbon storage, and the repair of insect killer-spoiled soil), its long-term ecological durability is not fully acknowledged. (Malghani et al. 2013). The permanence of the pyrolyzed biochar depends on its feedstocks and temperatures. According to certain research, some biochar can degrade fast in specific soils depending on the pyrolysis and feedstock conditions used to create it. According to research, improved hydrolysis would result in more durable biochar (Smith et al. 2010). The hardness of biochar may be improved by increasing the pyrolysis temperature, though.

5.3 The pH of Biochar

Biochar has a pH that varies from neutral to alkaline. Biochar can elevate the pH of the soil when its pH is privileged than the pH of polluted soils (Rasse et al. 2017). It happened as a consequence of the base cations' transformation into metal oxides during the pyrolysis process, which includes the bases magnesium, calcium, potassium, and silicon. As a result, biochar having highest ash content and pH is produced from natural sources with the highest compositions of minerals. (Schulze et al. 2016). They are thus expected to cause the step up elevation in pH of the soil after treatment. As the soil pH increases, less metal cations are soluble in the solution of soil. As a result, precipitation of metals take place out from this solution, largely as phosphates (Zeng et al. 2018). Biochar produced from sources like manure products has a high mineral ash composition that makes it excellent at obstructing high levels of metal (Wu et al. 2016). A unique quality of biochar for agricultural usage, particularly in soil amendment, is its pH value. It is one of the features as

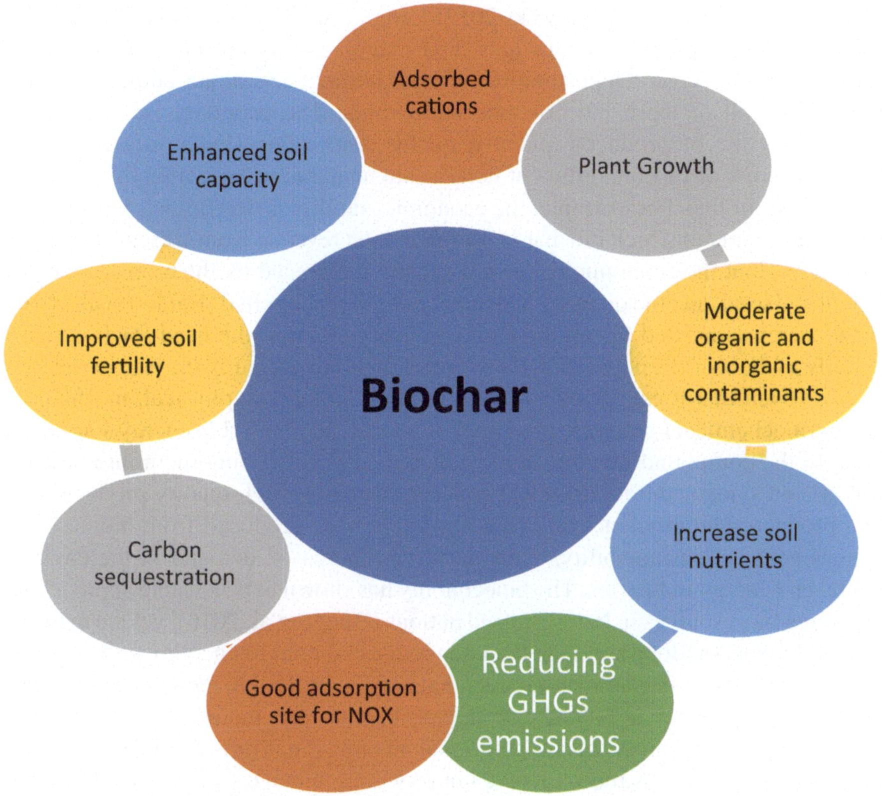

Fig. 1 Environmental significance of biochar application

well, and in such situation, chars produced by pyrolysis differ greatly from chars produced through hydrothermal carbonisation. Inherently either somewhat acidic or basic, raw biomass has a pH range of 5.2–7.3. (Wu et al. 2016). The process also raises the basic ash content. As a result, the degree of carbonisation increases with an increase in PH value (Beesley and Marmiroli 2011).The various pivotal roles played by biochar are depicted in Fig. 1.

6 Biochar and Bioeconomy

Utilising biotechnology to produce new, economically valuable bioproducts is what is meant by the term "bioeconomy", which denotes the research and exploitation of bioresources (Bugge et al. 2016). The feeds are the bioresources in this situation, and biochar, a crucial component of the bioeconomy, is the bioproduct. Any adaptable bioresource can provide a variety of bioproducts to quickly increase the resource's value (Budzianowski 2017; Leng et al. 2012). In general, commercialisation, sales, publicity campaigns, and manufacturing are all essential to the bioeconomy's survival. Opportunities for both indirect and direct employment are provided by all of these bioeconomic endeavours. The economy is significantly impacted by the bioproduct's quantity, quality, and safety (Fuentes-Saguar et al. 2017). In terms of biochar, efficient hefty-scale manufacturing will help agronomy and the bottom line. For example, the economic stability is determined by an acquiesce of the plants to which biochar is useful and the revenue from increased harvest. This shows that the economic balance is greatly influenced by the crop chosen, the kind of soil used, and the quantity and quality of biochar applied (Baidoo et al. 2016). Since energy is needed for all industries in every country, the need for it is rising globally as the world's population rises. Fossil fuels are currently the primary energy source. Given that it can be transformed through biological, physical, mechanical, and thermochemical processes, bioenergy is a practical renewable energy source. By severing the compound bond of crude materials and converting them into biochar, bio-oil, and syngas, thermochemical conversion creates high-quality products with high product yield efficiency. Biochar has lately been produced from biomass due to its benefits in sustainability, economics, and increased demand in the environmental and energy industries. The bioeconomy has shown to be a viable approach for upgrading bioresources in industrialised nations (Bugge et al. 2016), yet the majority of impoverished nations still lack the knowledge essential to encourage the production of bioproducts. A viable bioproduct that may be utilised in industry, agriculture, and the energy sector is biochar. Producing biochar can therefore assist improve soils and opening up prospects for additional income. It has further effects on associated companies, such as those involved in providing biochar equipment and carbon sequesters. An economy, particularly those involved in forestry and agriculture, may be impacted by the fabrication of biochar from forest biomass and its use as a soil improvement. According to certain research, the carbon dioxide storage subsidies can be quite significant for the financial viability of biochar. Biochar may be promoted by creating a CO_2 sequestration payment programme. Employing biochar for the interim production of high-yield cash crops may prove advantageous for farmers. The potential application of biochar for enhancing soil fertility and boosting agricultural productivity appears promising. However, biochar help the soil and water to be retained, which lowers the need for fertilisers and irrigation and restores depleted soils. As opposed to agricultural fertilisers, biochar as a soil addition lasts a long time and does not need to be applied every year, making it a cost-effective option. Biomass pyrolysis yields oil and syngas as its two major byproducts. Harsono et al. (2013) showed the economic assumption results for the production of biochar in Selangor at 532.00 US$/year, and the total revenue from biochar sale was 8012 US$/year. These products could be utilised as biofuel to provide green energy. Syngas is generally used as a component in the pyrolysis method, acting like a heat cause for the aeration or furnace stages. Oil may generate significant amounts of particles and black carbon while being used as a standalone product and derived from biomass. It is essential to convert the oil to biodiesel for use as carrying fuel (Rojas-Jiménez et al. 2023).

7 Biochar Benefits

Biochar helps to prevent nutrient leakage when applied to soil. Several investigations have shown that the inclusion of biochar often reduces N_2O emissions by 83%. This also holds true for the advantages of atmospheric Nitrogen loss, especially those connected to soil amendments and bio-fertilisers, which boost soil quality, biological growth, pH level, recycling of nutrient elements, water retention capacity, soil pollution, and other positive effects (Nguyen et al. 2014). Moreover, enhanced soil properties include greater porosity, the ability to hold more water and a reduction in bulk density. The hydrophilic, basic, hydrophobic, acidic, etc. characteristics of biochar's surface might result in the biochar absorbing solute molecules out from soil solution that can affect nutrient retention. A wood-based biochar produced at 450 °C absorbed between 250 and 520 mg P/g (McBeath et al. 2011; Zimmerman et al. 2011). Under atmospheric circumstances, biochar is employed as a bio-storage method (Roy and McDonald 2015). Agricultural soil conditioning with biochar works well. Agricultural practices only cause organic carbon to get worse every day (Xia et al. 2017). The presence of organic carbon in soil, combined with nutrients and water, especially N, P and K that provides the home for soil microorganisms which improve structure of soil, is extremely important for the maintenance of agricultural yields. Toxic contaminants can be removed from contaminated soils using biochar as an adsorbent. Depending on the material source, biochar has an organic carbon content that can reach 90%, and such substance has improved storage of carbon implementation processes (Ippolito et al. 2017).

8 Future Prospect and Conclusion

The use of biochar to remove contaminants from the soil as a cheaper alternative to pricey activated carbon is gaining popularity quickly. Future research is necessary to create biochar for soil improvement since the quality of biochar and how it performs depend on the kind of feedstock and the pyrolysis conditions (Li et al. 2017). The increase in storage of carbon and fertility of soil have received significant study attention. Efforts should continue to be made to maximise the usage of biochar to lower emissions of CO_2 and other greenhouse gases into the atmosphere. It is necessary to minimise CO_2 pollution in agricultural land to combat temperature alteration (Wang et al. 2018). Biochar is created from a wide range of biomass-based raw materials that are pyrolyzed under various circumstances, producing great diversity in physical and chemical properties which can improve carbon storage, reduce greenhouse gas emissions, stop soil erosion, lessen the island effect, and have other beneficial effects on the environment. Biochar may lessen the bioavailability and efficacy of pollutants in soil, both organic and inorganic. The kind of feedstock used and the pyrolysis conditions affect the amounts and properties of biochar. As their adsorption locations are loaded with organic matter out from native soil and contending contaminants over time, biochar's ability to immobilise pollutants may either increase, diminish, or alter, hence it is crucial to understand this before adding biochar to soils. Though several parts of biochar technology remain underdeveloped. The word "bioeconomy", which also refers to the application of biotechnology for the manufacture of new, valuable bioproducts, involves the exploitation and exploration of biological reserves or resources. A valuable bioproduct, biochar may be used to make biogas and bio-oil and is useful in industry, agricultural land and the energy sector. Making biochar can therefore help to enhance land surface and create increase in economy. Biochar is more affordable, lighter, and lasts longer once applied to the soil than fertiliser. Biochar might be encouraged by establishing a carbon storage payment programme.

References

Al-Wabel MI, Al-Omran A, El-Naggar AH, Nadeem M, Usman AR (2013) Pyrolysis temperature induced changes in characteristics and chemical composition of biochar produced from conocarpus wastes. Bioresour Technol 131:374–379

Ameloot N, Graber ER, Verheijen FG, De Neve S (2013) Interactions between biochar stability and soil organisms: review and research needs. Eur J Soil Sci 64(4):379–390

Awad YM, Ok YS, Abrigata J, Beiyuan J, Beckers F, Tsang DC, Rinklebe J (2018) Pine sawdust biomass and biochars at different pyrolysis temperatures change soil redox processes. Sci Total Env 1(625):147–154

Baidoo I, Sarpong DB, Bolwig S, Ninson D (2016) Biochar amended soils and crop productivity: a critical and meta-analysis of literature. Int J Develop Sustain 5(9):414–432

Beesley L, Marmiroli M (2011) The immobilisation and retention of soluble arsenic, cadmium and zinc by biochar. Environ Pollut 159(2):474–480

Brewer CE, Chuang VJ, Masiello CA, Gonnermann H, Gao X, Dugan B et al (2014) New approaches to measuring biochar density and porosity. Biomass Bioenergy 66:176–185

Budzianowski WM (2017) High-value low-volume bioproducts coupled to bioenergies with potential to enhance business development of sustainable biorefineries. Renew Sustain Energy Rev 70:793–804

Bugge MM, Hansen T, Klitkou A (2016) What is the bioeconomy? A review of the literature. Sustainability 8(7):691

Cao X, Ma L, Gao B, Harris W (2009) Dairy-manure derived biochar effectively sorbs lead and atrazine. Environ Sci Technol 43(9):3285–3291

Cetin E, Gupta R, Moghtaderi B (2005) Effect of pyrolysis pressure and heating rate on radiata pine char structure and apparent gasification reactivity. Fuel 84(10):1328–1334

El-Naggar A, Lee SS, Awad YM, Yang X, Ryu C, Rizwan M et al (2018) Influence of soil properties and feedstocks on biochar potential for carbon mineralization and improvement of infertile soils. Geoderma 332:100–108

Freddo A, Cai C, Reid BJ (2012) Environmental contextualisation of potential toxic elements and polycyclic aromatic hydrocarbons in biochar. Environ Pollut 171:18–24

Fuentes-Saguar PD, Mainar-Causapé AJ, Ferrari E (2017) The role of bioeconomy sectors and natural resources in EU economies: a social accounting matrix-based analysis approach. Sustainability 9(12):2383

Gurwick NP, Moore LA, Kelly C, Elias P (2013) A systematic review of biochar research, with a focus on its stability in situ and its promise as a climate mitigation strategy. PLoS ONE 8(9):e75932

Hale L, Luth M, Crowley D (2015) Biochar characteristics relate to its utility as an alternative soil inoculum carrier to peat and vermiculite. Soil Bio Biochem 81:228–235

Harsono SS, Grundman P, Lau LH, Hansen A, Salleh MAM, Meyer-Aurich A (2013) Energy balances, greenhouse gas emissions and economics of biochar production from palm oil empty fruit bunches. Resour Conserv Recycl 77(2013):10–15

Harvey OR, Kuo LJ, Zimmerman AR, Louchouarn P, Amonette JE, Herbert BE (2012) An index-based approach to assessing recalcitrance and soil carbon sequestration potential of engineered black carbons (biochars). Environ Sci Technol 46(3):1415–1421

Herath HMSK, Camps-Arbestain M, Hedley MJ, Kirschbaum MUF, Wang T, Van Hale R (2015) Experimental evidence for sequestering C with biochar by avoidance of CO_2 emissions from original feedstock and protection of native soil organic matter. GCB Bioenergy 7(3):512–526

Ippolito JA, Bjorneberg D, Stott D, Karlen D (2017) Soil quality improvement through conversion to sprinkler irrigation. Soil Sci Soc Am J 81(6):1505–1516

Joshi S, Ramola S, Singh B, Anerao P, Singh L (2022) Waste to wealth: types of raw materials for preparation of biochar and their characteristics. Engineered biochar: fundamentals, preparation, characterization and applications. Springer, Singapore, pp 21–33

Kim JH, Ok YS, Choi GH, Park BJ (2015) Residual perfluorochemicals in the biochar from sewage sludge. Chemosphere 134:435–437

Laird DA, Fleming P, Davis DD, Horton R, Wang B, Karlen DL (2010) Impact of biochar amendments on the quality of a typical Midwestern agricultural soil. Geoderma 158(3–4):443–449

Lehmann J, Gaunt J, Rondon M (2006) Bio-char sequestration in terrestrial ecosystems: a review. Mitig Adapt Strat Global Change 11(2):403–427

Lehmann J, Joseph S (2015) Biochar for environmental management: an introduction. In: Biochar for environmental management. Routledge, pp 1–13

Lehmann J, Rillig MC, Thies J, Masiello CA, Hockaday WC, Crowley D (2011) Biochar effects on soil biota–a review. Soil Biol Biochem 43(9):1812–1836

Leng L, Huang H (2018) An overview of the effect of pyrolysis process parameters on biochar stability. Bioresour Technol 270:627–642

Leng RA, Inthapanya S, Preston TR (2012) Biochar lowers net methane production from rumen fluid in vitro. Livestock Res Rural Develop 24(6):103

Li H, Dong X, da Silva EB, de Oliveira LM, Chen Y, Ma LQ (2017) Mechanisms of metal sorption by biochars: biochar characteristics and modifications. Chemosphere 178:466–478

Lu W, Ding W, Zhang J, Li Y, Luo J, Bolan N, Xie Z (2014) Biochar suppressed the decomposition of organic carbon in a cultivated sandy loam soil: a negative priming effect. Soil Biol Biochem 76:12–21

Luo Y, Durenkamp M, De Nobili M, Lin Q, Brookes PC (2011) Short term soil priming effects and the mineralisation of biochar following its incorporation to soils of different pH. Soil Biol Biochem 43(11):2304–2314

Ma R, Levard C, Judy JD, Unrine JM, Durenkamp M, Martin B et al (2014) Fate of zinc oxide and silver nanoparticles in a pilot wastewater treatment plant and in processed biosolids. Environ Sci Technol 48(1):104–112

Malghani S, Gleixner G, Trumbore SE (2013) Chars produced by slow pyrolysis and hydrothermal carbonization vary in carbon sequestration potential and greenhouse gases emissions. Soil Biol Biochem 62:137–146

Mašek O, Buss W, Roy-Poirier A, Lowe W, Peters C, Brownsort P et al (2018) Consistency of biochar properties over time and production scales: a characterisation of standard materials. J Anal Appl Pyrol 132:200–210

McBeath AV, Smernik RJ, Schneider MP, Schmidt MW, Plant EL (2011) Determination of the aromaticity and the degree of aromatic condensation of a thermosequence of wood charcoal using NMR. Org Geochem 42(10):1194–1202

Nguyen BT, Koide RT, Dell C, Drohan P, Skinner H, Adler PR, Nord A (2014) Turnover of soil carbon following addition of switchgrass-derived biochar to four soils. Soil Sci Soc Am J 78(2):531–537

Niraj KC, Mahat B (2024) A review article on the effect of biochar on soil properties. J Wastes Biomass Manag 6(2):77–81

Novak JM, Busscher WJ, Laird DL, Ahmedna M, Watts DW, Niandou MA (2009) Impact of biochar amendment on fertility of a southeastern coastal plain soil. Soil Sci 174(2):105–112

Novak JM, Busscher WJ, Watts DW, Laird DA, Ahmedna MA, Niandou MA (2010) Short-term CO_2 mineralization after additions of biochar and switchgrass to a Typic Kandiudult. Geoderma 154(3–4):281–288

O'Toole A, Knoth de Zarruk K, Steffens M, Rasse DP (2013) Characterization, stability, and plant effects of kiln-produced wheat straw biochar. J Environ Qual 42(2):429–436

Ok YS, Chang SX, Gao B, Chung HJ (2015) SMART biochar technology—a shifting paradigm towards advanced materials and healthcare research. Environ Technol Innov 4:206–209

Oleszczuk P, Jośko I, Kuśmierz M (2013) Biochar properties regarding to contaminants content and ecotoxicological assessment. J Hazard Mater 260:375–382

Quilliam RS, Glanville HC, Wade SC, Jones DL (2013) Life in the 'charosphere'–does biochar in agricultural soil provide a significant habitat for microorganisms? Soil Biol Biochem 65:287–293

Ramola S, Belwal T, Li CJ et al (2021a) Preparation and application of novel rice husk biocharcalcite composites for phosphate removal from aqueous medium. J Clean Prod 299:126802. https://doi.org/10.1016/j.jclepro.2021.126802

Ramola S, Rawat N, Shankhwar AK, Srivastava RK (2021b) Fixed bed adsorption of Pb and Cu by ironmodified bamboo, bagasse and tyre biochar. Sustain Chem Pharm 21:100486. https://doi.org/10.1016/j.scp.2021.100486

Rasse DP, Budai A, O'Toole A, Ma X, Rumpel C, Abiven S (2017) Persistence in soil of Miscanthus biochar in laboratory and field conditions. PLoS ONE 12(9):e0184383

Rhodes AH, Carlin A, Semple KT (2008) Impact of black carbon in the extraction and mineralization of phenanthrene in soil. Environ Sci Technol 42(3):740–745

Rojas-Jiménez HH, Peñaloza-Farfán L, Monroy S, Figueroa MÁ (2023) Bioeconomía y desarrollo territorial en Nariño, Tolima, Guajira y Bogotá-Región: aproximación desde la triple hélice. In: Gómez FC, Gómez LM, Valencia D, Gómez S, López JM, Villota-Paz JM (eds) Avances y desafíos en las ciencias y la ingeniería: nuevos conocimientos para un futuro sostenible. Editorial UNIMAR, pp 9–37. https://doi.org/10.31948/editorialunimar.208.c350

Roy M, McDonald LM (2015) Metal uptake in plants and health risk assessments in metal-contaminated smelter soils. Land Degrad Develop 26(8):785–792

Saxena M, Maity S, Sarkar S (2014) Carbon nanoparticles in 'biochar' boost wheat (*Triticum aestivum*) plant growth. RSC Adv 4(75):39948–39954

Schulze M, Mumme J, Funke A, Kern J (2016) Effects of selected process conditions on the stability of hydrochar in low-carbon sandy soil. Geoderma 267:137–145

Smith JL, Collins HP, Bailey VL (2010) The effect of young biochar on soil respiration. Soil Biol Biochem 42(12):2345–2347

Spokas KA (2010) Review of the stability of biochar in soils: predictability of O: C molar ratios. Carbon Manag 1(2):289–303

Sun Y, Gao B, Yao Y, Fang J, Zhang M, Zhou Y, Chen H, Yang L (2014) Effects of feedstock type, production method, and pyrolysis temperature on biochar and hydrochar properties. Chem Eng J 240:574–578

Uchimiya M, Wartelle LH, Klasson KT, Fortier CA, Lima IM (2011) Influence of pyrolysis temperature on biochar property and function as a heavy metal sorbent in soil. J Agricult Food Chem 59(6):2501–2510

Varma AK, Shankar R, Mondal P (2018) A review on pyrolysis of biomass and the impacts of operating conditions on product yield, quality, and upgradation. Recent Adv Biofuels Bioenergy Utilizat 14:227–259

Wang H, Gao B, Fang J, Ok YS, Xue Y, Yang K, Cao X (2018) Engineered biochar derived from eggshell-treated biomass for removal of aqueous lead. Ecol Eng 121:124–129

Wild E, Jones KC (2009) Novel method for the direct visualization of in vivo nanomaterials and chemical interactions in plants. Environ Sci Technol 43(14):5290–5294

Wu M, Han X, Zhong T, Yuan M, Wu W (2016) Soil organic carbon content affects the stability of biochar in paddy soil. Agricult Ecosyst Environ 223:59–66

Xia T, Qi Y, Liu J, Qi Z, Chen W, Wiesner MR (2017) Cation-inhibited transport of graphene oxide nanomaterials in saturated porous media: the Hofmeister effects. Environ Sci Technol 51(2):828–837

Yin YF, He XH, Ren GAO, Yang YS (2014) Effects of rice straw and its biochar addition on soil labile carbon and soil organic carbon. J Integr Agricult 13(3):491–498

Zeng X, Xiao Z, Zhang G, Wang A, Li Z, Liu Y et al (2018) Speciation and bioavailability of heavy metals in pyrolytic biochar of swine and goat manures. J Anal Appl Pyrol 132:82–93

Zhu L, Xiao Q, Cheng H, Shi B, Shen S (2017) Seasonal dynamics of soil microbial activity after biochar addition in a dryland maize field in North-Western China. Elsevier, Amsterdam, pp 141–149. https://doi.org/10.1016/j.ecoleng.2017.04.026

Zimmerman AR, Gao B, Ahn MY (2011) Positive and negative carbon mineralization priming effects among a variety of biochar-amended soils. Soil Biol Biochem 43(6):1169–1179

The manufacturer's authorised representative in the EU is Springer Nature Customer Service Centre GmbH, Europaplatz 3, 69115 Heidelberg, Germany. If you have any concerns regarding our products, please contact ProductSafety@springernature.com

Printed and bound by CPI Group (UK) Ltd, Croydon, CR0 4YY
30/09/2026
02225562-0011